Data Fusion

Harvey B. Mitchell

# Data Fusion

## Concepts, Ideas and Deep Learning

Third Edition

Springer

Harvey B. Mitchell
Mazkaret Batya, Israel

ISBN 978-3-662-71022-7 ISBN 978-3-662-71023-4 (eBook)
https://doi.org/10.1007/978-3-662-71023-4



This Springer imprint is published by the registered company Springer-Verlag GmbH, DE, part of Springer Nature.
The registered company address is: Heidelberger Platz 3, 14197 Berlin, Germany

*This book is dedicated to our children*

*DANIEL ASHER MITCHELL AND NOA SOPHIA MITCHELL*

*How much better is it to get wisdom than Gold!*

*and to get understanding rather to be chosen than Silver!*

*Proverbs 16:16*

# Preface

The purpose of this book is to provide an introduction to the concepts and ideas of multi-sensor data fusion, and includes a comprehensive discussion of deep learning techniques. It is an extensively revised and updated third edition of the author's book *Data Fusion: Concepts and Ideas*, which was originally published by Springer in 2007 (first edition) and 2012 (second edition). As in the previous editions, the present work has been designed as a textbook for a one-semester graduate course in multi-sensor data fusion. It should, however, also be useful to advanced undergraduates in electrical engineering or computer science who are studying data fusion for the first time and to practicing engineers who wish to use the concepts of data fusion in practical applications.

Since the appearance of the second edition in 2012, the field of data fusion has changed considerably. A large part of this is due to the increasing use of deep learning techniques in data fusion. As a result we have extensively revised and updated the book as follows:

*New Material.* We have added a new chapter on Deep Learning (Chap. 5) and have included significant amounts of new material in Chaps. 1, 4, 6–10, 12, 13, and 15 (Introduction, Common Representational Format, Spatial Alignment, Temporal Alignment, Semantic Alignment, Radiometric Normalization, Bayesian Inference, Robust Statistics, Sequential Bayesian Inference and Kalman Filters, and Ensemble Learning). The remaining chapters have been updated with new examples and references and have been rewritten. Altogether the book has more than 100 new examples and sections and more than 200 new references.

More than half of the text is completely new or has been rewritten for added clarity. At the same time, material which is out-of-date, or has been superseded by new developments, has been eliminated.

*Software Code.* Where appropriate we have given details of both MATLAB and Python code which may be downloaded from the internet. In several places in the text, we have also included pseudo-code.

*Figures.* More than 40 new figures have been added to the text. Of the remaining figures, most have been redrawn for added clarity.

*Typography.* The layout and typography of the book has been extensively revised. All examples appear on a gray-background for easy identification. Advanced material has been marked with an asterisk *. This material may be left for a second reading.

*Problems.* The problems at the end of each chapter originally appeared on the author's website. Their purpose serves to highlight new concepts or ideas which were introduced in the chapter.

*Index.* The index contains new headings and new entries. A considerable effort has been made to make the index comprehensive and useful to the reader.

Data fusion relies on a diverse collection of theories and techniques drawn from a wide range of subjects, including convolutional neural networks, signal processing, statistical estimation as well as computer vision and machine learning. We have attempted to integrate this diverse collection of th]eories and techniques by adopting a Bayesian statistical framework. This has also enabled us to keep the book to a reasonable size while treating many new and important techniques in depth.

As with any other branch of engineering, data fusion is a pragmatic activity which is driven by practicalities. It is therefore important that the student is able to experiment with different techniques presented in the book. For this purpose software code, written in MATLAB or Python, is particularly convenient. In this edition, we have included details of relevant MATLAB and Python code which may be downloaded from the internet. For the professional engineer, we have both illustrated the theory with real-life applications and have provided him with an extensive list of up-to-date references.

The book is based on seminars, lectures, and courses on data fusion which have been taught over several years. The structure and content of the book is based on material gathered and ideas exchanged with my colleagues. Particular thanks are extended to Prof. Jaakko Suutala who made a detailed list of suggestions for the new edition and to my son, Daniel Mitchell, who generously reviewed several chapters in the manuscript. In previous editions, Prof. Brian Cowan, Prof. Stanley Rotman, Dr. Paul Schaefer, Dr. Ruth Rotman, and Mr. Michael Avraham were kind enough to read and comment on the book.

I am indebted to Dr. Tom Ditzinger who accepted the original manuscript for publication in 2007 and has encouraged the publication of the subsequent second and third editions. I am also indebted to my wife and daughter for the support and patience they have shown me while the book was being written.

Finally to the reader. We hope you enjoy reading this book and that it will prove a useful addition to the increasingly important and expanding field of data fusion.

Mazkaret Batya, Israel
December 2024

Harvey B. Mitchell

# Contents

# Chapter 1
# Introduction

**Abstract** In this chapter we provide a brief overview of the theories and techniques of (multi-sensor) data fusion. We describe several common data fusion strategies and conceptual frameworks. We then describe a new unified data fusion model around which we categorize the different data fusion systems and algorithms described in the book. Finally we briefly mention the drawbacks to using data fusion and in particular the problem of catastrophic fusion. We show briefly how catastrophic fusion may be avoided by employing additional auxiliary sensors or by using various consistency tests.

## 1.1 Definition

The subject of this book is *data fusion* which we define as "the theory, techniques and tools which are used for combining sensor data, or data derived from sensory data, into a common representational format". In performing data fusion, our aim is to improve the quality of the information, so that it is, in some sense, *better* than would be possible if the data sources were used individually.

The above definition implies that the sensor data, or the data derived from the sensory data, consists of multiple measurements which have to be combined. The multiple measurements may, of course, be produced by multiple sensors. However, the definition also includes multiple measurements, produced at different time instants, by a single sensor.

The general concept of data fusion is analogous to the manner in which humans and animals use a combination of multiple senses, experience and the ability to reason to improve their chances of survival.

The basic problem of data fusion is one of determining the best procedure for combining the different data inputs. The view adopted in this book is that combining multiple sources of information with a priori information is best handled within a statistical framework. The main advantage of a statistical approach is that explicit probabilistic models are employed to describe the various relationships between sensors and sources of information taking into account the underlying uncertainties.



H. B. Mitchell, *Data Fusion*, https://doi.org/10.1007/978-3-662-71023-4_1

In particular we restrict ourselves to the Bayesian methodology which provides us with a useful way to formulate the data fusion problem in mathematical terms and which yields an assessment of the uncertainty in all unknowns in the problem.

## 1.2 Synergy

The principal motivation for data fusion is to improve the quality of the information output in a process known as *synergy*. Strictly speaking, synergy does not require the use of multiple sensors. The reason being that the synergistic effect may be obtained on a temporal sequence of data generated by a single sensor. However, employing more than one sensor may enhance the synergistic effect in several ways, including: increased spatial and temporal coverage, increased robustness to sensor and algorithmic failures, better noise suppression and increased estimation accuracy.

**Example 1.1** *Multi-modal Biometric Systems* [8, 9]. Biometric systems that rely on a *single* biometric trait for recognition are often characterized by high error rates. This is due to lack of completeness or *universality* in most biometric traits. For example, fingerprints are not truly universal since it is not possible to obtain a good quality fingerprint from people with hand-related disabilities, manual workers with many cuts and bruises on their fingertips or people with very oily or very dry fingers. Multi-modal biometric sensor systems solve the problem of non-universality by fusing evidence obtained from multiple traits.

**Example 1.2** *Multiple Camera Surveillance Systems* [14, 17]. The increasing demand for security by society has led to a growing need for surveillance activities in many environments. For example, the surveillance of a wide-area urban site may be provided by periodically scanning the area with a single narrow field-of-view camera. The temporal coverage is, however, limited by the time required for the camera to execute one scan. By using multiple cameras we reduce the mean time between scans and thereby increase the temporal coverage.

Broadly speaking, data fusion may improve the performance of the system in four different ways:

**Representation.** The information obtained during, or at the end, of the fusion process has an abstract level ("granuality"), higher than each input data set. The new abstract level or granuality provides a richer semantic on the data than each initial source of information.

**Certainty.** If $V$ is the sensor data before fusion and $p(V)$ is the a priori probability of the data before fusion, then the gain in certainty is the growth in $p(V)$ after fusion. If $\widetilde{V}$ denotes data after fusion, then we expect $p(\widetilde{V}) > p(V)$.

**Accuracy.** The variance of the data after the fusion process is smaller than the variance provided directly by the sources. If data is noisy or erroneous, the fusion process tries to reduce or eliminate noise and errors. In general, the gain in accuracy and the gain in certainty are correlated.

**Completeness.** Bringing new information to the current knowledge on an environment allows a more complete the view on this environment. In general, if the information is redundant and concordant, we could also have a gain in accuracy.

**Example 1.3** *Multi-modal Medical Imaging: Gain in Completeness.* We consider images obtained through Magnetic Resonance Imaging (MRI), Computed Tomography (CT) and Positron Emission Tomography (PET). The data fusion of all three images allows a surgeon to view "soft tissue" information (MRI) in the context of "skeleton" or "bone information" (CT) and in the context of "functional" or "physiological information" (PET).

## 1.3 Data Fusion Strategies

As the above examples show, data fusion is a wide-ranging subject with many different facets. In order to understand it better, and to become familiar with its terminology, we shall consider it from several different points of view [2, 6, 11].

### Fusion Type

In this scheme we classify a data fusion system according to its fusion type, i. e. according to what aspect of the system is fused.

**Fusion across sensors.** In this situation, a number of sensors nominally measure the same property. For example, several temperature sensors measuring the temperature of an object.

**Fusion across attributes.** In this situation, a number of sensors measure different quantities associated with the same experimental situation. E.g. measuring air temperature, pressure and humidity to determine air refractive index.

**Fusion across domains.** In this situation, a number of sensors measure the same attribute over different ranges or domains. This arises, for example, in the definition of a temperature scale.

**Fusion across time.** In this situation, current measurements are fused with historical information, for example, from an earlier calibration. Often the current information is not sufficient to determine the system accurately and historical information has to be incorporated to determine the system accurately.

**Example 1.4** *Flood Forecasting* [16]. Water companies are under constant pressure to reduce frequency of sewer overflows to natural water courses from urban drainage systems (UDS). The management of storm-water through the UDS requires accurate real-time estimates of the rainfall. One way water companies have done this is to fuse measurements of the rainfall made by ground-based rain gauges and a weather radar system. This is "fusion across sensors" since, nominally, the two sensors measure the same property.

### Sensor Configuration

In this scheme we classify a data fusion system according to its sensor configuration. There are three basic configurations:

**Complementary.** A sensor configuration is called complementary if the sensors do not directly depend on each other, but can be combined in order to give a more complete image of the phenomena under observation. Complementary sensors help resolve the problem of *incompleteness*.

**Competitive.** A sensor configuration is competitive if each sensor delivers an independent measurement of the same property. The aim of competitive fusion is to reduce the effects of *uncertain* and *erroneous* measurements.

**Cooperative.** A cooperative sensor configuration network uses the information provided by two, or more, independent sensors to derive information that would not be available from the single sensors.

The following example illustrates the complementary fusion of an infra-red and a visible-light image.

**Example 1.5** *Bilateral Cross-Modal Infra-red and Visible Image fusion* [5, 18]. The goal of infrared-visible image fusion is to retain both the abundant texture details from the visible image (VI) and the salient target detection from the infrared image (IR). By fusing together the two images we can avoid problems caused by the low resolution and noise of IR and simultaneously overcome the limitations of VI in harsh environments such as rain, snow, fog and low illumination.

The following example illustrates noise suppression by competitive fusion of several noisy inputs.

**Example 1.6** *Competitive Noise Suppression.* We capture an image $I$ of a given a scene $S$. We model $I$ as an $M \times N$ matrix $\mathbf{I}$, whose elements are:

$$I(m,n) = S(m,n) + N(m,n),$$

where $S(m,n)$ denotes the $(m,n)$th element of a deterministic signal which represents the scene properties and $N(m,n)$ denotes the image noise at the $(m,n)$th element which we model as independent and identically distributed (iid) zero-mean Gaussian noise: $N(m,n) \sim \mathcal{N}(0,\sigma^2)$.

By averaging $K$ such images we obtain a fused image $\overline{I}$ whose $(m,n)$th element is

$$\overline{I}(m,n) = S(m,n) + \overline{N}(m,n),$$

and whose noise variance is reduced by a factor of $K$: $\overline{N}(m,n) \sim \mathcal{N}(0,\sigma^2/K)$.

The following example shows cooperative fusion in the field of image fusion.

**Example 1.7** *Cooperative Stereo Vision for Autonomous Driving* [4]. Autonomous car systems enable self-driving cars to navigate in complicated environments without human driver intervention. An important part of the autonomous system is the perception module which analyzes the raw data collected by the different sensors (RGB cameras, LIDAR and radar) and outputs a 3D map of the environment. An important component of the perception module is stereo vision which is used to extract depth information from multiple 2D images of the same scene.

### Input/Output Characteristics

In this scheme (see Table 1.1) we classify a data fusion system according to its joint input/output characteristics. In this scheme the input/output combinations are restricted to combinations in which the output and input are at the same semantic level, or the output is one semantic level higher than the input.

**Table 1.1** Input/Output data fusion model

| Symbol | Name | Description |
|---|---|---|
| DaI-DaO | Data input-data output | Input data is smoothed/filtered |
| DaI-FeO | Data input-feature output | Generate multiple features from input data |
| FeI-FeO | Feature input-feature output | Reduce # input features or generate new features |
| FeI-DeO | Feature input-decision output | Fuse multiple input features |
| DeI-DeO | Decision input-decision output | Fuse multiple input decisions |

**Example 1.8** *Multi-modal Face and Palmprint Fusion* [1, 15]. Biometric recognition systems often use face images or palmprint images. In general, the recognition rates obtained using only face or palmprint images is relatively low. However, significantly better recognition rates are obtained by separately extracting features from the face and palmprint images which are then fused together. The final recognition is then performed on the fused features. This is fusion in the FeI-FeO domain.

Alternatively, preliminary recognition (decision) is performed separately on the face and palmprint features. The final recognition is then performed by fusing the face and palmprint decisions. This is fusion in the DeI-DeO domain.

### Early, Intermediate and Late Fusion

In this scheme we classify a data fusion system according to its output characteristics. This scheme is widely used when the fusion process is performed using a deep-learning multi-layer neural network. In this case, fusion may be performed in any given layer. Fusion performed in the first few layers is classified as early fusion and corresponds to the fusion of lightly processed raw sensor data, while fusion performed in the last fully-connected layers is classified as late fusion and often corresponds to the fusion of scores or decisions. Fusion performed in the remaining layers is classified as intermediate fusion and corresponds to fusion of features.

**Example 1.9** *Intermediate Fusion of RGB and Depth Images* [19]. Detecting weeds in a wheat field by analyzing the corresponding color (RGB) image of the wheat field has low performance. In an effort to improve the weed detection performance, the RGB image is *fused* with the corresponding depth image.

In the early layers of the network, color, position, texture and depth features are extracted from the RGB and Depth images. Then in an *intermediate layer* the color, position, texture and depth features are fused together ("intermediate-level" fusion). The fused features are processed further by the remaining layers. The final output of the network is a binary image $B$, where $B(i, j) = 1$ signifies that pixel $(i, j)$ is a weed; otherwise $B(i, j) = 0$.

**Example 1.10** *Late Feature Fusion** [12]. We describe a convolutional neural network (CNN) (see Chap. 5) for late fusion of fingerprints and online signatures. Input to the network is a fingerprint image $F$ and an image of an online signature $S$. The network is divided into two parts which separately extract low, intermediate and high-level features from $F$ and $S$ (Fig. 1.1). The low and intermediate-level features are extracted using a series of convolutional and max-pooling layers while the high-level features are extracted using two fully-connected layers. The high-level fingerprint and signature features are then fused together by concatenation. This is an example of "late fusion". The final output of the network is a binary decision $D$, where $D = 1$ signifies that the fingerprint $F$ and signature $S$ match; otherwise $D = 0$.

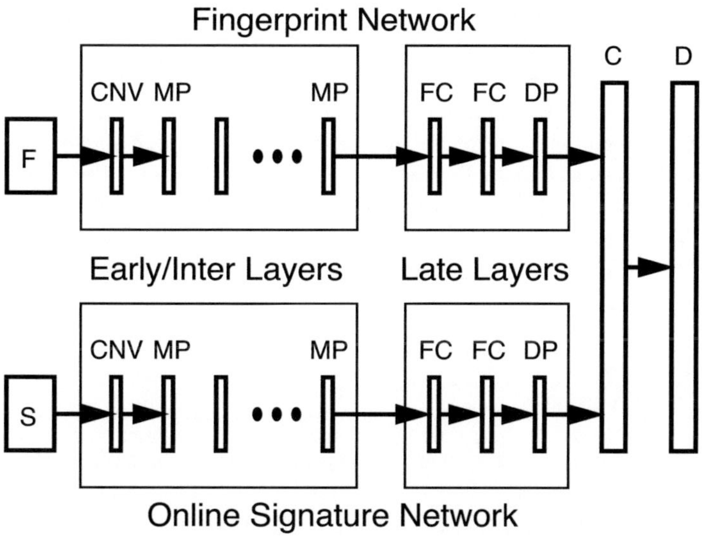


**Fig. 1.1** Shows deep learning convolutional neural network (CNN) for late feature fusion. The network consists of a fingerprint network and a signature network. The fingerprint network extracts low and intermediate-level features using a CNN with 5 convolutional (CNV) layers and 5 max-pooling (MP) layers. The signature network extracts low and intermediate-level features using a CNN with 4 CNV layers and 2 MP layers. High-level fingerprint and signature features are separately extracted using 2 fully-connected (FC) layers. The high-level features are fused together by concatenation (C). To prevent overfitting the network has dropout (DP) layers. The final output of the network is a decision $D$, where $D = 1$ signifies that $F$ and $S$ match; otherwise $D = 0$

**Table 1.2** Unified data fusion model

| Level | Name | Description |
|---|---|---|
| I-A | Early-DaO | Early fusion with data output (≡ DaI-DaO) |
| I-B | Early-FeO | Early fusion with feature output (≡ DaI-FeO) |
| II-A | Inter-FeO | Intermediate fusion with feature output (≡ FeI-FeO) |
| II-B | Inter-DeO | Intermediate fusion with decision output (≡ FeI-DeO) |
| III-A | Late-DeO | Late fusion with decision output (≡ DeI-DeO) |

DeO can be further divided into "soft" (DsO) and "hard" (DhO) decisions

### Unified Fusion Model

In this book we shall classify all data fusion algorithms using a hierarchical scheme (see Table 1.2) which combines the input/output and the early/intermediate/late fusion models.

## 1.4 Formal Framework

Data fusion systems are often of a size and a complexity that requires the use of a formal framework around which we organize the system. One class of formal frameworks is the functional model which gives an overview of the main processes, or functions, in the system. Table 1.3 lists some of the functional models which are used today. For smaller systems an architectural model which shows the individual modules in the system and the flow of data between modules is often more useful. In this book we have opted for the architectual model of Luo-Kay [13] which consists of a *distributed* network of autonomous modules, in which each module represents a separate function in the data fusion system.

**Table 1.3** Data fusion frameworks

| Name | Description |
|---|---|
| JDL | Consists of 5 levels at which fusion takes place: Pre-Processing (Level 0); Object Refinement (1); Situation Assessment (2); Threat Assessment (3); Process Refinement (4) |
| Thompoulos | Consists of 3 modules: Signal, Evidence, Dynamics. Each module integrates input data at a different level |
| Pau | Consists of a sequence of modules that must be completed before final output is established. Modules are: Feature Extraction; Association; Sensor Attribute Fusion; Analysis; Aggregation & Representation |
| Waterfall | Similar to Pau except the system is continuously updated with feedback information from the decision-making module |
| Omnibus | Consists of 4 modules: Signal Processing; Pattern Recognition & Feature Extraction; Decision Making; Context Processing & Control. Modules operate in a closed loop |

**Example 1.11** *Physiological Measurements.* We consider two physiological measurements made on a patient: temperature and blood pressure. The measurements are provided by two sensors: a thermometer and a tensiometer. The data sources are *distributed* and *complementary*. In forming a diagnosis, the physician fuses together the two measurements. In the physiological space of the patient the physician is said to perform "fusion across attributes".

Apart from providing us with a structured framework, the modular design decreases the complexity involved in designing a data fusion system by compartmentalizing the design problem. Modularity also helps in the construction and maintenance of an operational data fusion system.

In analyzing a data fusion system it is useful to divide the system into three parts: the physical, informative and cognitive domains and to determine the flow of data between these parts [3] (Fig. 1.2).

**Physical Domain: Hardware.** The physical domain contains the sensor modules each of which represents a sensor which physically interacts with the external environment. Each module contains a *sensor model* which provides us with a description of the measurements made by the sensor and of the local environment within which the sensor operates. In some applications we may wish to physically change the external environment. In this case the physical domain will also contain actuators which are able to modify the external environment.

**Informative Domain: Software.** The informative domain constitutes the heart of a data fusion system. It contains three blocks which are responsible for data fusion, control application/resource management and human-machine interface

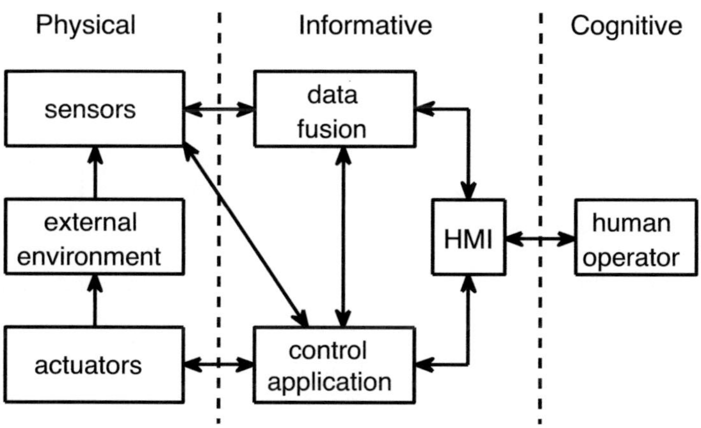


**Fig. 1.2** Shows the division of a generic data fusion system into three parts or domains: physical, informative and cognitive and the flow between the parts. The figure is divided into three panels which correspond to the physical domain (leftmost panel), informative domain (middle panel) and cognitive domain (rightmost panel)

(HMI). The data fusion block is constructed as an autonomous network of "fusion" modules. This network is responsible for combining all the sensor data into a unified view of the environment in the form of an "environmental image". The control application/resource management block is constructed as autonomous networks of "control" modules. This network is responsible for all decisions which are made on the basis of the environmental image. In many applications the decisions are fed back to the sensor block. In this case the process is known as "sensor management".

**Cognitive Domain: Human User.** In many multi-sensor data fusion applications the human user is the final arbiter or decision maker. In this case it is important to design the system in such a way that all the information which is transmitted to the human user is transformed into a form which is intuitively usable by the user for his decision-making process.

We now consider in a little more detail the data fusion block and the control loop which appear in Fig. 1.2.

### Data Fusion Block

In the Luo and Kay model we make a clear distinction between data fusion and data integration [13]. Signal processing blocks which perform data fusion combine input data into a "common representational format". On the other hand, blocks which perform multi-sensor integration may assist in a particular task but do *not* convert the input data into a common format. Thus, in the block labeled "data fusion" only a few of the autonomous modules actually perform "data fusion" as defined above; the remaining modules performing auxiliary tasks. Figure 1.3a illustrates a data fusion block in which we simultaneously have data integration and data fusion.

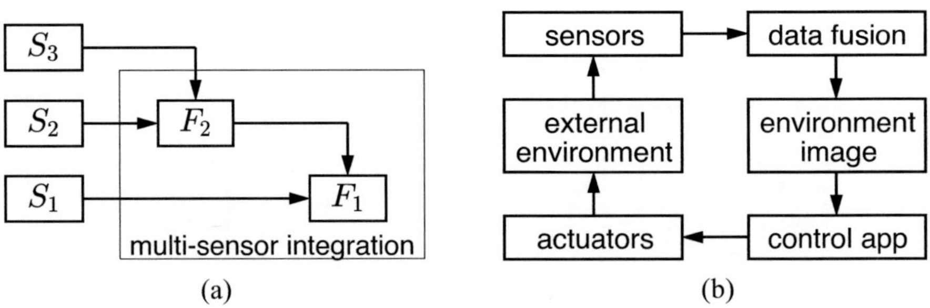


**Fig. 1.3** Shows a generic data fusion system with data fusion block and control loop. **a** Shows data fusion block which performs both data fusion and integration. $F_2$ performs data fusion on the data supplied by sensors $S_2$ and $S_3$ while $F_1$ performs data fusion on data supplied by $S_1$ and $F_2$. Integration is performed on the outputs from $F_1$ and $F_2$. **b** Shows a simple control loop built from four blocks: sensors, actuators, data fusion and control applications. A feedback mechanism is included by allowing the control application to act on the external environment via an actuator block

### Control Loop

The sensor, actuators, data fusion and control applications in Fig. 1.2 form a closed "control loop" [3]. In the control loop the (external) environment is observed with one or more sensors. The corresponding sensor observations are passed to the data fusion block where they are combined to form a unified view of the environment ("environmental image"), which in turn, is passed to the control application block. The loop is closed by feeding the decisions made by the control application block back to the environment. See Fig. 1.3b.

## 1.5 Catastrophic Fusion

The unsuspecting reader may conclude, on the basis of what has been presented so far, that a data fusion system is *always* better than a single sensor system. This conclusion is, however, mistaken: Often the performance of a data fusion system $F$ is below that of the individual sensors. This phenomena is known as *catastrophic fusion* and clearly should be avoided at all times.

Let $p_F$ denote the performance of the system $F$, then formally catastrophic fusion occurs when

$$p_F \ll \min_m(p_m) ,$$

where $p_m$ is the performance of the $m$th sensor $S_m$. In general, each sensor $S_m, m \in \{1, 2, \ldots, M\}$, is designed to operate correctly only under specific conditions, or environment, $E_m$. Let $E_F$ denote an environment in which all the sensors $S_m, m \in \{1, 2, \ldots, M\}$, are operating correctly, then $E_F$ is the "intersection" of the $E_m$:

$$E_F = E_1 \wedge E_2 \wedge \ldots E_M . \tag{1.1}$$

Sometimes, however, $F$ is used in an environment, which is inconsistent with one of the $E_m$, say $E^*$. When this happens, the signal from the corresponding sensor, $S^*$, may dominant the fused output with catastrophic results. To prevent this happening, we may employ *secondary classifiers* $C_m, m \in \{1, 2, \ldots, M\}$, which monitor the performance of the primary sensors $S_m$.

**Example 1.12** *Preventing Catastrophic Fusion in Automatic Speech Recognition (ASR)* [10]. In ideal conditions ASR systems perform very well using a single audio sensor $S_A$. However, there is substantial reduction in performance when background noise, channel distortion or reverberation are present. This has led to the development of audio-visual (AV) systems which employ an audio sensor $S_A$ and a visual sensor $S_V$. Both sensors are *complementary*: speech characteristics that are visually confusable are often acoustically distinct, and visual characteristics that are acoustically confusable are often visually distinct. The AV system works as follows: Given a finite set of utterances $U_i, i \in \{1, 2, \ldots, N\}$, we identify an unknown utterance, $U$, as

$$U^* = \arg\max_i \big(P(U_i | y_A, y_V)\big) , \tag{1.2}$$

where $P(U_i|y_A, y_V)$ is the conditional probability of $U_i$ given an observation $(y_A, y_V)$. To a first approximation, the audio and visual features are independent. In this case, Eq. (1.2) becomes:

$$U^* = \arg\max_i \big(P(U_i|y_A) \times P(U_i|y_V)\big) .$$

where the probabilities $P(U_i|y_A)$ and $P(U_i|y_V)$ are calculated off-line using a training set $\mathcal{D}$.

If $\mathcal{D}$ matches the speech data that is being tested, then $U^*$ is optimal. However, when the (test) speech is contaminated with noise, this assumption is no longer valid and the probability estimates are incorrect. In such cases, we use a weighted integration scheme:

$$U^* = \arg\max_i P(U_i|y_A)^{\alpha_A} \times P(U_i|y_V)^{\alpha_V} ,$$

where $\alpha_A + \alpha_V = 1$. The effect of the weights $\alpha_m, m \in \{A, V\}$, is to "neutralize" a sensor $S_m$, whenever $C_m$ determines that $S_m$ is operating outside its specified environment $E_m$.

### Consistency Test

In some applications we avoid catastrophic fusion by using a consistency test to determine which subset of sensor modalities can be safely fused.

**Example 1.13** *Non-parametric Consistency Test for Multiple Sensor Data Fusion* [7]. Consider a multi-modal systems, in which one sensor is trusted more than the others. For example, for autonomous driving we often employ a radar and a laser sensor. As a rule, the radar sensor is more reliable since it is generally immune to varying environmental conditions while the laser sensor is sensitive to airborne contaminants such as smoke and dust. Nevertheless the use of the laser data may be beneficial due to its higher resolution. In this case, we choose the radar sensor as a *reference*. Fusion of the radar and laser data is then *only* performed if a consistency test shows that the laser data statistically improves the reference radar model.

## 1.6 Organization

Although we shall discuss the physical, information and cognitive domains in a multi-sensor data fusion system, the emphasis will be on the information domain, and specifically on the data fusion block. The book is organized into three parts:

**Part I: Basic Concepts.** This consists of three chapters. In Chap. 1 we provide a general introduction to data fusion. This is followed by Chaps. 2 and 3 which provide overviews on sensors and data fusion architectures.

**Part II: Common Representational Format.** This consists of Chaps. 4–9. In Chap. 4 we provide an overview of the basic concept of a common representational format. This is followed by Chap. 5 which provides an overview of deep learning techniques. Then Chaps. 6–9 describe the different techniques used to create a common representational format.

**Part III: Data Fusion.** This consists of Chaps. 10–15. In Chap. 10 we give an overview of the Bayesian approach to data fusion and the different techniques involved. This is followed by Chaps. 11–13 in which we deal with parameter estimation theory including sequential estimation theory. This is followed by Bayesian decision theory in Chap. 14 and Ensemble learning in Chap. 15. The book ends with a short summary and postscript.

## Problems

**1.1** (a) What is the motivation for data fusion? (b) What are the merits of using a Bayesian approach for data fusion?

**1.2** When sensor measurements are transformed into a common representational format, their statistical distribution may change. Give examples.

**1.3** The single fusion cell (see Sect. 3.3) has three inputs: input data, auxiliary data and external knowledge. (a) Explain the difference between these inputs. (b) Describe the triangulation algorithm in Example 4.2 using the fusion cell. Identify corresponding input data, auxiliary data and external knowledge.

**1.4** (a) Explain why catastrophic fusion occurs. (b) Explain how auxiliary sensors $C_A$ and $C_V$ in Example 1.12 prevent catastrophic fusion.

## References

1. Ahmad MI, Woo WL, Dlay S (2016) Non-stationary feature fusion of face and palmprint multimodal biometrics. Neurocomputing 177:49–61
2. Bloch I, Hunter A, Appriou A et al (2001) Fusion: general concepts and characteristics. Int J Intell Syst 16:1107–1134

3. Elmenreich W (2002) Sensor fusion in time-triggered systems PhD thesis, Institut fur Technische Informatik, Technischen Universitat Wien
4. Fan R, Wang L, Bocus MJ et al (2023) Computer stereo vision for autonomous driving. In: Recent advances in computer vision applications, pp 41–70
5. Gao X, Liu S (2024) Bcmfifuse: bilateral cross-modal feature interaction-based network for infrared and visible image fusion. Remote Sens 16:3136
6. Gaw N, Yousefi S, Gahrooei MR (2022) Multimodal data fusion for systems improvement: a review. IISE Trans 54:1098–1116
7. Gerardo-Castro MP, Peynot T, Ramos F, Fitch R (2015) Non-parametric consistency test for multiple-sensing-modality data fusion. International conference on information fusion(FUSION), pp 443–451
8. Jain AK, Nandakumar K, Ross AA (2016) 50 years of biometric research: accomplishments, challenges and opportunities. Patt Recogn 79:80–105
9. Jagadiswary D, Saraswady D (2016) Biometric authentication using fused multimodal biometric. Procedia Comp Sci 85:109–116
10. Katsaggelos AK, Bahaadini S, Molina R (2015) Audiovisual fusion: challenges and new approaches. Proc IEEE 103:1635–1653
11. Lahat D, Adali T, Jutten C (2015) Multimodal data fusion: an overview of methods, challenges and prospects. Proc IEEE 103:1449–1477
12. Leghari M, Memon S, Dhomeja LD et al (2021) Deep feature fusion of fingerprint and online signature for multimodal biometrics. Computers 10:21
13. Luo RC, Yih C-C, Su KL (2002) Multisensor fusion and integration: approaches, applications and future research directions. IEEE Sens J 2:107–119
14. Mihailescu R-C, Davidsson P, Eklund U et al (2018) A survey and taxonomy on intelligent surveillance from a system perspective. Know Eng Rev 33:1–25
15. Raghavendra R, Dorizzi B, Rao A, Kumar GH (2011) Designing efficient fusion schemes for multimodal biometric systems using face and palmprint. Patt Recogn 44:1076–1088
16. Todini E (2001) A Bayesian technique for conditioning radar precipitation estimates to rain-gauge measurements. Hydrology Earth Sys Sci 5:187–199
17. Valera M, Velastin SA (2005) Intelligent distributed surveillance systems: a review. IEE Proc - Vis Image Signal Proc 152:192–204
18. Wang H, Li L, Li C et al (2023) Infrared and visible image fusion based on autoencoder composed of CNN-transformer. IEEE Access 11:78956–78969
19. Xu K, Li H, Cao W, Zhu Y, Chen R (2020) Recognition of wheat fields based on the fusion of RGB images and depth images. IEEE Access 8:110362–110370

# Chapter 2
# Sensors

**Abstract** In this chapter we consider sensors. These are special devices which interact directly with the environment and are ultimately the source of all the input data in a multi-sensor data fusion system. We describe the concept of a smart sensor and how it communicates with the outside world. This includes a brief discussion on different interface file systems and universal timing schemes. The problem of faulty sensors and errors in the sensor measurements are presented using a Bayesian sensor model.

## 2.1 Introduction

In this chapter we consider the sensors. These are special devices which interact *directly* with the environment and which are ultimately the source of all the input data in a data fusion system [7]. The physical element which interacts with the environment is known as the *sensor element* and may be any device which is capable of perceiving a physical property, or environmental attribute, such as heat, light, sound, pressure, magnetism or motion. To be useful, the sensor must map the value of the property or attribute to a quantitative measurement in a *consistent* and *predictable* manner.

In Chap. 1 we introduced a formal framework in which we represented a sensor fusion system as a distributed system of autonomous modules. To support such a scheme, the sensors must not only measure a physical property, but must also perform additional functions. These functions can be described in terms of compensation, information processing, communication and integration:

**Compensation.** This refers to the ability of a sensor to detect and respond to changes in the environment through self-diagnostic tests, self-calibration and adaption.

**Information Processing.** This refers to processes such as signal conditioning, data reduction, event detection and decision-making, which enhance the information content of the raw sensor measurements.



H. B. Mitchell, *Data Fusion*, https://doi.org/10.1007/978-3-662-71023-4_2

**Communication.** This refers to the use of a standardized interface and a standardized communication protocol for the transmission of information between the sensor and the outside world.

**Integration.** This refers to the coupling of the sensing and computation processes on the same silicon chip. Often this is implemented using microelectro-mechanical systems (MEMS) technology.

A practical implementation of such a sensor is known as a "smart" sensor [3, 13, 17].

## 2.2 Smart Sensor

A *smart sensor* is a hardware/software device that comprises in a compact small unit a sensor element, a micro-controller, a communication controller and the associated software for signal conditioning, calibration, diagnostic tests and communication. The smart sensor transforms the raw sensor signal to a standardized digital representation, checks and calibrates the signal, and transmits this digital signal to the outside world via a standardized interface using a standardized communication protocol [5, 7].

Figure 2.1 shows the measurement of a physical property by a smart sensor.

The transfer of information between a smart sensor and the outside world is achieved by reading (writing) the information from (to) an interface file system (IFS) which is encapsulated in the smart sensor as shown in Fig. 2.2.

**Example 2.1** *Smart Image Sensor* [1]. When a sensor is packaged together with the processing resources shown in Fig. 2.1 it is called a "smart sensor". Although more tightly packaged, the smart sensor is rarely smarter than its conventional counterpart. In fact, the "smart sensor" only becomes really "smart" when the tight integration of the sensing and the processing results in a sensor that can react to changing environmental conditions and consistently deliver useful measurements even under the harshest conditions.

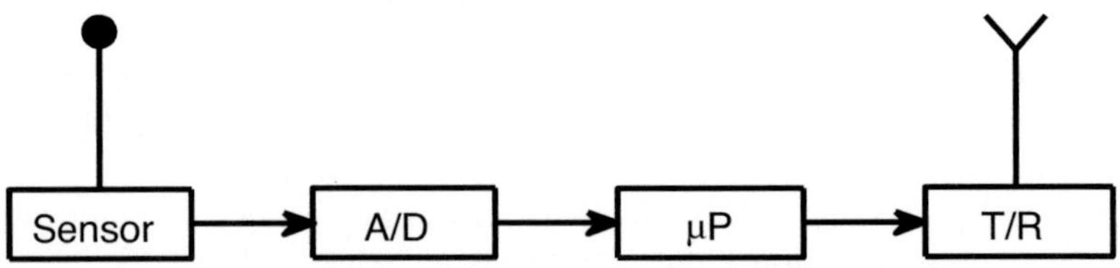


**Fig. 2.1** The sensor element measures a physical property and outputs an analog signal which is amplified, filtered and then converted to a digital signal by the analog-to-digital unit (A/D). The digital signal is processed by the microprocessor ($\mu$P), where it is temporally stored before being transmitted by the transmitter/receiver (T/R). The smart sensor may also receive command instructions via T/R

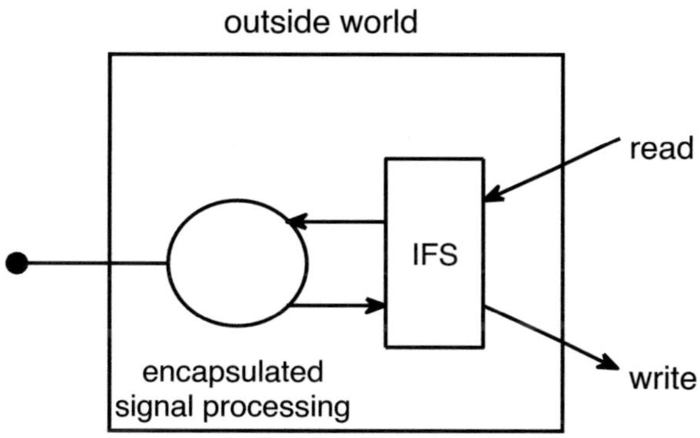


**Fig. 2.2** Shows a smart sensor with a sensor element and the encapsulated signal processing functions and the encapsulated interface file system (IFS)

## 2.3 Logical Sensors

A *logical sensor* is defined as any device which functions as a source of information for a multi-sensor data fusion node. Thus a logical sensor encompass both physical sensors and any fusion node whose output is subsequently fed into another fusion node. Unless stated otherwise, from now on the term "sensor" will refer to a logical sensor or a source-of-information. The following example illustrates a multi-sensor data fusion application which has one "physical" sensor but $N > 1$ "logical" sensors.

**Example 2.2** *Lesion Localization in Dermoscopy Images* [2]. Dermoscopy is a major tool used in the diagnosis of melanoma and other pigmented skin lesions. Due to the difficulty of human interpretation, automated analysis of dermoscopy images has become an important research area. In [2] an accurate automatic lesion localization method is presented which relies on thresholding the dermoscopy image $I_0$ using $K$ global thresholding algorithms $\mathcal{A}_k, k \in \{1, 2, \ldots, K\}$. The thresholded images $I_k, k \in \{1, 2, \ldots, K\}$, are fused together to give lesion localization which is more accurate than any of the individual images $I_k$.

The camera which records the dermoscopy image is the "physical" sensor $S_0$. The algorithms $\mathcal{A}_k, k \in \{1, 2, \ldots, K\}$, constitute the "logical" sensors $S_k, l \in \{1, 2, \ldots, K\}$. The thresholded images $I_k, k \in \{1, 2, \ldots, K\}$ are the inputs to the data fusion block $F$ where a more accurate localization of the lesions is calculated (Fig. 2.3).

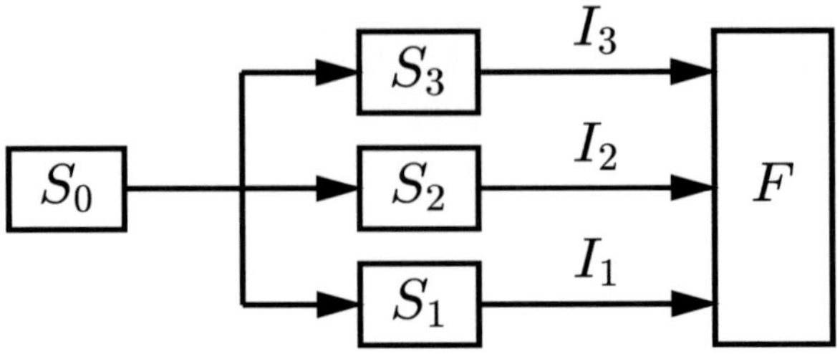


**Fig. 2.3** Shows the accurate localization of a skin lesion. $I_0$ is an image of the lesion captured by a camera (sensor) $S_0$. The logical sensors $S_k$, $k \in \{1, 2, \ldots, K\}$, are the different algorithms $\mathcal{A}_k$ which generate thresholded images $I_k$. The images $I_k$, $k \in \{1, 2, \ldots, K\}$, are fused together in $F$ to give an accurate lesion localization

## 2.4 Interface File System (IFS)

The IFS provides a structured space which is used for communicating information between a smart sensor and the outside world [6]. The following example neatly illustrates the concept of an IFS.

**Example 2.3** *Brake Pedal in a Car: An Interface File System* [11]. We consider a driver in a car. For the purposes of braking, the *brake pedal* acts as an IFS between the driver and the brakes. At this interface there are two relevant variables. The first variable is $P$: the pressure applied to the brake pedal by the driver and the second variable is $R$: the resistance provided by the brakes back to the driver. The relative position of the brake pedal uniquely identifies the values of $P$ and $R$ to both the driver and the brakes. The temporal association between sensing the information (e.g. by the driver) and receiving the information (e.g. by the brakes) is implicit, because of the mechanical connection between the driver and the brakes.

A record of the IFS can be accessed, both from the sensor and from the outside world (Fig. 2.2). Whenever one of the internal processes of a smart sensor requires information from the outside world or produces information for the outside world, it accesses the appropriate records of the IFS and reads (writes) the information to (from) this record. The internal processes of a smart sensor are thus not visible to the outside world.

Often we implement the IFS so that it acts as a temporal firewall between the sensors and the outside world. In this case, the IFS uses *local* interface file systems that can be written by the sensor and read by the outside world, without having direct communication between the sensor and the outside world (Fig. 2.4).

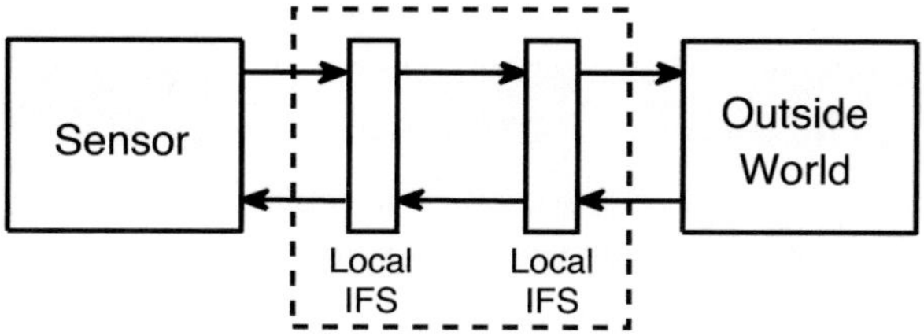


**Fig. 2.4** Shows the Interface file system (IFS) built as a temporal firewall. The IFS contains two separate local interface file systems: one is accessed by the sensor and the other is accessed by the outside world

### *2.4.1 Interface Types*

In the smart sensor model we distinguish between three interface types: the real-time service interface, the diagnostic and maintenance interface and the configuration and planning interface. All information that is exchanged across these interfaces is stored in files of the IFS.

**Real-time Service (RS) Interface.** This interface provides time sensitive information to the outside world. The communicated data is sensor observations made by a sensor on real-time entities in the environment. The RS accesses the real-time IFS files within the smart sensor which is usually time-critical.

**Diagnostic and Management (DM) Interface.** This interface establishes a connection to each smart sensor. Most sensors need both parameterization and calibration at start-up and the periodic collection of diagnostic information to support maintenance activities. The DM interface accesses the diagnostic IFS files within the smart sensor which is not usually time-critical.

**Configuration and Planning (CP) Interface.** This interface is used to configure a smart sensor for a given application. This includes the integration and set-up of newly connected sensor modules. The CP interface accesses the configuration IFS files within the smart sensor which is not usually time-critical.

### *2.4.2 Timing*

For reading (writing) information from (to) the different interfaces we could use any possible communication and computational schedule [4]. In distributed systems two different types of scheduling are used:

**Event Triggered.** In this schedule, all communication and processing activities are initiated whenever a significant change of state occurs.

**Time Triggered.** In this schedule, all communication and processing activities are initiated at pre-determined times.

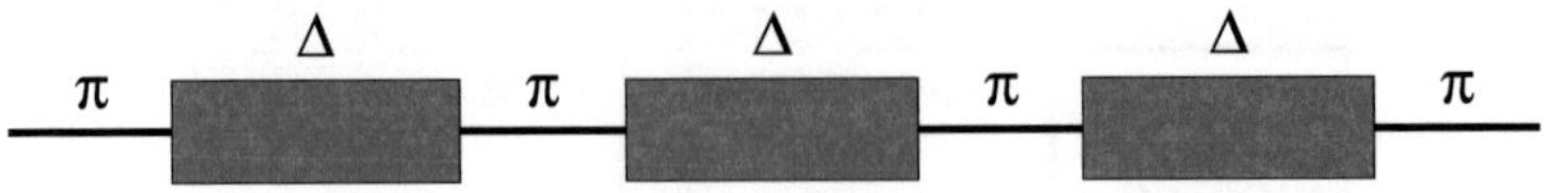


**Fig. 2.5** Shows the sparse time based used in the time-triggered architecture. $\pi$ denotes the intervals of activity and $\Delta$ denotes the intervals of silence. If $\delta$ is the precision of the clock synchronization, then we require $\Delta > \pi > \delta$

For many safety-critical real-time applications the *time-triggered architecture* (TTA) [10] is preferred because its schedule is deterministic. The TTA relies on the synchronization [14] of all local clocks[1] to a global clock: in which all measurements, communication and processing activities are initiated at *pre-determined* time instants which are *a priori* known to all nodes.

The use of a global clock does not, in and of itself, ensure that in a distributed system we are able to *consistently* order events according to their global time. This is illustrated in the following example.

**Example 2.4** *Ordering Events in a Distributed Network* [10]. Consider two local clocks $c$ and $C$. Suppose the following sequence of events occurs: local clock $c$ ticks, event $e$ occurs, local clock $C$ ticks. In such a situation the event $e$ is time-stamped by the two clocks with a difference of one tick. The finite precision of the global time-base makes it impossible, in a distributed system, to order events consistently on the basis of their global time stamps.

The TTA solves this problem by using a *sparse time base* in which we partition the continuum of time into an infinite sequence of alternating durations of *activity* (duration $\pi$) and *silence* (duration $\Delta$) (Fig. 2.5). The architecture must ensure that significant events, such as the sending of a message or the observation of an event, occur only during an interval of activity. Events occuring during the same segment of activity are considered to have happened at the same time. Events that are separated by at least one segment of silence can be consistently assigned to different time-stamps for all clocks in the system. This ensures that temporal ordering of events is always maintained.

[1] A clock is defined as a physical device for measuring time. It contains a counter and a physical oscillation mechanism that periodically generates an event which increases the counter by one. This event is the "tick" of the clock. Whenever the sensor perceives the occurrence of an event $e$ it will instantaneously record the current tick number as the time of the occurrence of the event $e$.

## 2.5 Sensor Observation

We may regard a sensor as a small "window" through which we are able to view a physical property which is characteristic of the outside world. In general the physical property is evolving continuously in time and value. However, the sensor only provides us with a "snapshot" of the process: Often the output of a sensor is reduced to a single scalar value.

The output of a sensor is known as a *sensor observation* and includes:

**Entity-Name** $E$. This is the name of the physical property which was measured by the sensor and the units in which it is measured. The units are often defined implicitly in the way the system processes the measurement value.

**Spatial Location** $\mathbf{x}$. This is the position in space to which the measured physical property refers. In many cases the spatial location is not given explicitly and is instead defined implicitly as the position of the sensor element.

**Time Instant** $t$. This is the time when the physical property was measured. In real-time systems the time $t$ is often as important as the value itself.

**Measurement** $\mathbf{y}$. This is the value of the physical property as measured by the sensor element. We write the physical property as a vector $\mathbf{y}$ since it may be multi-dimensional.

**Uncertainty** $\Delta\mathbf{y}$. This is a generic term and includes many different types of errors in $\mathbf{y}$, including measurement, calibration, loading and environmental errors. Some of these errors are defined *a priori* in the sensors data sheet and others may be calculated internally if the sensor is capable of validating its own measurements.

Symbolically we represent a sensor observation using the following five-tuple

$$O = \langle E, \mathbf{x}, t, \mathbf{y}, \Delta\mathbf{y} \rangle. \tag{2.1}$$

Sometimes not all the fields in Eq. (2.1) are present. In this case, we represent the missing fields by an asterix ($*$). For example, if the spatial location $\mathbf{x}$ is missing from the sensor observation, then we write the corresponding observation as

$$O = \langle E, *, t, \mathbf{y}, \Delta\mathbf{y} \rangle.$$

**Example 2.5** *Time-of-Flight Ultrasonic Sensor* [9]. In a time-of-flight (ToF) ultrasonic sensor, the sonar transducer emits a short pulse of sound and then listens for the echo. The measured quantity is the time between the transmission of the pulse and reception of the echo. This is converted to a *range reading* $r$ which is one-half of the distance traveled by the sound pulse. In this case, $\mathbf{y}$ is the range reading and the corresponding sensor observation is

$$O = \langle ToF, *, t, r, \Delta r \rangle,$$

where $\Delta r$ is the uncertainty in $r$.

**Example 2.6** *A Digital RGB (Color) Image.* A digital RGB (color) image $I$ is characterized by an array of pixels. At each pixel location, $\mathbf{x}$, there are a set of three color values $R(\mathbf{x})$, $G(\mathbf{x})$ and $B(\mathbf{x})$. Let $\mathbf{X}$ denote the relative positions of the array of pixels, then the corresponding sensor observation is

$$O = \langle I, \mathbf{X}, t, \boldsymbol{\Theta}, * \rangle,$$

where $\boldsymbol{\Theta}$ denotes the array of $(R, G, B)$ values.

## 2.6 Sensor Uncertainty

Sensor measurements are uncertain, which means that they can only give an estimate of the measured physical property. The uncertainty can manifest itself in several different ways:

**Random Errors.** These errors are characterized by a lack of repeatability in the output of the sensor. They are caused by measurement noise. For example, they may be due to fluctuations in the capacity of the resistance of the electrical circuits in the sensor, or due to the limited resolution of the sensor.

**Systematic Errors.** These errors are characterized by being consistent and repeatable. There are many different types of systematic errors:

*Calibration errors.* These are a result of error in the calibration process, and are often due to linearization of the calibration process for devices exhibiting non-linear characteristics.

*Loading errors.* These arise if the sensor is *intrusive* which, through its operation, alters the measurand.

*Environmental Errors.* These arise from the sensor being affected by environmental factors which have not been taken into account.

*Common Representation Format Errors.* These occur when we transform from the original sensor space to a common representational format (see Chap. 4).

**Spurious Readings.** These errors are non-systematic measurement errors. An example of such an error occurs when a sensor detects an obstacle at a given location **x** when, in fact, there is no obstacle at **x**.

**Example 2.7** *Systematic Errors in a Tipping-Bucket Rain Gauge* [16]. Rain-gauges, and in particular the most commonly used tipping-bucket rain-gauge (Fig. 2.6), are relatively accurate instruments. However, at high rain intensities, they suffer from "undercatchment": part of the water is not "weighed" by the tipping-bucket mechanism. This is a *systematic error* which, to a good approximation, may be corrected using the following calibration curve:

$$y = \alpha y_G^{\beta},$$

where $y$ is the actual rain intensity, $y_G$ is the rain-gauge measured intensity and $\alpha$ and $\beta$ are two calibration parameters [16].

**Example 2.8** *Spurious Readings in Time-of-Flight Ultrasonic Sensor*. In a time-of-flight ultrasonic sensor (Example 2.5), the sensor calculates the distance $r$ to the nearest obstacle from the measured time between the transmission

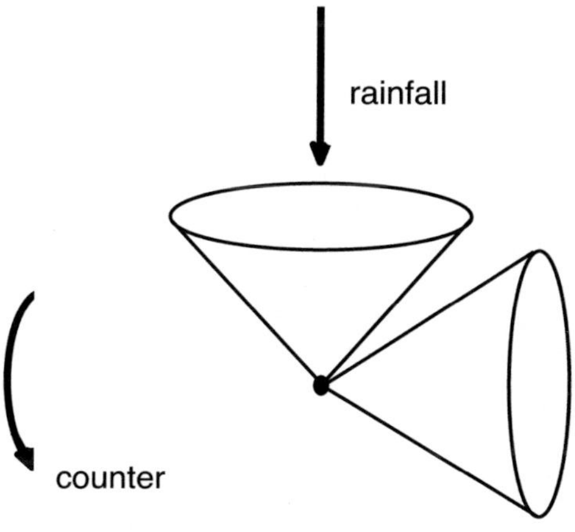


**Fig. 2.6** Shows a tipping-bucket rain-gauge consisting of buckets which tip when the weight of 0.01 inches of rain falls into them. When one bucket tips, the other bucket quickly moves into place to catch the rain. Each time a bucket tips, an electronic signal is sent to a recorder. The tipping-bucket rain-gauge is good at measuring drizzle and very light rainfall. However, it tends to underestimate during very heavy rain events, such as thunderstorms

of a pulse and the reception of its echo. Sometimes the sensor will detect an obstacle at a given range $r$ when, in fact, there is no object at $r$. This can happen if the sensor receives a pulse emitted by second sensor and interprets the pulse as if it were its own pulse. This is a *spurious reading*.

## 2.7 Sensor Characteristics

In selecting an appropriate sensor, or set of sensors, we need to consider the characteristics of the individual sensors. For this purpose, it is convenient to group the sensor characteristics into the following categories:

**State.** The sensors are classified as being internal or external. Internal sensors measure "internal" system parameters, e. g. position, velocity and acceleration. Such sensors include potentiometers, tachometers, accelerometers and optical encoders. External sensors monitor the system's geometric and/or dynamic relation to its tasks and environment. Such sensors include proximity devices, strain gauges and ultrasonic, pressure and electromagnetic sensors.

**Function.** The sensors are classified in terms of their functions, i.e. in terms of the parameters, or *measurands*, which they measure. In the mechanical domain the measurands include displacement, velocity, acceleration, mass and force.

**Performance.** The sensors are classified according to their performance measures which include: accuracy, repeatability, linearity, sensitivity, resolution, reliability and range.

**Output.** The sensors are classified according to the nature of their output signal: analog , digital , frequency (use of output signal's frequency) and coded (modulation of output signal's frequency, amplitude or pulse).

**Energy Type.** The sensors are classified according to the type of energy transferred to the sensor. For example, thermal energy involves temperature effects in materials including thermal capacity, latent heat and phase change, or electrical energy involves electrical parameters such as current, voltage and resistance.

For data fusion applications which involve multiple sensors we must also take into account the fusion type and sensor configuration as described in Sect. 1.3.

## 2.8 Sensor Model*

As we explained in Sect. 2.2, the smart sensor smoothes, checks and calibrates the sensor signal before it is transmitted to the outside world. In order to perform these functions, we require a rich *sensor model* which will provide us with a coherent description of the sensors ability to extract information from its surroundings and

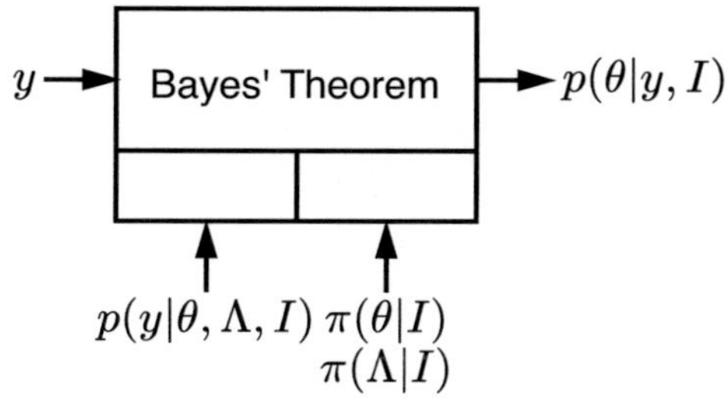


**Fig. 2.7** Shows the faulty sensor model. Input to the model is: *a priori* pdf, $\pi(\boldsymbol{\theta}|I)$, likelihood function, $p(\mathbf{y}|\boldsymbol{\theta}, \Lambda, I)$, and discrete sensor reliability $\pi(\boldsymbol{\Lambda}|I)$. The model uses Bayes' theorem to calculate the joint pdf $p(\boldsymbol{\theta}, \boldsymbol{\Lambda}|\mathbf{y}, I) \sim p(\mathbf{y}|\boldsymbol{\theta}, \boldsymbol{\Lambda}, I)\pi(\boldsymbol{\theta}|I)\pi(\boldsymbol{\Lambda}|I)$ and then eliminates $\boldsymbol{\Lambda}$ by marginalization. Output of the model is *a posterior* pdf $p(\boldsymbol{\theta}|\mathbf{y}, I) \sim \pi(\boldsymbol{\theta}|I) \int p(\mathbf{y}|\boldsymbol{\theta}, \boldsymbol{\Lambda}, I)\pi(\boldsymbol{\Lambda}|I)d\boldsymbol{\Lambda}$

make meaningful sensor observations. This information is also required when we consider the fusion of multi-sensor input data. In that case we often include the sensor model within the general “background information” $I$.

We begin our development of the sensor model by distinguishing between the variable $\boldsymbol{\Theta}$ in which we are interested, and a sensor measurement $\mathbf{y}$. We directly observe $N$ *raw* sensor measurements $\mathbf{y}_i, i \in \{1, 2, \ldots, N\}$, while the variable $\boldsymbol{\Theta}$ is not directly observed and must be *inferred*. In mathematical terms we interpret the task of inferring $\boldsymbol{\Theta}$ as estimating the *a posteriori* probability, $P(\boldsymbol{\Theta} = \boldsymbol{\theta}|\mathbf{y}, I)$, where $\boldsymbol{\theta}$ represents the true value of the variable of $\boldsymbol{\Theta}$ and $\mathbf{y} = \left(\mathbf{y}_1^T, \mathbf{y}_2^T, \ldots, \mathbf{y}_N^T\right)^T$ denotes the vector of $N$ sensor measurements.[2] This is in accordance with the Bayesian viewpoint (see Chap. 10) which assumes that all the available information concerning $\boldsymbol{\Theta}$ is contained in $p(\boldsymbol{\Theta} = \boldsymbol{\theta}|\mathbf{y}, I)$.[3]

Figure 2.7 shows a simple, but useful sensor model. Input to the model are three probability distributions:

**A Priori pdf $\pi(\boldsymbol{\theta}|I)$.** This is a continuous pdf which describes our *a priori* beliefs about $\boldsymbol{\theta}$. In the absence of any further information we may model the pdf using a histogram of historical data, or we may construct it from *a priori* information we have concerning typical $\boldsymbol{\theta}$ values. For computational convenience we often assume that $\pi(\boldsymbol{\theta}|I)$ is a Gaussian pdf with mean, covariance parameters $(\boldsymbol{\mu}, \boldsymbol{\Sigma})$:

$$\pi(\boldsymbol{\theta}|I) = \mathcal{N}(\boldsymbol{\theta}|\boldsymbol{\mu}, \boldsymbol{\Sigma}), \tag{2.2}$$

---

[2] We use bold letters for the variable of interest $\boldsymbol{\Theta}$ and the sensor measurements $\mathbf{y}_i, i \in \{1, 2, \ldots, N\}$ to emphasize that $\boldsymbol{\Theta}$ and $\mathbf{y}_i$ may be multi-dimensional. Unless stated otherwise, we shall assume $\boldsymbol{\Theta}$ and $\mathbf{y}_i, i \in \{1, 2, \ldots, N\}$, are continuous. In this case, a probability which is a function of $\boldsymbol{\Theta}$ or $\mathbf{y}_i$ is interpreted as a pdf. To emphasize this, we denote pdf's with a small $p$.

[3] For variables with continuous values we do not normally distinguish between the random variable and its value. Thus from now on we shall write the pdf $p(\boldsymbol{\Theta} = \boldsymbol{\theta}|I)$ as $p(\boldsymbol{\theta}|I)$. However, in the case of variables with discrete values, this may cause some confusion. For these variables we shall therefore continue to write $P(\boldsymbol{\Lambda} = \boldsymbol{\lambda}|I)$.

**Sensor Reliability** $\pi(\mathbf{\Lambda}|I)$. This is a *discrete* pdf which specifies the *a priori* reliability of the sensor. In the simplest model, there are two states: $\Lambda = \{\lambda_0, \lambda_1\}$ where $\lambda_0$ denotes fault-free operation and $\lambda_1$ denotes faulty operation, where ordinarily $\pi(\Lambda = \lambda_0|I) \approx 1$. Since the status of the sensor may change from measurement to measurement, we write the sensor status as $\mathbf{\Lambda} = (\Lambda_1, \Lambda_2, \ldots, \Lambda_N)^T$, where $\Lambda_i$ denotes the status of the sensor when it makes the $i$th measurement $\mathbf{y}_i$.

**Likelihood** $p(\mathbf{y}|\boldsymbol{\theta}, \mathbf{\Lambda}, I)$. This is a continuous function which describes how the raw sensor measurements $\mathbf{y}$ depend on the true value $\boldsymbol{\theta}$, the background information $I$ and the sensor status $\mathbf{\Lambda}$.

We now illustrate the fusion of multiple measurements when one or more of the sensor measurements may be spurious. For this purpose we use the Kumar-Garg-Zachery sensor model [12, 18]. See Chap. 12 for a full discussion on how to deal with spurious measurements in data fusion.

### Kumar-Garg-Zachery (KGZ) Model*

Given $K$ scalar measurements $y_k, k \in \{1, 2, \ldots, K\}$, of a given parameter $\Theta = \theta$, we assume a Gaussian sensor model:

$$p(y_k|\theta, \Lambda_k = \lambda_0, I) = \frac{1}{\sigma_k\sqrt{2\pi}} \exp -\frac{1}{2}\left(\frac{\theta - y_k}{\sigma_k}\right)^2 .$$

The probability that the measurement from the $k$th sensor $S_k$ is *not* spurious given the true parameter value $\theta$ and given the sensor measurement $y_k$ is:

$$p(\Lambda_k = \lambda_0|\theta, y_k, I) \propto \exp -\left(\frac{\theta - y_k}{\alpha_k}\right)^2 ,$$

where the parameter $\alpha_k$ jointly depends on the variance $\sigma_k^2$ of the $k$th measurement *and* the distance between $y_k$ and the other measurements $y_l, l \neq k, l \in \{1, 2, \ldots, K\}$. The idea is that as the probability that $y_k$ is a spurious measurement, or outlier, increases so the variance $\alpha_k$ should also increase. In the KSG model, $\alpha_k^2$ is defined as

$$\alpha_k^2 = \frac{2\sigma_k^2 m}{\prod_{\substack{l=1\\ l\neq k}}^{K}(y_k - y_l)^2} . \tag{2.3}$$

where $m = \max_k \left(\prod_{\substack{l=1\\ l\neq k}}^{K}(y_k - y_l)^2\right)$. However, this may give inconsistent results when some of the $y_k$ are very close to each other. Instead, the following formula may be used:

$$\alpha_k^2 = \frac{2\sigma_k^2 m'}{\prod_{\substack{l=1 \\ l \neq k}}^{K} (\max(|y_k - y_l|, \sigma_k))^2}, \tag{2.4}$$

where $m' = \max_k \left( \prod_{\substack{l=1 \\ l \neq k}}^{K} \left( \max(|y_k - y_l|, \sigma_k) \right)^2 \right)$.

If $y_{1:K} = \{y_1, y_2, \ldots, y_K\}$, then the *a posteriori* pdf $p(\theta|y_{1:K}, I)$ is given by

$$p(\theta|y_{1:K}, I) = \frac{\pi(\theta|I)}{p(y_{1:K}|I)} \prod_{k=1}^{K} \pi(\Lambda_k = \lambda_0|I) \times \frac{p(y_k|\theta, \Lambda_k = \lambda_0, I)}{\pi(\Lambda_k = \lambda_0|\theta, y_k, I)}.$$

Substituting Eq. (2.4) into $p(\theta|y_{1:K}, I)$ gives

$$\begin{aligned} p(\theta| \ y_{1:K}, I) = & \frac{\pi(\theta|I)}{p(y_{1:K}|I)} \prod_{k=1}^{K} \frac{\pi(\Lambda_k = \lambda_0|I)}{\sigma_k \sqrt{2\pi}} \\ & \times \exp -\Big[ \frac{(\theta - y_k)^2}{2\sigma_k^2} \Big( 1 - \frac{1}{m'} \prod_{\substack{l=1 \\ l \neq k}}^{K} \left( (\max |y_k - y_l|, \sigma_k) \right)^2 \Big) \Big]. \end{aligned} \tag{2.5}$$

The following two examples illustrate using the KSG model (Eq. 2.5) to fuse a set of experimental measurements. In the first example, one of the measurements is clearly spurious.

**Example 2.9** *Fusion of Experimental Neutron Fission Cross-Section Measurements.* Table 2.1 lists a set of experimental neutron fission cross-section measurements $(y_k, \sigma_k), k \in \{1, 2, \ldots, K\}$. The corresponding Gaussian likelihoods $p(y_k|\theta, I) \sim \mathcal{N}(y_k, \sigma_k^2)$ are shown in Fig. 2.8a.

Data point #5 is clearly an outlier being eight standard deviations away from the mean of the other four points. Assuming a flat prior on the cross-section $\theta$, the corresponding KSG *a posteriori* pdf is shown in Fig. 2.8b. The result is a Gaussian curve which is very close to the product of $\mathcal{N}(\theta|y_k, \sigma_k^2)$ for $k \in \{1, 2, 3, 4\}$, which is what we might expect if we believe that point #5 is an outlier that should be discounted.

In the second example, there are no measurements which are clearly spurious but the measurements have wildly varying variances.

**Table 2.1** Experimental neutron fission cross-section measurements [8]

| Parameter | Experimental values | | | | |
|---|---|---|---|---|---|
| $y_i$ | 2.385 | 2.521 | 2.449 | 2.420 | 2.670 |
| $\sigma_i$ | 0.027 | 0.027 | 0.027 | 0.027 | 0.027 |

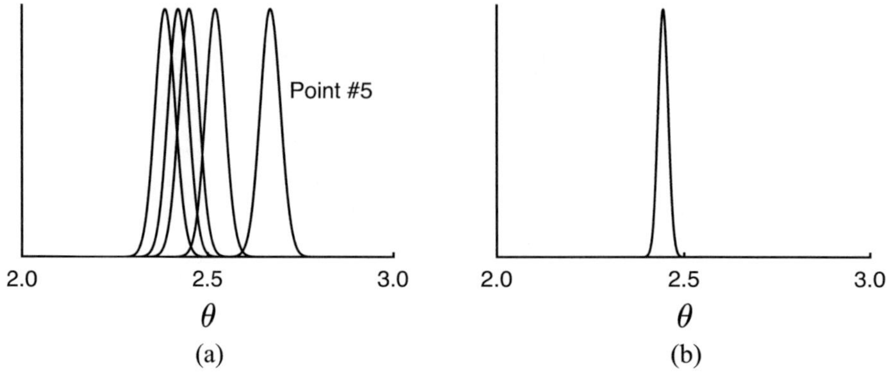


**Fig. 2.8** Shows experimental neutron fission cross-section measurements. **a** Shows the Gaussian likelihoods $\mathcal{N}(y_k, \sigma_k^2), k \in \{1, 2, \ldots, K\}$ for the cross sections $y_k$ and error bars $\sigma_k$. **b** Shows the *a posteriori* pdf $p(\theta|y_{1:K}, I)$ calculated using the KSG model

**Table 2.2** Experimental measurements for hubble's constant

| Parameter | Experimental values | | | | | | | | | | | | |
|---|---|---|---|---|---|---|---|---|---|---|---|---|---|
| $y_i$ | 49 | 51 | 55 | 61 | 89 | 68 | 67 | 80 | 82 | 76 | 81 | 73 | 80 |
| $\sigma_i$ | 10 | 13 | 25 | 19 | 10 | 22 | 10 | 25 | 18 | 19 | 17 | 10 | 12 |

The values $y_k$ and the error bars $\sigma_i$ were taken from Fig. 2 in [15]. The values are given in appropriate units

**Example 2.10** *Fusion of Experimental Hubble Constant Measurements.* Table 2.2 lists $N = 13$ experimental measurements $\mathbf{y} = (y_1, y_2, \ldots, y_N)^T$ and error bars $\boldsymbol{\sigma} = (\sigma_1, \sigma_2, \ldots, \sigma_N)^T$ of Hubble's constant $\theta$. The measurements include a wide variety of different techniques and are assumed to be independent. Assuming a flat prior on Hubble's constant $\theta$, the corresponding KSG *a posteriori* pdf is shown in Fig. 2.9. The result is a Gaussian curve which is close to the product of $\mathcal{N}(\theta|y_k, \sigma_k^2)$ for all $N = 13$ measurements.

In the above example we followed standard practice and used a Gaussian function for $p(\mathbf{y}|\boldsymbol{\theta}, I)$. However, there are applications which require a different shaped pdf. E.g. an asymmetric or non-Gaussian likelihood function.

**Example 2.11** *Konolige Model for a Time-of-Flight (ToF) Ultrasonic Sensor* [9]. Consider a ToF ultrasonic sensor (see Example 2.5). Suppose we obtain a range reading equal to $R$. Then the fault-free (i.e. $\Lambda = \lambda_0$) likelihood function is $p(r \circ R|r_0)$, where $r \circ R$ denotes that the *first* detected echo corresponds to a distance $R$ ("$r = R$") and that no return less than $R$ was received ("$r \nless R$").

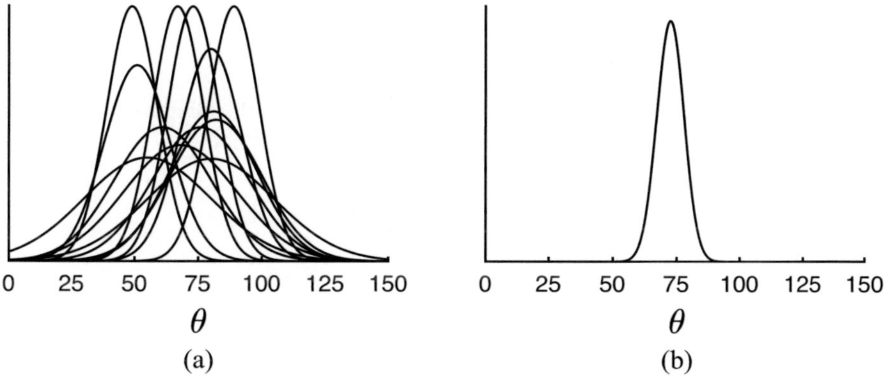


**Fig. 2.9** Shows experimental measurements of Hubble's Constant. **a** Shows the Gaussian likelihoods $\mathcal{N}(y_k, \sigma_k^2)$, $k \in \{1, 2, \ldots, K\}$ for Hubble's constant $y_k$ and error bars $\sigma_k$. **b** Shows the *a posteriori* pdf $p(\theta|y_{1:K}, I)$ calculated using the KSG model

Let $p(r = R|r_0)$ and $p(r \not< R|r_0)$ denote the conditional pdfs corresponding to $r = R$ and $r \not< R$, then the likelihood function for a ToF sensor is

$$p(r \circ R|r_0) = p(r = R|r_0) \times p(r \not< R|r_0),$$

where $p(r \not< R|r_0) = 1 - \int_0^R p(r = x|r_0)dx$. To a good approximation,

$$p(r = R|r_0) \propto \frac{1}{\sqrt{2\pi\sigma^2}} \exp\left[-\frac{1}{2}\left(\frac{r - r_0}{\sigma}\right)^2\right] + F,$$

where $F$ is a small constant which takes into account multiple targets which may reflect the sonar beam, in addition to the target at $r_0$ (see Sect. 12.3.4). In practice, we observe the range error gets larger and the probability of detection gets smaller at increasing range. Incorporating these effects into the likelihood function gives a new (Konolige) likelihood function:

$$p(r \circ R|r_0) = \gamma\left[\frac{\alpha(r)}{\sqrt{2\pi\sigma^2(r)}} \exp -\frac{1}{2}\left(\frac{r - r_0}{\sigma(r)}\right)^2 + F\right]\left(1 - \int_0^R p(r = x|r_0)dx\right),$$

where $\gamma$ is a normalization constant, $\alpha(r)$ describes the attenuation of the detection rate with increasing distance and $\sigma(r)$ describes the increase in the range variance with increasing distance.

## Problems

**2.1** Compare and contrast the Real-time Service interface, the Diagnostic and management interface and the Configuration and Planning interface.

**2.2** Measurement uncertainty may be broadly divided into random errors, systematic errors and spurious readings. Explain and define these errors.

**2.3** The following are all systematic errors: calibration errors, loading errors, environmental errors and common representational errors. Compare and contrast these errors. Give examples of each type of error.

## References

1. Brajovic V, Kanade T (2004) When are the smart vision sensors smart? an example of an illumination-adaptive image sensor. Sensor Rev 24:156–166
2. Celebi ME, Hwang S, Iyatomi H et al (2010) Robust border detection in dermoscopy images using threshold fusion. In: Proceedings of IEEE international conference on image processing, pp 2541–2544
3. Dodia A, Shah P, Sekhar R et al (2023) Smart sensors in Industry 4.0. In: 14th International conference emerging technology, pp 1–6
4. Eidson JC (2010) Measurement, control and communication using IEEE 1588. Published Springer
5. Elmenreich W (2002) Sensor fusion in time-triggered systems PhD thesis, Institut fur Technische Informatik, Technischen Universitat Wien
6. Elmenreich W, Haidinger W, Kopetz H (2001) Interface design for smart transducers. In: IEEE instrumentation and measurement technology conference (IMTC), vol 3, pp 1643–1647
7. Fowler KR, Schmalzel JL (2004) Sensors: the first stage in the measurement chain. In: IEEE instrumentation and measurement magazine, pp 60–65
8. Hanson KM (2005) Bayesian analysis of inconsistent measurements of neutron cross sections. In: AIP conference proceedings of the Bayesian inference and maximum entropy methods in science and engineering
9. Konolige K (1997) Improved occupancy grids for map building. Auton Robot 4:351–367
10. Kopetz H, Bauer G (2003) The time-triggered architecture. Proc IEEE 91:112–126
11. Kopetz H, Holzmann M, Elmenreich W (2001) A universal smart transducer interface: TTP/A. In: 3rd IEEE international symposium object-oriented real-time distributed computing
12. Kumar M, Garg DP, Zachery RA (2007) A method for the judicious fusion of inconsistent multiple sensor data. IEEE Sens J 7:723–733
13. Mukhopadhyay SC, Tyagi SKS, Suryadevara N et al (2021) Artificial intellegence-based sensors for next generation IoT applications: a review. IEEE Sens J 21:24920–24932
14. Paulitsch M (2002) Fault-tolerant clock synchronization for embedded distributed multi-cluster systems. PhD thesis, Institut fur Technische Informatik, Technischen Universitat Wien
15. Press WH (1996) Understanding data better with Baysian and global statistical methods. Unsolved problems in astrophysics. Published by Princeton University Press
16. Todini E (2001) A Bayesian technique for conditioning radar precipitation estimates to rain-gauge measurements. Hydrology Earth Sys Sci 5:187–199
17. Sehrawat D, Gill NS (2019) Smart sensors: analysis of different types of IoT sensors. In: Proceedings of the 3rd international conference trends electronics
18. Wang Z, Chiang L (2019) Monitoring chemical processes using judicious fusion of multi-rate sensor data. Sensors 19:2240

# Chapter 3
# Architecture

**Abstract** In this chapter we consider the architecture of a data fusion system. We start by considering the data fusion system and how it communicates with its environment in a fault-tolerant manner. Next we consider different fusion networks starting from the single fusion cell and moving onto serial, parallel and iterative networks. We then consider different network topologies—centralized, decentralized and hierarchical—and their relative advantages and disadvantages.

## 3.1 Introduction

In Chap. 1 we introduced a formal framework in which we represented a data fusion system as a distributed system of autonomous modules. In this chapter we shall consider the architecture of a data fusion system and, in particular, the architecture of the "data fusion block" (Sect. 1.4).

The modules in the data fusion block are commonly called *fusion nodes*. A communication system allows the transfer of information from one node to another node via an exchange of messages. An algorithmic description of the fusion block is provided by the software which is embedded in the nodes and which determines the behaviour of the block and coordinates its activities.

Although the data fusion block is represented as a distributed system we perceive it as a whole rather than as a collection of independent nodes. This phenomena is known as *transparency* [15] and is concerned with the "concealment" of nodes in a distributed system. There are many different forms of transparency which reflect on the different motivations and goals of the distributed systems. Two important forms of transparency are:

**Replication Transparency.** Here we increase the reliability and performance of the system by using multiple copies of the same sensor. However the user, or the application program, has no knowledge of the duplicate sensors [15, 17].

**Failure Transparency.** Here we conceal failures in the hardware or software components and thereby allow the users and application programs to complete their tasks. However the user, or the application program, has no knowledge of the failures [15].



H. B. Mitchell, *Data Fusion*, https://doi.org/10.1007/978-3-662-71023-4_3

The following example illustrates a *fault-tolerant framework* using replication transparency.

**Example 3.1** *Fault-Tolerant Framework* [1]. We consider the communication between a sensor node $S$ and a fusion node $F$ via an interface file system (IFS) (see Fig. 3.1). In general the environmental image produced by $F$ is sensitive to errors in the value $y$ received from $S$. We may reduce this sensitivity by using a fault-tolerant framework (Fig. 3.2): We replace the sensor $S$ by a set of $M$ redundant sensors $S_m$, $m \in \{1, 2, \ldots, M\}$. If $y_m$ denotes the $y$ value produced by $S_m$, then we insert an extra component into the fusion node $F$ in order to process the $y_m$, $m \in \{1, 2, \ldots, M\}$, and produce a new $y$ value which is less likely to be in error. Suppose $y_{\text{NEW}}$ denotes the new $y$ value, then the fusion algorithm remains unchanged: It operates on $y_{\text{NEW}}$ value and has no knowledge of the $y_m$.

The fault-tolerance processing in Fig. 3.2 can be of various kinds. For example, in a *voting algorithm* we select one observation out of the multiple input observations (see Chap. 15).

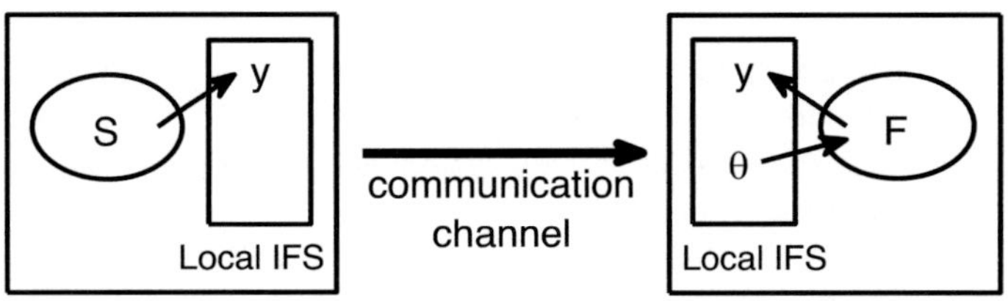


**Fig. 3.1** Shows the communication and processing of data. If the sensor $S$ provides a false value for $y$, then the fusion cell $F$ will be affected by the false value and will in turn produce a wrong value for its output $\theta$

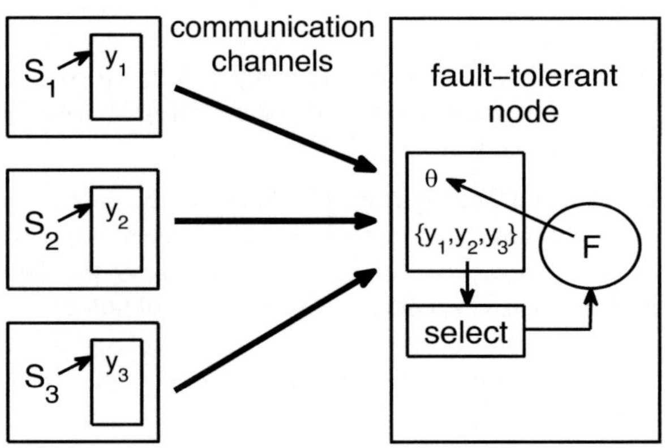


**Fig. 3.2** Shows fault-tolerant communication and processing of data. As long as not more than one of the sensor nodes $S_m$, $m \in \{1, 2, 3\}$, provides a false value for $y$, then the fusion cell $F$ will *not* be affected by the false value and will in turn produce a correct value for its output $\theta$

## 3.2 Fusion Node

We model the data fusion block as a distributed network of autonomous fusion cells or nodes. Each node receives one, or more, input *sensor observations*

$$O_i = \langle E_i, \mathbf{x}_i, t_i, \mathbf{y}_i, \Delta\mathbf{y}_i \rangle, \qquad i \in \{1, 2, \ldots\}. \tag{3.1}$$

If necessary, the observations and any a priori knowledge, are converted to a common representational format, which are then fused together to produce new output observations. The output observation may, or may not, differ in position ($\mathbf{x}$), time ($t$), value ($\mathbf{y}$) and uncertainty ($\Delta\mathbf{y}$) from the input observation(s). However, the output observations are *always* assigned new entity names.

**Example 3.2** *Image Smoothing.* We reconsider Example 2.6 and assume three input images $I_i, i \in \{1, 2, 3\}$, each characterized by an array of pixel gray-levels $\mathbf{G}_i$. The corresponding (censored) observations are $O_i = \langle I_i, \mathbf{X}_i, t_i, \mathbf{G}_i, * \rangle$. The fusion cell $F$ performs pixel-by-pixel smoothing and generates an output $\widetilde{O}$, where $\widetilde{O}$ is a *new* image (entity name $\widetilde{I}$) characterized by an array of pixels with relative positions $\widetilde{\mathbf{X}}$ and gray-levels $\widetilde{\mathbf{G}}$:

$$\widetilde{O} = \langle \widetilde{I}, \widetilde{\mathbf{X}}, \widetilde{t}, \widetilde{\mathbf{G}}, * \rangle. \tag{3.2}$$

Figure 3.3 shows a graphical representation of a single fusion cell or node [20]. The node may receive three distinct types of input data or information:

**Sensor Information.** This includes data which comes *directly* from the sensors $S_m, m \in \{1, 2, \ldots, M\}$, as well as input which has been produced by other fusion cells. Ordinarily, this constitutes most of the input data into the fusion cell.

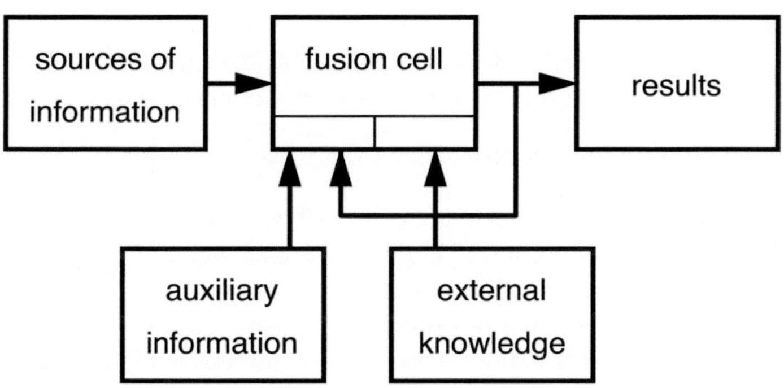


**Fig. 3.3** Shows a single fusion node and its three types of input data or information: sources of information, auxiliary information and a priori external knowledge and its output results. In the figure we show feedback of the results as auxiliary information

**Auxiliary Information.** This includes additional data which is derived by specific processing of the sensor information.

**External Knowledge.** This includes additional data and consists of all the elements of the a priori, or external, knowledge.

**Example 3.3** *Sensor Information Versus Auxiliary Information.* Consider a fusion cell $F$ which receives information directly from a physical sensor, $S$, and from a source-of-information $A$ in the form of data $R$. If $R$ is of the same "type" as $S$, or if the data rate of $R$ is similar to that of $S$, we classify $A$ as a logical sensor. Otherwise we classify $R$ as auxiliary information and $A$ as specific *algorithm* which generates $R$.

### Properties

The chief properties [5] of the fusion node are:

**Common Structure.** Each node has the same structure. This allows designers to easily implement and modify all nodes.

**Completeness.** The node possesses all the information that it requires to perform its tasks. Some of this information, such as the sensor observations, is received through a communications channel.

**Relevance.** Each node only has information relevant to the functions it is required to perform.

**Uncertainty.** Observation error and noise mean that the information in a node is uncertain to a greater, or lesser, extent. To ensure a reasonable degree of operational reliability, we must take this uncertainty into account when processing the information in a node.

**Communication.** Individual nodes are only able to communicate with other nodes that are directly connected to it. A node is not able to send a message to, or receive a message from, another node if it is not directly connected to it.

Other properties which are sometimes important are

**Self-Identification.** This refers to the ability of the node to automatically identify itself. This facility is important in large networks containing very many nodes.

**Self-Testing.** Self-testing is the process of inspection and verification which ensures the node is functioning properly. In case of a malfunction the node is able to inform the network about its condition and thus avoid corrupting the system with wrong data.

## 3.3 Simple Fusion Networks

Figure 3.3 is a graphical representation of the single fusion node or cell. It is the basic building block for all data fusion networks. We illustrate its versatility by using it to construct four different fusion networks.

### 3.3.1 Single Fusion Cell

The simplest data fusion network consists of a single fusion cell. This cell may be used to fuse together raw sensor measurements. In the network we have the option of introducing external knowledge, in which case, the network represents a *supervised* process.

**Example 3.4** *Exploiting camera meta-data for scene classification* [2]. Semantic scene classification based only on low-level vision cues (e.g. color, texture, etc.) has had limited success. To improve the classification performance we may use camera meta-data as follows:

**Scene brightness.** This includes exposure time, aperture, f-number, exposure time and shutter speed. Outdoor scenes tend to be brighter than indoor scenes, even under outcast skies. They therefore tend to have shorter exposure times and smaller aperture.

**Flash.** Because of the above lighting differences, camera flash is used on a much higher percentage of images of indoor scenes than of outdoor scenes.

**Subject Distance.** Most outdoor scenes and landscape images have a large subject distance.

**Focal length.** Each frame of a video zoom-in is more likely to be used for distant outdoor objects while zoom-out is more likely to be used for long-distance, outdoor scenes.

The above meta-data is auxiliary data. To exploit the meta-data, the fusion cell also requires external data in the form of statistical tables.

#### Single Deep-Learning Fusion Cell*

Increasingly, fusion cells $F$ are implemented using a deep-learning network. The following example illustrates a single deep-learning fusion cell used for video classification.

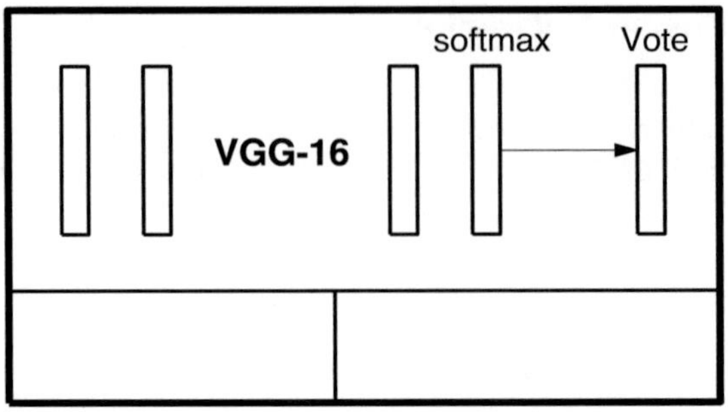


**Fig. 3.4** Shows video classification using single fusion cell consisting of a VGGNet-16 CNN and a vote counter. Input to the cell is a series of video frames. For each frame $i$, output from softmax layer (last layer in VGGNet-16) is a vector $\mathbf{s}_i = (s_i(1), s_i(2), \ldots, s_i(K))^T$, where $s_i(k)$ is the a posteriori probability (score) that frame $i$ belongs to class $c_k$, $k \in \{1, 2, \ldots, K\}$. The vectors $\mathbf{s}_i$ are fed to a vote counter which outputs the classification of the *entire* video: $c^* = c_{k^*}$, where $c_{k^*}$ is the class with the highest vote

**Example 3.5** *Video Classification Using a Single Deep-Learning Fusion Cell* [8]. Video classification requires analyzing the content of videos. The classification is widely used by suppliers of videos to manage advertisements and video contents and to suggest interest-based videos to consumers. Figure 3.4 shows the internal structure of single fusion cell which consists of a pre-trained VGGNet-16 convolutional neural network (CNN) and a simple vote counter.

For each frame $i$ of a video the softmax layer of the network generates a score $s_i(k)$ for each possible video class $k$, $k \in \{1, 2, \ldots, K\}$. The classification of the entire video is $c^* = c_{k^*}$, where $k^*$ is the index with the maximum voter score:

$$k^* = \max_k \Big( \sum_{i=1} V(i, k) \Big),$$

and

$$V(i, k) = \begin{cases} 1 & \text{if } \; s_i(k) = \max(s_i(1), s_i(2), \ldots, s_i(K)) \\ 0 & \text{otherwise.} \end{cases}$$

### 3.3.2 *Parallel Network*

After the single fusion cell is the parallel network which represents the next higher level of complexity. In the parallel network (Fig. 3.5a) each fusion cell $F_m$, $m \in \{1, 2, \ldots, M\}$, separately processes the input data it receives from the sensor $S_m$ and delivers its results $R_m$ to a fusion process $\widetilde{F}$ which generates the final result $\widetilde{R}$.

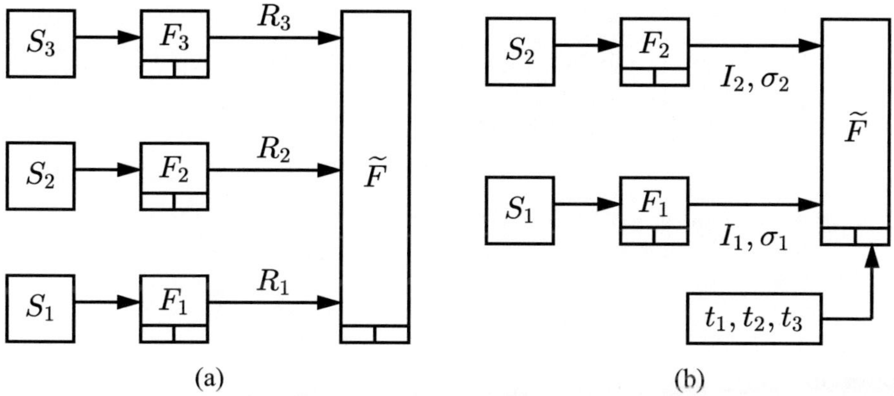


**Fig. 3.5** Shows a parallel network. **a** Shows a generic network consisting of three fusion cells $F_m, m \in \{1, 2, 3\}$. Each cell $F_m$ acts as a virtual sensor whose output is $R_m$. The $R_m, m \in \{1, 2, 3\}$, are then fused together by $\widetilde{F}$. **b** Shows the parallel architecture of the tire pressure monitoring system. Sensor $S_1$ sends resonance frequency measurements to $F_1$ which outputs variables $(I_1, \sigma_1)$ and sensor $S_2$ sends wheel radii measurements to $F_2$ which outputs variables $(I_2, \sigma_2)$. The cell $\widetilde{F}$ receives $(I_1, \sigma_1)$ and $(I_2, \sigma_2)$ and generates intermediate variables $(I_3, \sigma_3)$. It outputs a warning if $\max\big(I_1(\sigma_1 - t_1), I_2(\sigma_2 - t_2), I_3(\sigma_3 - t_3)\big) > 0$. The thresholds $t_1, t_2, t_3$ (regarded as "external knowledge") are calculated using a set of training samples $\mathcal{D}$

Typically, the intermediate result $R_m, m \in \{1, 2, \ldots, M\}$, are *redundant* and the fusion process $\widetilde{F}$ is used to increase reliability and accuracy of the final result $\widetilde{R}$.

**Example 3.6** *Tire Pressure Monitoring Device* [16, 21]. Maintaining the correct tire pressure for road vehicles is very important since incorrect pressure may adversely affect vehicle handling, fuel consumption and tire lifetime. The simplest way to monitor the tire pressure is to directly monitor the pressure in each tire using a separate pressure sensor mounted at the rim of each tire. However, this is very expensive and instead we measure the tire pressure *indirectly* using existing sensors. Two sensors used for this purpose are:

**Vibration Sensor $\mathbf{S_1}$.** This sensor relies on the fact that the tire acts like a spring when excited by road roughness. The idea is to monitor the resonance frequency which in turn is correlated with the tire pressure.

**Wheel Radius Sensor $\mathbf{S_2}$.** This sensor relies on the fact that the tire pressure affects the rolling radius of the tire. The idea is to monitor the difference in wheel speeds, which should be close to zero when the tires are equally large.

Each fusion cell $F_m, m \in \{1, 2\}$, carries information about the current tire using a pair of variables $(I_m, \sigma_m)$, where $I_m = 1$ indicates under-inflation; otherwise $I_m = 0$ and $\sigma_m$ is the expected confidence in $I_m$.

If the two pairs of variables are independent, then we issue a warning when:

$$\text{Warning} \quad \text{if } \max\big((I_1(\sigma_1 - t_1), I_2(\sigma_2 - t_2)\big) > 0,$$

where $t_1$ and $t_2$ are given thresholds.

We may enhance the robustness and sensitivity of the system by *fusing* $(I_1, \sigma_1)$, $(I_2, \sigma_2)$ and $((I_1 + I_2)/2, (\sigma_1 + \sigma_2)/2)$ in a simple voting scheme (Fig. 3.5b). In this case, a warning is issued when

$$\text{Warning} \quad \text{if } \max\big(I_1(\sigma_1 - t_1), I_2(\sigma_2 - t_2), I_3(\sigma_3 - t_3)\big) > 0,$$

where $I_3 = (I_1 + I_2)/2$, $\sigma_3 = (\sigma_1 + \sigma_2)/2$ and $t_3 < \min(t_1, t_2)$.

**Deep-Learning Parallel Network***

The following example illustrates a general deep-learning parallel network for image fusion. This network has been used to fuse various types of *spatially-aligned* images including multi-focus, infrared-visual, multi-modal medical and multi-exposure images.

**Example 3.7** *General Deep Learning Image Fusion Block* [22]. Figure 3.6 shows a general purpose deep-learning CNN image fusion block which is used to fuse together two spatially-aligned input images $I'$ and $I''$. The early convolutional layers (CNV1 and CNV2) are arranged in two parallel branches each of which is used to extract salient image features from a different input image. If $F'_2$, $F''_2$ denote the features extracted from $I'$ and $I''$, then $F'_2$, $F''_2$ are fused together *pixel-by-pixel* (see Sect. 5.5). The fused features $\widetilde{F}_3$ are then reconstructed by convolutional layers CNV3 and CNV4. Output of CNV4 is the fused image $\widetilde{I}$. The model in Fig. 3.6 is fully convolutional and may be trained end-to-end.

## 3.3.3 Serial Network

The serial network (Fig. 3.7) contains $M$ fusion cells, each cell $F_m, m \in \{1, 2, \ldots, M-1\}$, receives one, or more, observations from the sensor $S_m$ and it sends its results $R_m$ to the next cell $F_{m+1}$.

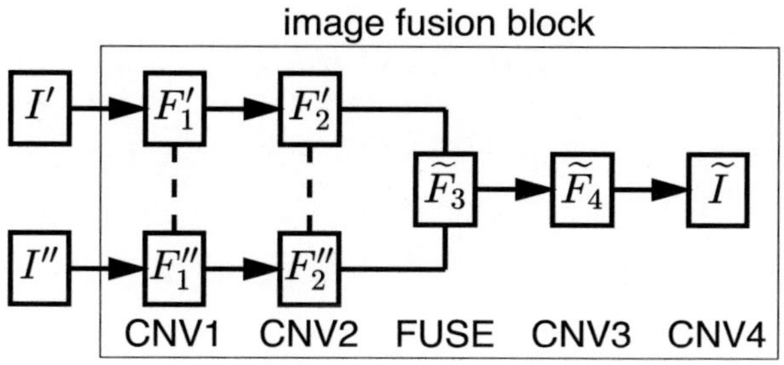


**Fig. 3.6** Shows a general deep-learning image fusion framework. Input to the network are two $H \times W \times 3$ *spatially-aligned* color (RGB) images $I'$ and $I''$ (see Chap. 6). Feature extraction is performed by the (tied) convolutional layers CNV1 and CNV2. After feature extraction we have two feature maps $F_2'$, $F_2''$, which are fused together to give a fused feature map $\tilde{F}_3$. After further convolution by CNV3 and CNV4, we obtain the fused output image $\widetilde{I}$ of size $H \times W \times 3$.

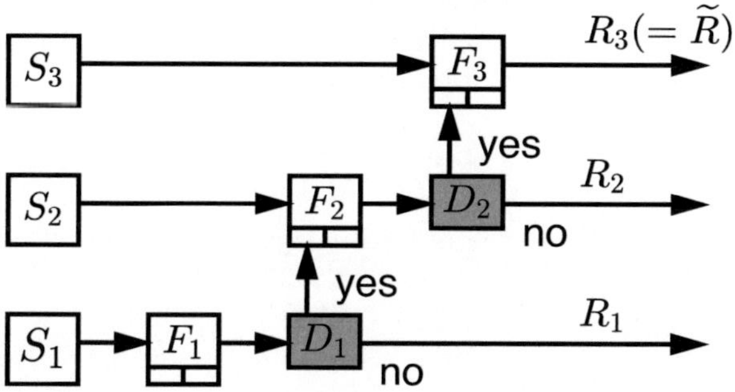


**Fig. 3.7** Shows the serial arrangement of three fusion cells $F_1$, $F_2$ and $F_3$ configured as a cascade network. The network includes two decision boxes $D_m$, $m \in \{1, 2\}$ whose purpose is to determine whether an additional sensor $S_{m+1}$ is required: If $D_1$ = no, only sensor $S_1$ is used; if $D_1$ = yes and $D_2$ = no, then sensors $S_1$ and $S_2$ are used and if $D_1$ = yes and $D_2$ = yes, all three sensors $S_1$, $S_2$, $S_3$ are used

### Cascaded Network

The serial network is often configured as a cascade network. In this case, we add decision modules $D_m$, $m \in \{1, 2, \ldots, M-1\}$, to the network. The purpose of each decision node $D_m$ is to determine whether we can rely on the intermediate result $R_m$. If yes we stop processing and output $R_m$; otherwise we continue processing and add the next sensor $S_{m+1}$ (see Fig. 3.7). Cascaded networks are often used in identification schemes as described in the following example.

**Example 3.8** *Multi-modal Biometric Identification Scheme* [14]. Figure 3.7 shows a multi-modal biometric identification scheme using cascaded architecture. If after processing the first $m$ biometric features, we have sufficient confidence concerning the identity of the user, we stop the process and the user is not required to provide any more biometric features.

### *3.3.4 Iterative Network*

Figure 3.8 represents an iterative network in which we re-introduce the result $R$ as auxiliary knowledge into the fusion cell $F$. This network is generally used in applications which possess dynamical data sources. In that case, $S$ consists of data changing through time. This scheme integrates many applications such as temporal tracking of objects in an image sequence, or the navigation of a mobile robot. $F$ could be, for example, a Kalman filter (KF) (see Chap. 13).

**Example 3.9** *Heart-Rate Estimation Using a Kalman Filter (KF)* [18]. Analysis of heart rate variability requires the calculation of the mean heart rate which we adaptively estimate using a KF (see Chap. 13). Let $\mu_{k|k}$ denote the estimated mean heart rate at time step $k$ given the sequence of measured heart-rates $y_{1:k} = (y_1, y_2, \ldots, y_k)$. Then we assume $\mu_{k|k}$ follows a random walk process:

$$\mu_{k|k} = \mu_{k-1|k-1} + v_{k-1},$$

where $v_{k-1} \sim \mathcal{N}(0, Q)$ represents the process noise at time step $k-1$. We suppose the measured heart-rate at time step $k$ is $y_k$:

$$y_k = \mu_k + w_k,$$

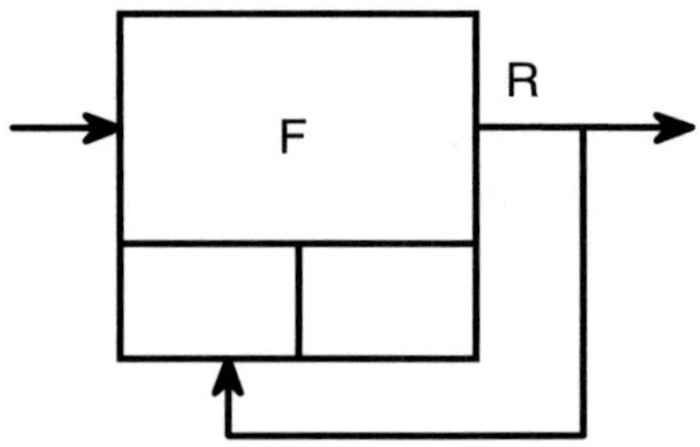


**Fig. 3.8** Shows single fusion cell operating in an iterative mode. The result $R$ is re-introduced as auxiliary knowledge into $F$

where $w_k \sim \mathcal{N}(0, R)$ represents the measurement noise. The KF estimates $\mu_{k|k}$ in a predictor/corrector algorithm: In the prediction phase we project the filtered heart rate $\mu_{k-1|k-1}$ and its covariance matrix $\Sigma_{k-1|k-1}$ to the next time step $k$. Let $\mu_{k|k-1}$ and $\Sigma_{k|k-1}$ be the corresponding *predicted* values, then

$$\mu_{k|k-1} = \mu_{k-1|k-1} \quad \text{and} \quad \Sigma_{k|k-1} = Q + \Sigma_{k-1|k-1}.$$

In the correction phase we fuse the predicted values with the measurement $y_k$ made at the $k$th time step:

$$\mu_{k|k} = \mu_{k|k-1} + K_k(y_k - \mu_{k|k-1}) \quad \text{and} \quad \Sigma_{k|k} = (I - K_k)\Sigma_{k|k-1},$$

where $I$ is a unit matrix and $K_k = \Sigma_{k|k-1}(\Sigma_{k|k-1} + R)^{-1}$.

## 3.4 Network Topology

Fundamentally the fusion nodes may be arranged in three different ways (topolgies), or architectures: *centralized*, *decentralized* and *hierarchical*. In Table 3.1 we list several multi-sensor data fusion applications and their architectures. We now briefly consider each of these architectures in turn.

**Table 3.1** Data fusion applications: networks and architectures

| Network | | Application |
|---|---|---|
| Single (C) | Example 3.10 | Satellite imaging of the earth |
| Single (C) | Example 11.6 | Nonintrusive speech quality estimation using GMM's |
| Single (C) | Example 13.1 | INS/Radar altimeter: a hybrid navigation principle |
| Serial (H) | Example 3.8 | Multi-modal biometric identification scheme |
| Iterative (C) | Example 3.9 | Heart-rate estimation using a Kalman filter |
| Parallel (D) | Example 3.11 | Covariance intersection in a decentralized network |
| Parallel/Serial (H) | Example 3.6 | Tire pressure monitoring device |
| Parallel/Serial (H) | Example 3.14 | Data incest |
| Parallel/Serial (H) | Example 3.15 | Audio-visual speech recognition system |
| Parallel/Serial (H) | Example 9.19 | Fusing similarity coefficients in virtual screening |

C, D, H refer, respectively, to centralized, decentralized, hierarchical network topologies

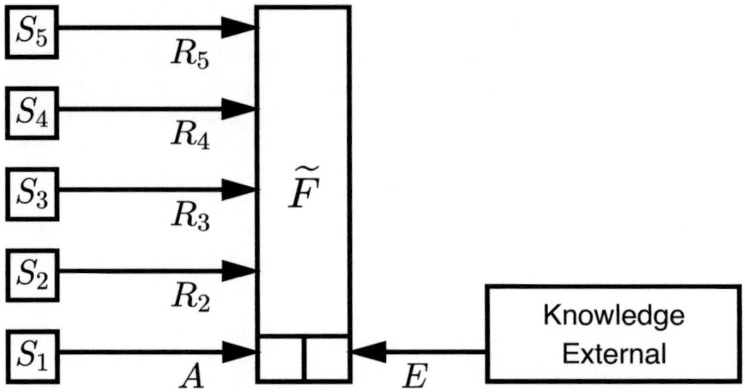


**Fig. 3.9** Shows the operation of a centralized node $\widetilde{F}$. Input to $\widetilde{F}$ are produced by sources-of-information, $S_m, m \in \{1, 2, \ldots, 5\}$, where $S_2 - S_5$ supply their information $R_2 - R_5$ directly to $\widetilde{F}$ while $S_1$ supplies its information $A$ in the form of auxiliary information. Additionally $\widetilde{F}$ receives information $E$ in the form of external knowledge

### 3.4.1 Centralized

In a centralized system, the sensor fusion unit is treated as a central processor or a node that collects *all* information from the different sources of information $S_m, m \in \{1, 2, \ldots, M\}$. Note that some of this information may be supplied in the form "auxiliary" knowledge. In addition the node may receive information in the form of "external knowledge" (see Fig. 3.9). All decisions are made at this node and instructions, or task assignments, are given out to the respective sensors.

Theoretically, a single fusion node has the best performance if all the sensors are accurately aligned.[1] However, small errors in the correct alignment of the sensors may cause substantial reduction in the performance of the fusion algorithm. In addition the centralized architecture suffers from the following drawbacks:

**Communication.** All sensors communicate directly with the centralized fusion node. This may cause a communication bottleneck.

**Inflexibility.** If the application changes, the centralized system may not easily adapt. For example, the central processor might not be powerful enough to handle the new application.

**Vulnerability.** If there is a failure in the central processor, or the central communications facility, the entire system fails.

**Non-Modularity.** If more sensors are added to a centralized system, some reprogramming is generally required: The communications bandwidth and computer power would need to increase (assuming a fixed total processing time constraint) in proportion to the amount of extra data.

[1] Technically a centralized architecture requires a single common representational format in which all sensors are accurately located. See Chap. 4.

**Example 3.10** *Satellite Imaging of the Earth* [20]. We consider the mapping of the earth from five different satellite images: $S_1$ = Texture map, $S_2$ = MSI (multi-spectral image), $S_3$ = Radar image, $S_4$ = GEO (geographical map) and $S_5$ = HRI (high resolution image) (Fig. 3.9). The texture information is supplied as *auxiliary information* and is itself derived from the high-resolution image. In this example, $\widetilde{F}$ is a classifier which uses pre-calculated classification codes supplied as external knowledge.

### *3.4.2 Decentralized*

In a decentralized system, the sensor measurements, or information, is fused locally using a set of local fusion nodes rather than by using a single central fusion node. The main advantage of a decentralized architecture is the lack of sensitivity regarding the correct alignment of the sensors. Other advantages of a decentralized architecture are:

**Communication.** The architecture is relatively immune to communication and signal processing bottlenecks. This is because the information flow between nodes is reduced, and the data processing load is distributed.

**Scalable.** The decentralized architecture is scalable in structure without being constrained by centralized computational bottleneck or communication bandwidth limitations.

**Robustness.** The decentralized architecture is robust against loss of sensing nodes and any changes in the network.

**Modular Design.** The decentralized architecture is modular in the design and implementation of the fusion nodes.

However, the decentralized network suffers from effects of *redundant* information [19]. For example, suppose a node $A$ receives a piece of information from a node $B$, the topology of the network may be such that $B$ is passing along the information it originally received from $A$. Thus if $A$ were to fuse the "new" information from $B$ with the "old" information it already has under the assumption of independence, then it would *wrongly* decrease the uncertainty in the fused information.

Theoretically we could solve the problem of redundant information by keeping track of all the information in the network. This is not, however, practical. Instead, fusion algorithms have been developed which are robust against redundant information. One such algorithm is the method of *covariance intersection* (CI) [3, 7].

**Example 3.11** *Covariance Intersection (CI) in a Decentralized Network* [3, 6]. We consider a decentralized network in which a given node $F$ receives the *same* piece of information from $m$ different sources of information $S_m, m \in \{1, 2, \ldots, M\}$. We suppose the piece of information sent by $S_m$ is a scalar measurement $y_m$ with an uncertainty $\sigma_m^2$. If we regard the measurements and uncertainties as independent, then the corresponding fused maximum likelihood (ML) measurement value and uncertainty (see Chap. 10) are given by:

$$\widetilde{y}_{\text{ML}} = \sum_m \alpha_m^2 y_m \Big/ \sum_m \alpha_m^2 \quad \text{and} \quad \widetilde{\sigma}_{\text{ML}}^2 = 1 \Big/ \sum_m \alpha_m^2,$$

where $\alpha_m = 1/\sigma_m$. Since the sources $S_m$ all supply the same piece of information, then

$$y_1 = y = y_2 = \cdots = y_M \quad \text{and} \quad \sigma_1 = \sigma = \sigma_2 = \cdots = \sigma_M,$$

and thus

$$\widetilde{y}_{\text{ML}} = y \quad \text{and} \quad \widetilde{\sigma}_{\text{ML}}^2 = \sigma^2/M.$$

The $1/M$ reduction in uncertainty is clearly incorrect: it occurs because the node $F$ assumes that each $(y_m, \sigma_m^2), m \in \{1, 2, \ldots, M\}$, represents an *independent* piece of information.

In the method of CI we solve this problem by fusing the measurements $y_m$, using a *weighted* ML estimate. In this case, the corresponding fused estimate and uncertainty are

$$\widetilde{y}_{\text{CI}} = \sum_m \Omega_m \alpha_m^2 y_m / \sum_m \Omega_m \alpha_m^2 \quad \text{and} \quad \widetilde{\sigma}_{\text{CI}}^2 = 1 \Big/ \sum_m \Omega_m \alpha_m^2.$$

where the weights $\Omega_m$, are chosen to minimize $\widetilde{\sigma}_{\text{CI}}^2$ subject to the constraints: $\Omega_m \geq 0$ and $\sum_{m=1}^{M} \Omega_m = 1$ [6]. If we assume all $\Omega_m$ are equal to $1/M$, then the CI estimates are:

$$\widetilde{y}_{\text{CI}} = y \quad \text{and} \quad \widetilde{\sigma}_{\text{CI}}^2 = \sigma^2.$$

We observe that by using CI we no longer obtain a reduction in uncertainty.

In the general case we have $M$ multi-dimensional measurements $\mathbf{y}_m, m \in \{1, 2, \ldots, M\}$, with covariance matrices $\mathbf{\Sigma}_m$. The CI estimate is $(\widetilde{\mathbf{y}}_{\text{CI}}, \widetilde{\mathbf{\Sigma}}_{\text{CI}})$, where

$$\widetilde{\mathbf{y}}_{\text{CI}} = \sum_m \Omega_m \widetilde{\mathbf{\Sigma}}_{\text{CI}} \mathbf{\Sigma}_m^{-1} \mathbf{y}_m \quad \text{and} \quad \widetilde{\mathbf{\Sigma}}_{\text{CI}}^{-1} = \sum_m \Omega_m \mathbf{\Sigma}_m^{-1}. \tag{3.3}$$

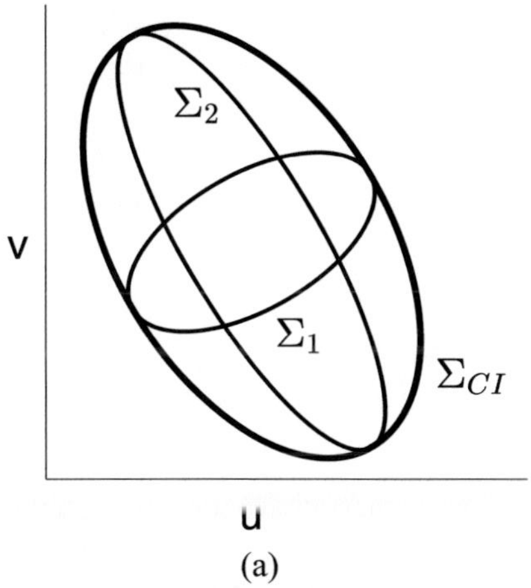


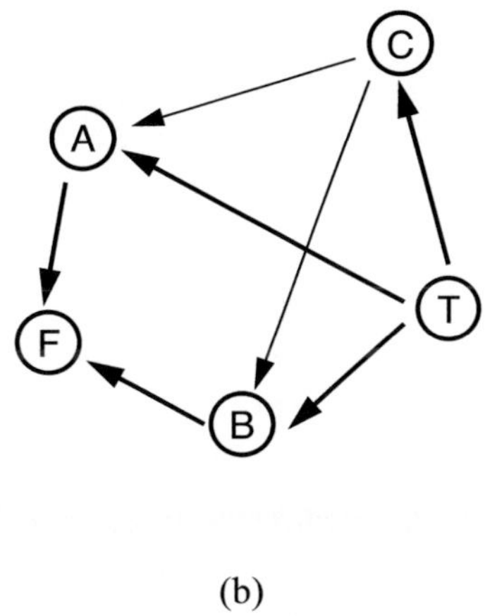


**Fig. 3.10** Shows problems arising from redundant information. **a** Shows the covariances $\mathbf{\Sigma}_1$ and $\mathbf{\Sigma}_2$, of two coincident sensor observations $O_1 = (\mathbf{y}_1, \mathbf{\Sigma}_1)$ and $O_2 = (\mathbf{y}_2, \mathbf{\Sigma}_2)$ in 2D. The covariance of the CI, $\widetilde{\mathbf{\Sigma}}_{\text{CI}}$, is the ellipse drawn with a thick solid line. **b** Illustrates data incest when sensor $C$ successfully sends information regarding target $T$ to $A$ and $B$ (via communication channels $C \rightarrow A$ and $C \rightarrow B$) which in turn forward the information onto fusion cell $F$. Unreliable communication channels are drawn with thin arrows

**Example 3.12** *Blood Pressure Measurements: Fusion Using CI* [11] Non-invasive blood pressure (BP) measurement relies on monitoring physiological signals such as electrocardiology ECG. The main drawback to using these measurements are their large measurement uncertainty. To reduce the uncertainty, we use CI to fuse several BP estimates $(y_m, \Sigma_m), k \in \{1, 2, \ldots, K\}$. The fused estimate $(\widetilde{y}_{\text{CI}}, \widetilde{\Sigma}_{\text{CI}})$ is then given by Eq. (3.3).

**Example 3.13** *CI of Two 2D Observations* [7]. Figure 3.10a shows two 2D observations $O_1 = (\mathbf{y}_1, \mathbf{\Sigma}_1)$ and $O_2 = (\mathbf{y}_2, \mathbf{\Sigma}_2)$ and their CI $\widetilde{O}_{\text{CI}} = (\widetilde{\mathbf{y}}_{\text{CI}}, \widetilde{\mathbf{\Sigma}}_{\text{CI}})$.

### *3.4.3 Hierarchical*

This design can be thought of as a hybrid architecture in which we mix together the centralized and decentralized systems. The hierarchical architecture combines the advantages of the centralized and decentralized architectures without some of their disadvantages. For example, the performance of the hierarchical architecture is

**Table 3.2** A Posteriori pdf at $F$

| $C \rightarrow A$ | $C \rightarrow B$ | A Posteriori pdf |
|---|---|---|
| No | No | $p(T\|A, B)$ |
| Yes | No | $p(T\|A, B, C)$ |
| No | Yes | $p(T\|A, B, C)$ |
| Yes | Yes | $p(T\|A, B, C, C)$ |

relatively insensitive to any errors in the correct alignment of the sensors. However, the hierarchical network may still suffer from the effects of redundant information, or *data incest*, as the following example shows.

**Example 3.14** *Data Incest* [12]. Three mobile sensors $A, B, C$ send information concerning a target $T$ to a fusion cell $F$ (Fig. 3.10b). $C$ is always out of range of $F$ and it can only relay its information to $F$ via $A$ or $B$. However $C$ is not always in range of $A$ and $B$. To increase the chances of its information reaching $F$, $C$ broadcasts its information to *both* $A$ and $B$. Assuming $F$ fuses the pieces of information it receives as if they were independent, then data incest occurs when the information from $C$ reaches both $A$ and $B$ and is used twice over by $F$. Table 3.2 shows the a posteriori probability density function (pdf) that $F$ estimates assuming $C$ transmits its information to $A$ and $B$ via channels $C \rightarrow A$) and $C \rightarrow B$. Data incest occurs when both $C \rightarrow A$ and $C \rightarrow B$ are successful and the corresponding a posteriori pdf is $p(T|A, B, C, C)$.

In the hierarchical architecture there are often several levels where the top level contains a single centralized fusion node and last level is made up of several decentralized (local) fusion nodes. Each of local fusion node may receive input from a small group of sensors or from an individual sensor. The following example illustrate the implementation of an audio-visual speech recognition system using a hierarchical architecture.

**Example 3.15** *Audio-Visual Speech Recognition System* [13]. Figure 3.11 shows the implementation of the audio-visual speech recognition system using a hierarchical architecture (see Example 1.12).

### Hierarchical Deep-Learning Fusion*

The following example illustrates feature/decision fusion using a deep learning hierarchical architecture.

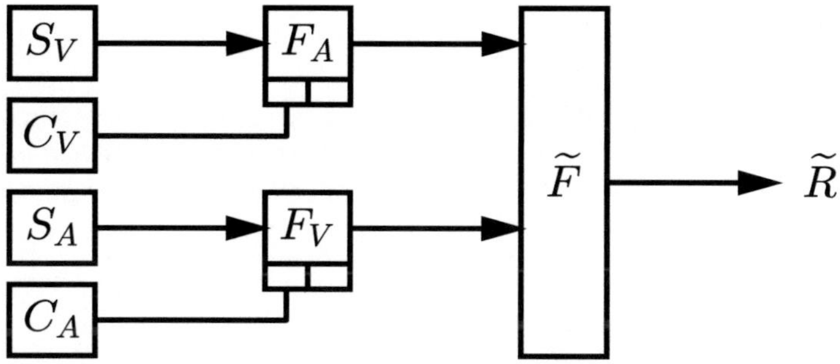


**Fig. 3.11** Shows the hierarchical architecture for an audio-visual speech recognition system. The network consists of a central fusion node $\widetilde{F}$ which receives input from audio fusion node $F_A$ and visual fusion node $F_V$. These nodes receive, *raw sensor data* from the audio and visual sensors $S_A$ and $S_V$ and *auxiliary data* from secondary classifiers $C_A$ and $C_V$

**Example 3.16** *Video Classification Using Hierarchical Deep-Learning Feature Fusion* [9]. Video classification described in Example 3.5 used a single deep-learning fusion cell (VGGNet-16 network). In a later work, the authors considered an enhanced video classification scheme using hierarchical deep-learning CNN. In the network the input video is pre-processed into three separate inputs:

**RGB Frames.** Set of individual RGB frames.
**Optical Flows.** Calculated optical flows (see Sect. 6.6) between pairs of video frames.
**Histogram of Oriented Gradients (HOG).** HOG (see Sect. 4.4.2) calculated for each video frame.

Each of these inputs are separately processed using a VGGNet-16 network. For each video frame $i$ the outputs from the VGGNet-16 networks are fused together using a multi-level feature/decision architecture (see Fig. 3.12).

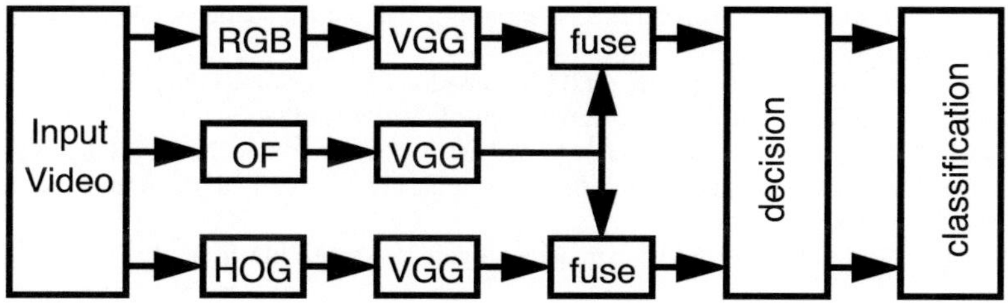


**Fig. 3.12** Shows the hierarchical architecture for deep learning video classification. The input video is decomposed into three separate inputs which are separately processed using a VGGNet-16 network. The outputs from the VGGNet-16 networks are then fused together in a multi-level feature/decision fusion architecture

## 3.5 Further Reading

General references on the different architectures used in multi-sensor data fusion are: [4, 10]. Covariance intersection is now widely used in decentralized and hierarchical systems. A modern and comprehensive review of covariance intersection is [6]. See also [7, 19].

## Problems

**3.1** Compare and contrast (a) Replication transparency and (b) Failure transparency in a decentralized multi-sensor data fusion network.

**3.2** List the relative advantages and disadvantages of the centralized, decentralized and hierarchical architectures.

**3.3** Give an example of a data fusion system which uses (a) A parallel network and (b) A serial network. What are the advantages/disadvantages of the two networks.

**3.4** Give an example of a data fusion system which uses an iterative network.

**3.5** Define data incest. Explain why it does not affect data fusion systems with centralized architectures.

**3.6** Define "covariance intersection". How does it "solve" problem of data incest?

**3.7** Suggest an alternative approach to prevent data incest.

## References

1. Bauer G (2001) Transparent fault tolerance in a time-triggered architecture. PhD thesis. Institut fur Technische Informatik, Technischen Universitat Wien
2. Boutell M, Luo J (2005) Beyond pixels: exploiting camera metadata for photo classification. Patt Recogn 38:935–946
3. Chen L, Arambel PO, Mehra RK (2002) Estimation under unknown correlation: covariance intersection revisited. IEEE Trans Autom Control 47:1879–1882
4. Delvai M, Eisenmann U, Elmenreich W (2003) A generic architecture for integrated smart transducers. In: Proceedings of the 13th international conference field programmable logic and applications, Lisbon, Portugal
5. Elmenreich W (2002) Sensor fusion in time-triggered systems PhD thesis, Technical University Wien
6. Forsling R, Noack B, Hendeby G (2024) A quarter century of covariance intersection: correlations still unknown? IEEE Control Sys Mag 44:81–105
7. Julier S, Uhlmann J (2001) General decentralized data fusion with covariance intersection (CI). In: Hall D, Llians J (eds) Handbook of multidimensional data fusion. CRC Press

8. Kiziltepe RS, Gan JQ, Escobar JJ (2019) Combining very deep convolutional neural networks and recurrent neural networks for video classification. LNCS 11507:811–822
9. Kiziltepe RS, Gan JQ, Escobar JJ (2024) Integration of feature and decision fusion with deep learning architectures for video classification. IEEE Access 12:19432
10. Kopetz H, Bauer G (2003) The time-triggered architecture. Proc IEEE 91:112–126
11. Landry C, Peterson SD, Arami A (2022) A fusion approach to improve accuracy and estimate uncertainty in cuffless blood pressure monitoring. Sci Rep 12:7948
12. McLaughlin S, Krishnamurthy V, Challa S (2003) Managing data incest in a distributed sensor network. In: Proceedings of the IEEE International Conference Acoustics, Speech Signal Processing, vol 5, pp V-269–V-272
13. Movellan JR, Mineiro P (1998) Robust sensor fusion: analysis and applications to audio-visual speech recognition. Mach Learn 32:85–100
14. Nandakumar K (2005) Integration of multiple cues in biometric systems. MSc thesis, Michagan State University
15. Paulitsch M (2002) Fault-tolerant clock synchronization for embedded distributed multi-cluster systems. PhD thesis, Institut fur Technische Informatik, Technischen Universitat Wien
16. Persson N, Gustafsson F, Drevo M (2002) Indirect tire pressure monitoring using sensor fusion. Proc SAE 2002
17. Poledna S (1997) Replica determinism and flexible scheduling in hard real-time dependable systems. Research Report Nr. 21/97 Nov 1997, Technishe Universitat Wien, Austria
18. Schlogl A, Fortin J, Habenbacher W, Akay M (2001) Adaptive mean and trend removal of heart rate variability using Kalman filtering. In: Proceedings of the 23rd international conference IEEE engineering in medicine and biology society
19. Uhlmann JK (2003) Covariance consistency methods for fault-tolerant distributed data fusion. Inform Fusion 4:201–215
20. Wald L (2002) Data fusion: definitions and architectures. Published by Les Presses de l'Ecole des Mines, Paris
21. Zhao J, Du J, Zhu B et al (2023) Indirect tire pressure monitoring method based on the fusion of time and frequency domain analysis. Meas: J Int Meas Conf 220:113282
22. Zhang Y, Liu Y, Sun P, Yan H, Zhao X, Zhang L (2020) IFCNN: a general image fusion framework based on convolutional neural network. Inf Fusion 54:99–118

# Chapter 4
# Common Representational Format

**Abstract** In this chapter we introduce the concept of a common representational format. This is a basic concept in all multi-sensor data fusion systems and is concerned with the conversion of all input signals to a common format. The process of converting sensor observations into a common representational format involves spatial, temporal and semantic alignments together with radiometric normalization. The chapter contains introductory discussion of these functions and includes more than fifteen extended examples of different common representational formats and their use in practical data fusion applications. The examples include the use of "off-the-shelf" deep-learning features.

## 4.1 Introduction

The subject of this chapter is the *common representational format*. Conversion of all sensor observations to a common format is a basic requirement for all data fusion systems. The reason for this is that only after conversion to a common format are the sensor observations compatible and sensor fusion may be performed. The following example, taken from the field of brain research, illustrates the concept of a common representational format.

**Example 4.1** *A Standardized Brain Atlas* [29]. In order to compare different brains and, to facilitate comparisons on a voxel-by-voxel basis, we use a standardized anatomically-based coordinate system or *brain atlas*. The idea is that, in the new coordinate system, all brains have the same orientation and size. The transformation to this coordinate system also allows us to enhance weak, or noisy, signals by averaging the transformed images. The standardized brain atlas allows us to catalogue the anatomical, metabolic, electrophysiological, and chemical architecture of different brains into the same coordinate systems. Table 4.1 lists several types of brain atlas which are in common use today.



H. B. Mitchell, *Data Fusion*, https://doi.org/10.1007/978-3-662-71023-4_4

**Table 4.1** Different brain atlases

| Name | Description |
|---|---|
| Adaptable | Brain atlas which can be individualized to anatomy of new subjects. This allows automated labeling of structures in new patients' scans |
| Probabilistic | Population-based atlas containing information on cross-subject variations in brain structure and function. The information is used to identify anomalies and label structures in new patients |
| Disease-specific | Brain atlas of patients suffering from a specific disease |
| Dynamic-4D | Atlas which incorporates probabilistic information on structural changes which take place during brain development, tumor growth or disease processes |

The common representational format may be usefully divided into two types [48]:

**Joint representation.** Here the common representational format consists of one latent space: All multimodal data is mapped into this single space or joint representation. In data fusion applications this is the most widely used format.

**Coordinated representation.** Here the common representational format consists of more than one latent space. A different latent space or representation is used for each modality. The different spaces are not independent but are constrained so that the data in one space can be compared to data in another space. An important example is Canonical Correlation Analysis (CCA) (Sect. 4.7.3).

In Eq. (2.1) we introduced the following notation for a sensor observation:

$$O = \langle E, \mathbf{x}, t, \mathbf{y}, \Delta\mathbf{y} \rangle, \tag{4.1}$$

where $E$ is the entity name which designates the physical property underlying the observation and $\mathbf{x}, t, \mathbf{y}$ and $\Delta\mathbf{y}$ are, respectively, the spatial position, the time, the value and the uncertainty of the physical property as measured by the sensor. Corresponding to the four types of information—entity, spatial, temporal and value—the process of converting the sensor observations into a common representational format involves the following four functions:

**Spatial Alignment.** Transformation of the local spatial positions $\mathbf{x}$ to a common coordinate system. The process involves geo-referencing the location and field-of-view of each sensor and is considered in Chap. 6.

**Temporal Alignment.** Transformation of the local times $t$ to a common time axis. In many applications, this is performed using a dynamic time warping (DTW) algorithm and is considered in Chap. 7.

**Semantic Alignment.** Conversion of the multiple inputs so they refer to the same objects or phenomena. In many applications the process is performed using an assignment algorithm and is considered in Chap. 8.

**Radiometric Normalization.** Calibration or normalization of the sensor values $\mathbf{y}$ and their uncertainties $\Delta\mathbf{y}$ to a common scale. The process is often performed using a statistical analysis and is considered in Chap. 9

In many data fusion applications, the construction of the common coordinate system is the primary fusion algorithm (see Table 4.2). Two examples are now given. The first example describes the fusion of $K$ direction-bearing measurements by creating a common Cartesian coordinate system (Fig. 4.1).

**Table 4.2** Applications in which construction of the common coordinate system is the primary fusion algorithm

| Class | | Application |
|---|---|---|
| Early-DaO | Example 4.1 | Standardized brain atlas |
| | Example 4.3 | Epipolar rectification |
| | Example 4.4 | Distributed surveillance system |
| | Example 4.5 | Lucky imaging |
| | Example 4.10 | Myocardial imaging |
| | Example 4.19 | Debiased cartesian coordinates |
| | Example 6.3 | Non-rigid spatial alignment in cardiac MR images |
| | Example 6.21 | Mosaic fingerprint image |
| | Example 7.1 | Cardiac MR image sequences |
| Early-FeO | Example 4.2 | Bayesian triangulation of direction-bearing measurements |
| | Example 4.11 | Green-based format for crop segmentation |
| | Example 4.13 | Off-the-shelf CNN features for iris recognition |
| | Example 4.14 | Background subtraction using local binary similarity pattern (LBSP) |
| | Example 4.16 | Facial expression recognition using hybrid HOG/CNN features |
| | Example 4.17 | Scene classification using bag-of-visual-words (BOVW) |
| | Example 4.21 | Principal Discriminant Analysis (PDA) for small-sample size problems |
| | Example 4.23 | Deep CCA: feature fusion for remote sensing classification |
| | Example 4.24 | Sparse face sketch synthesis (FSS) |
| | Example 7.2 | Visualization of the dysphagia (Disorders associated with swallowing) |

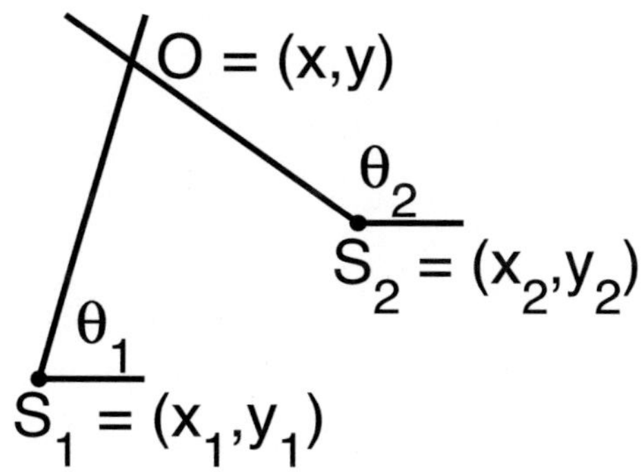


**Fig. 4.1** Shows the triangulation of two sensors $S_1$ and $S_2$. The sensors are located, respectively, and at $(X_1, Y_1)$ and $(X_2, Y_2)$, and only measure the bearings $\theta_1$ and $\theta_2$ to the object $O$. Fusion of the two sensor measurements gives an estimate $(\hat{x}, \hat{y})$ of the targets true location $(x, y)$

**Example 4.2** *Bayesian Triangulation of Direction-Bearing Measurements* [30]. Let $\phi_m, m \in \{1, 2, \ldots, M\}$, denote, respectively, the *true* bearings of the object $O$. Then

$$\phi_m = \tan^{-1} \frac{x - X_m}{y - Y_m},$$

where the *measured* bearings are distorted by random noise $w_m$:

$$\theta_m = \phi_m + w_m.$$

Assume $w_m$ are independent and identically distributed (iid) according to zero-mean Gaussian distribution: $w_m \sim \mathcal{N}(0, \sigma_m^2)$. Then, given the bearings $\phi_{1:M} = \{\phi_1, \phi_2, \ldots, \phi_M\}$, the *a posteriori* pdf is

$$p(\theta_{1:M}|\phi_{1:M}, I) = \prod_m \frac{1}{\sigma_m\sqrt{2\pi}} \exp -\left(\frac{\theta_m - \phi_m}{\sigma_m\sqrt{2}}\right)^2.$$

Use Bayes'theorem to invert this equation:

$$p(x, y|\theta_{1:M}) \sim \pi(x, y|I) \exp - \sum_m \frac{1}{2\sigma_m^2}\left(\theta_m - \tan^{-1}\frac{x - X_m}{y - Y_m}\right)^2,$$

where $\phi_m = \tan^{-1} \frac{x-X_m}{y-Y_m}$ and $\pi(x, y|I)$ is an *a priori* pdf for each $(x, y)$ given the background information $I$. The estimated location of object $O$ is given by the mean values:

$$\hat{x} = \int xp(x, y|\theta_{1:M})dxdy \quad \text{and} \quad \hat{y} = \int yp(x, y|\theta_{1:M})dxdy.$$

The second example involves epipolar rectification. This is a transformation of two stereo images to a common representational format.

**Example 4.3** *Epipolar Rectification* [16, 21]. Given two stereo images of a 3D scene, the distance (depth) of each point in the scene is found as follows: For each object in the 3D scene find *corresponding* 2D locations in each image (this is known as solving the "correspondence" problem). Then use triangulation between corresponding 2D locations to find depth of each object.

Solving the correspondence problem is a basic task in stereo vision. To simplify this task we transform the two images to a common representational format in a process known as *epipolar rectification*: Suppose the 3D scene is photographed by two pin-hole cameras $C_1$ and $C_2$ from different viewpoints.

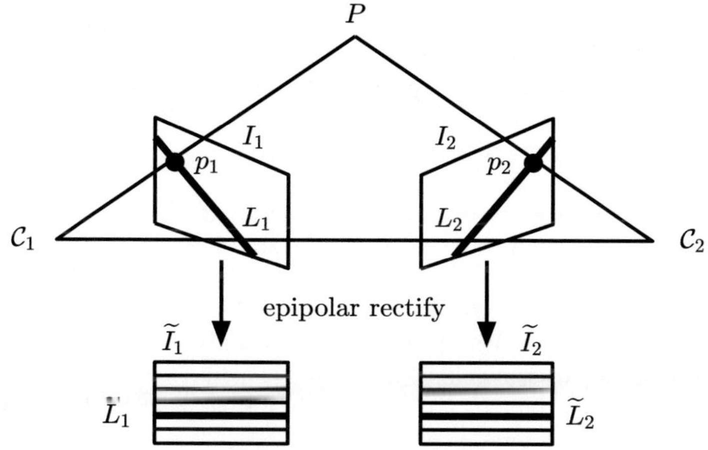


**Fig. 4.2** Shows epipolar rectification of two stereo images $I_1$ and $I_2$ captured by pinhole cameras $C_1$ and $C_2$. In the image $I_k$, $k \in \{1, 2\}$, the 3D point $P$ appears as a 2D point $p_k$ with coordinates $(i_k, j_k)$ which lies on the epipolar line $L_k$. In the corresponding rectified images $\widetilde{I}_k$, $k \in \{1, 2\}$, the points $p_k$ have coordinates $(\widetilde{i}_k, \widetilde{j}_k)$ and lie on epipolar lines $\widetilde{L}_k$, where $\widetilde{i}_1 = \widetilde{i}_2$ and $\widetilde{L}_1 = \widetilde{L}_2$

If a point $P$ in the 3D scene appears at the 2D position $p_1$ in image $I_1$, then corresponding point in $I_2$ must lie on the line $L_2$ (see Fig. 4.2). Similarly, if $P$ appears at $p_2$ in $I_2$, then corresponding point in $I_1$ must lie on the line $L_1$. The lines $L_1$ and $L_2$ are known as "epipolar lines". In epipolar rectification we separately transform $I_1$ and $I_2$ to a common representational format so that corresponding epipolar lines map onto the *same* horizontal line in the transformed images. If $\widetilde{I}_1$ and $\widetilde{I}_2$ denote the two transformed ("epipolar rectified") images, then, given any point $\widetilde{p}_1$ with coordinates $(\widetilde{i}_1, \widetilde{j}_1)$ in $\widetilde{I}_1$, the corresponding point $\widetilde{p}_2$ in $\widetilde{I}_2$ must lie on the same row, i.e. $\widetilde{i}_2$ must equal $\widetilde{i}_1$.

## 4.2 Spatial-Temporal Transformation

Let $\mathcal{T}$ denote a transformation which maps the position $\mathbf{x}$ and time $t$ of a sensor observation $O$ to a position $\mathbf{x}'$, and a time, $t'$, in a common representational format:

$$(\mathbf{x}', t') = \mathcal{T}(\mathbf{x}, t). \tag{4.2}$$

In general, $\mathcal{T}$ has the following form:

$$\mathcal{T}(\mathbf{x}, t) = \; (\mathcal{T}_{\mathbf{x}}(\mathbf{x}, t), \mathcal{T}_t(\mathbf{x}, t)), \tag{4.3}$$

where both the spatial transformation $\mathcal{T}_{\mathbf{x}}(\mathbf{x}, t)$ and the temporal transformation $\mathcal{T}_t(\mathbf{x}, t)$ are functions of $\mathbf{x}$ and $t$. However, in many fusion applications we approximate Eq. (4.3) with a "decoupled" spatial-temporal transformation:

$$\mathcal{T}(\mathbf{x}, t) = (\mathcal{T}_{\mathbf{x}}(\mathbf{x}), \mathcal{T}_t(t)). \tag{4.4}$$

Even when Eq. (4.4) is not very accurate we often find it convenient to use it as a first-order approximation. In some applications Eq. (4.4) is true by definition e. g. when the system under observation is not changing in time. The following example illustrates the construction of a common representational format for an environment which is essentially static and in which the sensors are all of the same type. In this case temporal and semantic alignment and radiometric calibration are not required and the construction of the common representational format reduces to the construction of a *common spatial coordinate system*.

**Example 4.4** *A Distributed Surveillance System* [26, 41]. The demand for surveillance activities for safety and security purposes has received particular attention for remote sensing in transportation applications (such as airports, railways, motorways) and in public places (such as banks, supermarkets, and parking lots). Such systems typically consist of several video-based television cameras located in multiple locations. Consider a sequence of $M$ narrow field-of-view "spot" images $I_m$, $m \in \{1, 2, \ldots, M\}$, taken of a wide surveillance area.

We establish a common coordinate system by building a panoramic or "mosaic" image $I^*$ from the sequence of images $I_m$ (Fig. 4.3). For each $I_m$, we

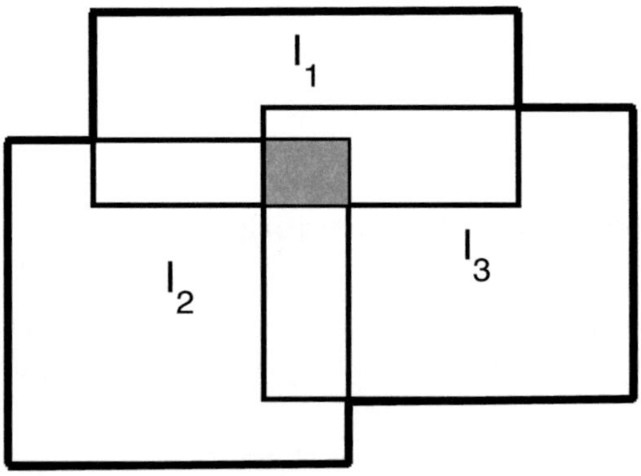


**Fig. 4.3** Shows the surveillance of a wide-area site with a sequence of "spot" images $I_m$, $m \in \{1, 2, 3\}$. Each spot image $I_m$ is transformed to a common coordinate system using a transformation $\mathcal{T}_m$. The region defined by the heavy border is the mosaic image $I^*$. The shaded area denotes a region where the transformed images $\widetilde{I}_1$, $\widetilde{I}_2$ and $\widetilde{I}_3$ overlap. In this region the mosaic image $I^*$ is formed by "stitching" $\widetilde{I}_1$, $\widetilde{I}_2$ and $\widetilde{I}_3$

find a geometric transformation $\mathcal{T}_m$ which maps the local "camera-centered" coordinate system of $I_m$ to the common "object-centered" coordinate system of $I^*$. If $\widetilde{I}_m$ denotes the transformed image, then we form the mosaic image $I^*$ by "stitching" together $\widetilde{I}_m$, $m \in \{1, 2, \ldots, M\}$. In this case, the aim of a stitching algorithm [49] is to produce a visually plausible mosaic image $I^*$ which is geometrically and photometrically as similar as possible to the transformed images $\widetilde{I}_m$ and the seams between the stitched images are invisible.

The following example describe the formation of a mosaic image using "lucky" imaging.

**Example 4.5** *Lucky Imaging* [19, 27]. Long-range imaging systems are heavily impacted by atmospheric turbulence which can cause severe distortions in the captured images or videos. This distortion is caused primarily by fluctuations in the refractive index of the air which lies between the camera and the scene being captured. The idea behind *lucky imaging* is that there is finite probability that in one of the frames the wavefront will pass undistorted from the scene to the camera. This is the so-called "lucky" frame.

If the scene is densely textured then we divide each frame $i$ of the input video into overlapping patches of size $w \times w$. Let $I_p(i)$ denote the $p$th patch in frame $i$. If $Q_p(i)$ denotes the quality of this patch, then the corresponding "lucky" patch is $\widetilde{I}_p \equiv I_p(i^*)$, where

$$i^* = \arg\max_i (Q_p(i)).$$

For the quality $Q_p(i)$ we convolve the $i$th frame with a $3 \times 3$ Laplacian [19] which we then sum up over all pixels in the patch $I_p(i)$. We establish a less distorted ("lucky") reference image $\widetilde{I}$ (i.e. common representational format), by stitching together the individual lucky patches $\widetilde{I}_p$.

In Table 4.3 we list two common types of stitching algorithms. The following example describes the transition smoothing, or feathering, of two input images $I_1$ and $I_2$.

**Table 4.3** Stitching algorithms

| Algorithm | Description |
|---|---|
| Optimal seam | Find curve in overlap region on which differences between the $\mathcal{T}_m(I_m)$ are minimal. Then copy each image to corresponding side of curve |
| Transition smoothing | Minimize seam artifacts by smoothing the transition region (small region which is spatially close to seam) [49] |

**Example 4.6** *Feathering* [49]. In feathering, the mosaic image $I^*$ is a weighted combination of the *transformed* images $\widetilde{I}_1$, $\widetilde{I}_2$, where the weighting coefficients vary as a function of the distance from the seam. In general, feathering works well as long as there is no significant misalignment. However, when the misalignments are significant, the mosaic image displays artifacts such as double edges. A modification of feathering which is less sensitive to misalignment errors, is to stitch the derivatives of the input images instead of the images themselves. Let $\partial \widetilde{I}_1/\partial x$, $\partial \widetilde{I}_1/\partial y$, $\partial \widetilde{I}_2/\partial x$ and $\partial \widetilde{I}_2/\partial y$ be the derivatives of the transformed images. If $F_x$ and $F_y$ denote the derivative images formed by feathering $\partial \widetilde{I}_1/\partial x$ and $\partial \widetilde{I}_2/\partial x$ and $\partial \widetilde{I}_1/\partial y$ and $\partial \widetilde{I}_2/\partial y$, then we choose the final mosaic image $I^*$ to be the image whose derivatives $\partial I^*/\partial x$ and $\partial I^*/\partial y$ are closest to $F_x$ and $F_y$.

## 4.3 Geographical Information System (GIS)*

An important example of a common representational format is a *Geographical Information System* (GIS). In a GIS we combine multiple images of the earth obtained from different sensors and maps, including demographic and infrastructure maps, into a common coordinate system.

**Example 4.7** *Detection and Tracking of Moving Objects in Geo-Coordinates* [23, 43]. The goal of video surveillance is to identify and track all relevant moving objects in a scene. This involves detecting moving objects, tracking them while they are visible and re-acquiring the objects once they emerge from occlusion. To track the objects from a moving camera, we need to describe the motion of the objects in a common coordinate system. Often we use absolute geo-coordinates, i.e. longitude and latitude, from a satellite image or map as the common coordinate system. By spatially aligning the images from the moving camera to a satellite image we generate the absolute geo-location of the moving objects. An important advantage of tracking in geo-coordinates is we obtain physically meaningful measurements of object motion.

In building a GIS we often need to convert a set of sensor measurements $y(\mathbf{u}_i)$, $i \in \{1, 2, \ldots, N\}$, made at specified spatial locations $\mathbf{u}_i$, into a continuous map. One way of doing this is known as *Kriging* [14, 33], where the term Kriging refers to a family of best (in the least squares sense) linear unbiased estimators (BLUE) (see Table 4.4). We interpret the $y_i = y(\mathbf{u}_i)$ as the outcome of a random field $y(\mathbf{u})$. If $y_i$, $i \in \{1, 2 \ldots, N\}$, denotes the sensor measurements which are located in the neighbor-

**Table 4.4** Kriging estimators

| Estimator | Description |
|---|---|
| Simple | Assumes known mean value $\mu(\mathbf{u}_0) = \mu_0$ in the local neighbourhood of $\mathbf{u}_0$ |
| Ordinary | Assumes unknown constant mean value $\mu(\mathbf{u}_0) = \mu_0$ in neighbourhood of $\mathbf{u}_0$ |
| Universal | Assumes mean value $\mu(\mathbf{u})$ varies smoothly in neighbourhood of $\mathbf{u}_0$. $\mu(\mathbf{u}_0)$ is modelled as $\mu(\mathbf{u}_0) = \sum_{l=1}^{L} \beta_l f_l(\mathbf{u}_0)$, where $f_l(\mathbf{u})$ are known basis functions. $\beta_l, l \in \{1, 2, \ldots, L\}$, found by solving appropriate Kriging Eqns. |
| Co-Kriging | Combines primary input measurements $y(\mathbf{u}_i)$ with $K$ auxiliary variables $z_1(\mathbf{u}), z_2(\mathbf{u}), \ldots, z_K(\mathbf{u})$. Often used when $z_k(\mathbf{u})$ are known at a few $\mathbf{u}_j$ |
| KED | Kriging with External Drift. Similar to Co-Kriging except auxiliary variables are known at all $\mathbf{u}$ including $\mathbf{u}_0$ |

hood of a given point $\mathbf{u}_0$, then the estimated sensor value at $\mathbf{u}_0$, is

$$\hat{y}(\mathbf{u}_0) = \mu(\mathbf{u}_0) + \sum_i \lambda_i (y_i - \mu(\mathbf{u}_0)), \tag{4.5}$$

where $\mu(\mathbf{u}_0)$ is the mean value at $\mathbf{u}_0$ and the weights $\lambda_i$, $i \in \{1, 2, \ldots, N\}$, are chosen to give the best unbiased solution. Mathematically, $\mu(\mathbf{u}_0)$ and $\lambda_i$, are solutions of a system of linear equations known as the *Kriging equations*.

**Example 4.8** *Ordinary Kriging* [14]. In ordinary Kriging the mean $\mu(\mathbf{u})$ is equal to an unknown constant $\mu_0$ in the local region of $\mathbf{u}_0$ and the weights $\lambda_i$ sum to 1. In this case, the estimated sensor value at $\mathbf{u}_0$ is

$$\hat{y}(\mathbf{u}_0) = \mu_0 + \sum_i \lambda_i (y_i - \mu_0) = \sum_i \lambda_i y_i .$$

Then the unknown mean $\mu_0$ and the weights $\lambda_i$ are given by the matrix equation:

$$\begin{pmatrix} \lambda_1 \\ \vdots \\ \lambda_N \\ \mu_0 \end{pmatrix} = \begin{pmatrix} \Sigma(\mathbf{u}_1, \mathbf{u}_1) & \ldots & \Sigma(\mathbf{u}_N, \mathbf{u}_1) & 1 \\ \vdots & \ddots & \vdots & \vdots \\ \Sigma(\mathbf{u}_1, \mathbf{u}_N) & \ldots & \Sigma(\mathbf{u}_N, \mathbf{u}_N) & 1 \\ 1 & \ldots & 1 & 0 \end{pmatrix}^{-1} \begin{pmatrix} \Sigma(\mathbf{u}, \mathbf{u}_1) \\ \vdots \\ \Sigma(\mathbf{u}, \mathbf{u}_N) \\ 1 \end{pmatrix},$$

where $\Sigma(\mathbf{u}_i, \mathbf{u}_j) = E\big((y(\mathbf{u}_i) - \mu(\mathbf{u}_i))(y(\mathbf{u}_j) - \mu(\mathbf{u}_j))^T\big)$ is the spatial covariance and $\mu(\mathbf{u}_i)$ is the mean sensor value in the neighbourhood of $\mathbf{u}_i$.

Apart from assuming a given model for the spatial mean $\mu(\mathbf{u})$, the only information required by the Kriging equations is the spatial covariance function $\Sigma(\mathbf{u}_i, \mathbf{u}_j)$.

**Table 4.5** Positive definite 2nd-order stationary univariate covariance functions

| Name | $\Sigma(h\|\boldsymbol{\theta})$ |
|---|---|
| Linear | $\alpha - \beta h$ |
| Exponential | $\alpha - \beta(1 - \exp -(\frac{h}{H}))$ |
| Gaussian | $\alpha - \beta(1 - \exp -(\frac{h}{H})^2)$ |

These formulas only apply for $h > 0$. By definition, $\Sigma(h = 0|\boldsymbol{\theta}) = 1$, where $\boldsymbol{\theta}$ denotes the parameters: $\{\alpha, \beta, H\}$

### Spatial Covariance Function

Conventionally we learn the spatial covariance, $\Sigma(\mathbf{u}_i, \mathbf{u}_j)$, from the input measurements $y_i$ subject to any *a priori* knowledge and given constraints. For example, we often require $\Sigma(\mathbf{u}_i, \mathbf{u}_j)$ to be both *positive definite* and *second-order stationary*.[1] One way of satisfying these constraints is to fit[2] the experimental covariance $\widetilde{\Sigma}(\mathbf{h})$, to a *parametric* curve $\Sigma(\mathbf{h}|\boldsymbol{\theta})$, which is known to be positive definite and second-order stationary. See Table 4.5 for a list of parametric univariate covariance functions which are positive-definite and second-order stationary.

Let $N(\mathbf{h})$ be the number of pairs of measurements $\big(y(\mathbf{u}_i), y(\mathbf{u}_j)\big)$ whose separation $(\mathbf{u}_i - \mathbf{u}_j)$ is close to $\mathbf{h}$. Then an experimental estimate of $\Sigma(\mathbf{h})$ is

$$\widetilde{\Sigma}(\mathbf{h}) \approx \frac{1}{N(\mathbf{h})} \sum_{(\mathbf{u}_i - \mathbf{u}_j) \approx \mathbf{h}} \big(y(\mathbf{u}_i) - \mu\big)\big(y(\mathbf{u}_j) - \mu\big)^T , \tag{4.6}$$

where $\mu = \frac{1}{N} \sum_{i=1}^{N} y(\mathbf{u}_i)$ is an estimate of the mean value of $y$.[3]

The following example illustrates how we may use $\widetilde{\boldsymbol{\Sigma}}(\mathbf{h})$ to estimate the standard deviation of the noise in an input image.

**Example 4.9** *Standard Deviation in an Input Image* [34]. Given an input image $I(\mathbf{u})$ we calculate the spatial covariance matrix $\widetilde{\boldsymbol{\Sigma}}(h)$ for different values of $h = |\mathbf{h}|$. We fit $\widetilde{\boldsymbol{\Sigma}}(h)$ to a parametric function $f(A, B, \ldots)$. The square root of the best fit at $h \to 0$ gives us the standard deviation of the noise in the image. Two recommended functions for $f(A, B, \ldots)$ are:

[1] Postive definiteness requires $\sum_i \sum_j a_i a_j \Sigma(\mathbf{u}_i, \mathbf{u}_j) \geq 0$ for all $\mathbf{u}_i$, $\mathbf{u}_j$ and real numbers $a_i$, $a_j$. Second-order stationarity requires $\mu(\mathbf{u})$ to be the same for all $\mathbf{u}$ and that $\Sigma(\mathbf{u}_i, \mathbf{u}_j)$ only depends on the difference $\mathbf{h} = (\mathbf{u}_i - \mathbf{u}_j)$.

[2] Example 11.8 shows an example of least square fitting of the experimental measured covariance $\widetilde{\Sigma}(\mathbf{h})$.

[3] Equation (4.6) is sensitive to outliers (see Chap. 12). If outliers are thought to be present, "robust" estimates should be used. Two such estimates are: $\widetilde{\Sigma}(\mathbf{h}) = \big(\sum a_{ij}\big)^4/(0.457N(\mathbf{h}) + 0.494)$ and $\widetilde{\Sigma}(\mathbf{h}) = \big(\mathrm{med}\{a_{ij}\}\big)^4/0.457$, where $a_{ij} = \sqrt{|(y(\mathbf{u}_i) - \mu)(y(\mathbf{u}_j) - \mu)^T|}$ and the summation and the median are taken over all pairs of points $(\mathbf{u}_i, \mathbf{u}_j)$ which satisfy: $(\mathbf{u}_i - \mathbf{u}_j) \approx \mathbf{h}$.

$$f(A, B, C) = A + B\big(1 - \exp(-h/C)^2\big) \quad \text{Gaussian function,}$$
$$f(A, B, C) = A + B\big(1 - \exp(-h/C)\big) \quad \text{Exponential function.}$$

### Gaussian Processes

The Kriging Eqs. were presented in the framework of a geographical information system. In fact the equations have a more general application where they are more generally known as Gaussian Processes. The very close similarities between Kriging and Gaussian Processes are discussed in detail in [25].

## 4.4 Common Representational Format

It is clear from the above that the common representational format plays a crucial role in a data fusion system. In fact in most cases the choice of common representational format will govern what type of fusion algorithm is used. The following example illustrates this situation when a person performs data fusion.

**Example 4.10** *Myocardial Imaging*. Polar maps ("bull's-eye" images), are a standard way of displaying myocardial functions. They are constructed by combining images from multiple planes so that information about the entire myocardium can be displayed in a single image. Each image plane forms a ring in the polar map. Although the rings may be divided into an arbitrary number of sectors, in practice, a clinician uses four (anterior, lateral, inferior and septal) or six (anterior, anterior-lateral, inferior-lateral, inferior, inferior-septal and anterior-septal) sectors for his visual interpretation of the image. Figure 4.4 shows the polar image representation of a left ventricle.

Choosing an appropriate common representational format is very dependent on the task at hand. For example, if the task were object recognition then we would require a common representational format which is insensitive to variations that can occur in an object. For detecting living plants, or crops, an "excess-green-index" common representational format is effective as the following example shows.

**Example 4.11** *Green-Based Format for Crop Segmentation* [22]. Transplanting rice seedlings is a preliminary stage for rice growing and production operations. To reduce costs automatic rice transplanters are often used. For the transplanter to operate successfully in a complex paddy field environment

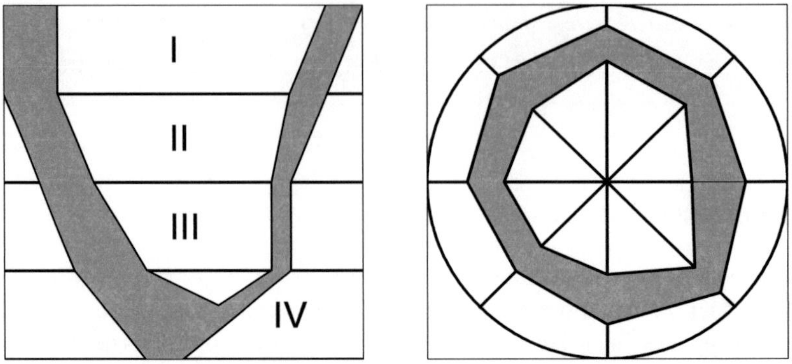


(a) (b)

**Fig. 4.4** Shows a polar image of a left ventricle. **a** Shows the left ventricle divided into four slices: I (Basal), II (Mid-Basal), III (Mid-Apical) and IV (Apical). **b** Shows a given slice divided into 8 sectors

accurate information regarding the location of the crop rows is required. To supply this information a color (RGB) image of the paddy field is segmented into two classes: living plants (crops) and background elements such as soil, stones and water.

For this purpose the RGB input image is converted into a *gray* image which highlights the plant [45], i.e. brighter pixels for pixels which belong to the plant and darker pixels for pixels which belong to the background. The plants, or crops, are then found by segmenting the gray image with a threshold $T$. The gray image is formed in two steps.

**Invariant Color Space.** The color (RGB) image is converted into an *illumination-invariant* $C_bC_rC_g$ color space [22]:

$$\begin{pmatrix} C_b \\ C_r \\ C_g \end{pmatrix} = \begin{pmatrix} -0.148 & -0.291 & 0.439 \\ 0.439 & -0.368 & -0.071 \\ 0.318 & 0.439 & -0.121 \end{pmatrix} \begin{pmatrix} R \\ G \\ B \end{pmatrix} + \begin{pmatrix} 128 \\ 128 \\ 128 \end{pmatrix} .$$

**Excess Green-Color Index.** The $C_bC_rC_g$ image is converted into a one-dimensional *excess green color index*, or gray, image:

$$\mathcal{G} = (2C_g - C_r - C_b).$$

The gray image $\mathcal{G}$ is then segmented using a threshold $T$:

$$S(i, j) = \begin{cases} 1 \text{ (``plant'')} & \text{if } \mathcal{G}(i, j) > T \\ 0 \text{ (``background'')} & \text{otherwise.} \end{cases}$$

*Note*: In this application, the number of pixels which belong to rice seedlings is much less than the number of pixels which belong to the background. As a result, the histogram of $\mathcal{G}$ tends to be unimodal which requires a special binarization threshold (see Sect. 9.6).

The following example illustrates a format which is invariant to translation, rotation, scaling and partially invariant to noise and occlusion. This format is useful when we perform object recognition by comparing *shapes* of objects.

**Example 4.12** *Shape Context and Inner Distance* [47]. Let $\mathbf{z}_i = (x_i, y_i), i \in \{1, 2, \ldots, N\}$, denote a set of $N$ sample points on a given contour (Fig. 4.5). Then the shape context $SC_i$ of the point $\mathbf{z}_i$ describes the Euclidean distance and orientation of the points $\mathbf{z}_j$, $j \neq i$, relative to $\mathbf{z}_i$. Let $r_{ij}$ and $\theta_{ij}$ denote the distance and orientation of $\mathbf{z}_j$ relative to $\mathbf{z}_i$. Divide the $(r, \theta)$ space into $L$ columns $\Theta_l, l \in \{1, 2, \ldots, L\}$, and $K$ rows $R_k, k \in \{1, 2, \ldots, K\}$. If $h(k, l)$ is the number of points $\mathbf{z}_j$, $j \neq i$, for which $r_{ij}$ lies in the $k$th row and $\theta_{ij}$ lies in the $l$th column, then

$$SC_i = \begin{pmatrix} h(1,1) & h(1,2) & \ldots & h(1,L) \\ h(2,1) & h(2,2) & \ldots & h(2,L) \\ \vdots & \vdots & \ddots & \vdots \\ h(K,1) & h(K,2) & \ldots & h(K,L) \end{pmatrix} .$$

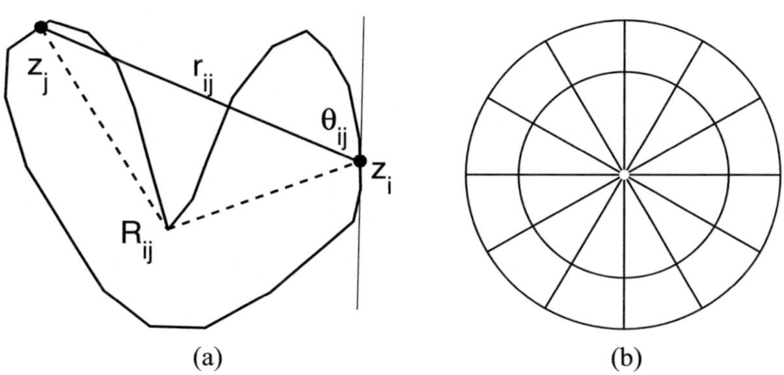


**Fig. 4.5** Shows calculation of the shape context. **a** Shows a closed contour defined by $N$ points $\mathbf{z}_i, i \in \{1, 2, \ldots, N\}$. The figure shows the Euclidean distance $r_{ij}$ and the inner distance $R_{ij}$ between the points $\mathbf{z}_j$ and $\mathbf{z}_i$. **b** Shows division of the $(r, \theta)$ space into $K \times L = 24$ bins, where $K = 2$ and $L = 12$

$SC_i$ is invariant to translation and rotation and approximately invariant to occlusion and to noise. It may be made scale invariant by dividing the distances $r_{ij}$ with $\bar{r}$, where $\bar{r}$ is value of $r_{ij}$ averaged over all $i$ and $j$, $j \neq i$.

For articulated objects the inner distance is often used instead of the shape context. This is identical to the shape context except we replace the pairwise distance $r_{ij}$ with $R_{ij}$, where $R_{ij}$ is the length of the shortest line which joins $\mathbf{z}_i$ and $\mathbf{z}_j$ and which lies within the given contour.

Recently shape-context has been used in matching time series. See Example 7.10.

Broadly speaking there are three different methods for forming a common representational format:

**Handcrafted features**. The common representational format consists of a set of features which are *manually* designed by the engineer or scientist to be invariant.

**Deep-learned features**. The common representational format consists of a set of features which are extracted *automatically* by training a deep neural network. The depth and architecture of the network and the variability of the training dataset determine the degree to which the features are invariant. Example 4.13 illustrates a common representational format extracted from a convolutional neural network (CNN).

**Hybrid features**. The common representational format consists of a set of features formed by fusing handcrafted features with deep-learned features. Example 4.16 illustrates a hybrid common representational format.

**Example 4.13** *Off-the-shelf CNN Features for Iris Recognition* [28, 32]. We consider a convolutional neural network (CNN) with many layers which has been pre-trained on a very large dataset. Then a feature vector extracted from one of the layers in the CNN can act as a *generic* "off-the-shelf" common representational format. In principle, the output of any layer can be used, although the early layers are generally not used. It is also possible to generate an ensemble of feature vectors by using more than one layer.

Once the CNN feature vectors are extracted they can be used directly in a wide range of tasks. For example, using off-the-shelf CNN feature vector for iris recognition [28]. The CNN vector, although originally trained for classifying generic objects, proved a powerful format for representing iris images.

Although deep-learned features are generally very effective, handcrafted features are sometimes preferred. Reasons for this include:

**Interpretability**. Unlike deep features, handcrafted features usually have a direct physical interpretation.

**Real-time**. Systems which operate in real-time often have limited computing power and memory. This may limit the use of deep features which generally require considerable computing power and memory

**Limited Data**. Unlike deep features, handcrafted features require little, or no, training data.

**Small or Low Resolution Images**. Handcrafted features are often effective in small or low-resolution images while deep learning features may fail due to the large size of their receptive fields.

**Generalization**. Handcrafted features often generalize better to new and unseen datasets.

For completeness we describe three handcrafted common representational formats LBP, HOG and SIFT which are widely used in different areas of machine learning and computer vision. They have also been used in conjunction with deep learning.

### *4.4.1 Local Binary Pattern (LBP)*

Local Binary Pattern (LBP) [39] is a common representational format which has been widely used for face recognition and in problems involving textures and their classification.

Let $G$ be an gray-scale input image. At each pixel $(i, j)$ we define a set of neighbouring pixels $n$, $n \in \{1, 2, \cdots, N\}$, which lie equally spaced on a radius $R$ centered at $(i, j)$. Each neighbouring pixel $n$ is labeled $\phi_n = 0$ if the center pixel has higher gray-scale value than the neighbouring pixel ($g_c > g_n$); otherwise $\phi_n = 1$. The LBP label for the center pixel $(i, j)$ is found by converting the vector $\mathbf{V} = (\phi_1, \phi_2, \ldots, \phi_N)^T$ into an integer in the range $[0, 2^N)$:

$$LBP(i, j) = \sum_n \phi_n 2^{(n-1)}.$$

Variations of the LBP include:

**Rotationally invariant LBP**.

$$LBP_{\text{ROT}}(i, j) = \min_{0 \leq k \leq N} \left( \sum_n \phi_n 2^{[(n+k-1) \mod N]} \right).$$

**Dominant LBP**. In dominant LBP we calculate the frequency of occurrence of all rotationally invariant LBP's. We then select a small number of the most frequently occurring patterns.

**Similarity LBP** Reference [36]. In similarity LBP we replace $\phi_n$ by

$$\phi_n = \begin{cases} 1 \text{ if } |g_n - g_c| \geq t, \\ 0 \text{ otherwise,} \end{cases}$$

where $t$ is an absolute threshold. In [36] the threshold is allowed to vary with the illumination by letting $t = \alpha \times g_c$ where $\alpha$ is a threshold *multiplier*.

The following examples illustrate the use of LBP in computer vision applications.

**Example 4.14** *Background Subtraction using LBP* [36]. Background subtraction (BS) is a widely used algorithm for moving object detection in a stationary video camera. BS aims to segment each input video frame into foreground and background, where foreground regions represent moving objects and background regions represent a static scene. BS is challenging when the background is changing because of swaying of trees, camera jitter etc. In *pixel-based* BS algorithms, each pixel is modeled by observing the change in intensity of a pixel over time to determine if a given pixel is an outlier (and so should be considered "foreground"). Traditional pixel-based BS algorithms do not include information regarding the spatial relations between neighboring pixels. In [36] this information is included by using a similarity LBP. The algorithm is made robust by allowing the threshold $t$ to vary with illumination: $t = \alpha \times g_c$, where $\alpha$ is a given threshold *multiplier*.

**Example 4.15** *Autonomous Crack Detection on Metallic Surfaces* [6]. Regular inspection of nuclear power plants is important to avoid accidents and ensure safe operation. For nuclear power plants crack detection in metallic surfaces is critical. In inspection videos, it is most important to distinguish cracks from other texture defects such as scratches, welds, and grind marks. In [6] LBP was successfully used for this purpose in a state-of-the-art autonomous crack detection algorithm (see Example 14.5).

### *4.4.2 Histogram of Oriented Gradients (HOG)*

Histogram of Oriented Gradients (HOG) [8] is a common representational format which is robust against local geometric and photometric transformations. It has been widely used for detecting and tracking humans in images. This is a challenging task owing to their variable appearance and wide range of poses that they can adopt.

HOG is calculated as follows. The input image is divided into small spatial regions ("cells"). In each cell we calculate a local histogram of gradient directions, or edge orientations, over the pixels in the cell. A HOG feature vector $\mathbf{f}(i, j)$ is created at $(i, j)$ by concatenating the histograms of several cells which are close to $(i, j)$.

The following example shows a deep learning network for facial expression recognition (FER) which uses both CNN features and handcrafted HOG features.

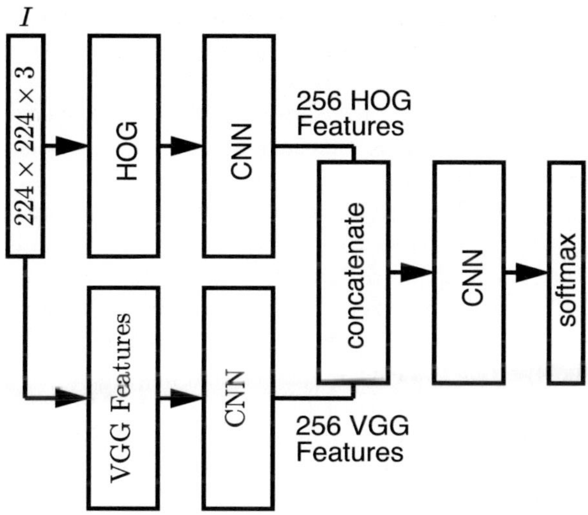


**Fig. 4.6** Shows neural network for FER using HOG and CNN features. HOG and CNN features are separately extracted from the input image $I$. These features are separately processed, reduced in number and fused together (concatenated). The 512 HOG-CNN features are processed. The final output (from softmax layer) is a probability score for each emotion class

**Example 4.16** *Facial Expression Recognition (FER) Using Hybrid HOG/CNN Features* [1]. Facial expressions are a prevalent way to recognize human emotions and automatic FER is a significant task in cognitive science, artificial intelligence and computer vision. Although conventional convolutional neural network (CNN) network have been used for FER their performance is limited due to their difficulty in distinguishing micro-facial expressions. HOG features count the localized gradient changes around each pixel. Intuitively they may be useful for modeling the structure of facial muscles and hence for distinguishing micro-facial expressions.

Given a $224 \times 224 \times 3$ input facial image $I$ we extract HOG features and CNN features (using VGG-Face). These features are separately processed, reduced in number and fused together (concatenated). The concatenated HOG-CNN features are then processed using a series of densely connected convolutional layers. The final output is a probability score for each emotion class. Figure 4.6 shows schematically the deep neural network used for FER.

### *4.4.3 Scale Invariant Feature Transform (SIFT) Descriptor*

Among the precursors of the HOG is the scale invariant feature transform (SIFT) operator [24]. Unlike LBP and HOG, the SIFT operator is both a *detector* and a *feature extractor*. As a detector it identifies a set of salient points in the input image (called "keypoints"). As a feature extractor, it extracts a distinct feature vector or descriptor from each keypoint in the image. SIFT was initially developed to spatially align two images by matching the feature vectors of keypoints in $I_1$ to the feature vectors of keypoints in $I_2$ (see Sect. 6.5).

The SIFT discriptor is invariant to changes in scale, rotation and illumination. It is also robust across a substantial range of affine distortion changes due to changes in camera viewpoint, occlusion and noise. The descriptor is also highly distinctive.

The SIFT descriptor is computed by partitioning the image region surrounding a given key-point into a $4 \times 4$ grid of cells, and computing a HOG of 8 bins in each cell. The grid is square with the $x-$axis oriented along the key-point gradient direction, and the width of the grid being approximately 12-times the detected scale of the key-point. For each key-point we form a 128-long SIFT feature vector by concatenating the $4 \times 4 = 16$ HOG's.

The following example describes Bag of Visual Words (BOVW) algorithm [2] for scene recognition. This algorithm originally used SIFT although other features, including convolutional neural network (CNN) features, may be used instead. See Example 5.9.

**Example 4.17** *Scene Classification Using Bag-of-Visual-Words (BOVW)* [35]. The idea underlying BOVW is to represent an image as a set of invariant features. Given a database $\mathcal{D}$ we apply SIFT to each image $I_i, i \in \{1, 2, \ldots, N\}$, in a $\mathcal{D}$. For each image we detect the keypoints and extract the corresponding descriptors. The keypoint descriptors are then clustered into $K$ groups using a $K$-means clustering algorithm (see Sect. 8.4). The representative descriptor of each cluster represents a "visual word" in a BOVW *dictionary*.

Given the BOVW dictionary, we encode the SIFT descriptors extracted from the test image $I_{\text{TEST}}$ into the nearest visual words and thus obtain a global representation of $I_{\text{TEST}}$ by forming a histogram $\mathcal{H}_{\text{TEST}}$ of the visual words, i.e. by counting the number of times each visual word is used to encode the SIFT descriptors which appear in $I_{\text{TEST}}$. Likewise, we find the BOVW representation $\mathcal{H}_i$ of each database image $I_i, i \in \{1, 2, \ldots, N\}$.

We then compare $\mathcal{H}_{\text{TEST}}$ with each $\mathcal{H}_i$ using an appropriate distance metric. We thus obtain a ranked retrieval list in increasing order of distance between $\mathcal{H}_{\text{TEST}}$ and $\mathcal{H}_i$.

For other examples of SIFT and its uses see:

**SIFT Registration**. Register two images $I_1$ and $I_2$ using SIFT. See Sect. 6.5.

**SIFT Image**. SIFT is applied to every pixel in a $M \times N$ gray-scale image. The output is a $M \times N \times 128$ "SIFT image". See Sect. 6.6.1.

**SIFT Flow**. Apply optical flow to a SIFT image. See Sect. 6.6.1.

**SIFT-CNN**. Train a CNN on a dataset of SIFT images. See Example 5.10.

**Table 4.6** Photometric invariant color space

| Color space | Uncertainty |
|---|---|
| Normalized $r$ and $g$ | $\sigma_r = \sqrt{(R^2(\sigma_B^2 + \sigma_G^2) + (G+B)^2\sigma_R^2)/(R+G+B)^2}$, |
| | $\sigma_g = \sqrt{(G^2(\sigma_B^2 + \sigma_R^2) + (R+B)^2\sigma_G^2)/(R+G+B)^2}$ |
| Opponent components | $\sigma_1 = \sqrt{(\sigma_G^2 + \sigma_R^2)/2}$, $\sigma_2 = \sqrt{(4\sigma_B^2 + \sigma_G^2 + \sigma_R^2)/3}$ |
| Hue $H$ | $\sigma_H = \sqrt{3(\sigma_B^2(G-R)^2 + \sigma_G^2(B-R)^2 + \sigma_R^2(B-G)^2)/(2\Delta)}$ |

$r = R/(R+G+B)$, $g = G/(R+G+B)$, $o_1 = (R-G)/2$, $o_2 = (2B-R-G)/\sqrt{3}$, $H = \arctan((\sqrt{3}(G-B)/(2R-G-B))$, $\Delta = B^2 + G^2 - GR + R^2 - B(G+R)^2$

### 4.4.4 *Uncertainty**

Sometimes what is important is how the sensor uncertainty propagates in a given common representational format. The following examples illustrate these concerns.

**Example 4.18** *Object Recognition Based on Photometric Color Invariants* [12]. A simple and effective scheme for 3D object recognition is to represent and match images on the basis of color histograms. For effective recognition we use a color space which is invariant to changes in viewing direction, object orientation and illumination.

In Table 4.6 we list common color spaces which are used for this purpose. We observe that measurement uncertainty is propagated differently in each space: normalized $rg$ is unstable around $R = G = B = 0$ ($\sigma_r, \sigma_g \to \infty$) and hue $H$ is unstable around $R = G = B$ ($\sigma_H \to \infty$) while the opponent space $o_1, o_2$ is relatively stable at all $RGB$ values [12].

**Example 4.19** *Debiased Cartesian Coordinates* [11]. We consider tracking an object $O$ in 2D using a range/bearing sensor measurements $(r_i, \theta_i)$. Let $(r_0, \theta_0)$ denote the true range/bearing of the object $O$, then we assume

$$r_i = r_0 + \delta r \quad \text{and} \quad \theta_i = \theta_0 + \delta\theta,$$

where $\delta r \sim \mathcal{N}(0, \sigma_r^2)$ and $\delta\theta \sim \mathcal{N}(0, \sigma_\theta^2)$ are iid random measurement errors.

For target tracking applications (see Chap. 13) it is convenient to use a Cartesian coordinates $(x, y)$ since this defines an inertial reference frame. If $(x_i, y_i)$ are the coordinates corresponding to $(r_i, \theta_i)$, then formally:

$$x_i = r_i \cos\theta_i \quad \text{and} \quad y_i = r_i \sin\theta_i .$$

However, $E(x_i) \neq 0$ and $E(y_i) \neq 0$. This means $x_i$ and $y_i$ are *not* unbiased estimates of the true Cartesian coordinates $x_0 = r_0 \cos\theta_0$ and $y_0 = r_0 \sin\theta_0$. To correct for these bias effects we use corrected, or "debiased", coordinates $(\widetilde{x}_i, \widetilde{y}_i)$. For range/bearing measurements, the corrected coordinates are found by dividing through by $\lambda = \exp(\sigma_\theta^2/2)$:

$$(\widetilde{x}_i, \widetilde{y}_i) \approx (x_i/\lambda, y_i/\lambda) .$$

## 4.5 Dimension Reduction

In many data fusion applications a paramount concern is to reduce the computational load and/or the storage requirements. These objectives may be achieved simultaneously by using a *low-dimensional* common representational format to represent the input data with minimal loss of information.

Some approaches which are used to produce a low-dimensional format are:

**Feature selection methods**. Here we optimally select a small number of the original parameters or features. These features are often chosen to be physically significant. Finding the optimal subset of features is, however, computationally intractable, so feature selection typically involves a heuristic search.

**Subspace techniques.** Here we represent the input data using a small number of new features which are specific combinations of the original input features. A disadvantage of this approach is the lack of interpretability as the new features are "artificial".

**Sparse methods.** Here we represent the input data using a sparse combination of features (i.e. combination of features in which only a small number of coefficients are non-zero).

**Auto-encoders.** Auto-encoders are discussed in Sect. 5.7.

## 4.6 Feature Selection

In feature selection a small subset of features are selected from the original features. Given a specified task, the subset of features may be found using the following iterative procedure. We start with an empty set of features. In each iteration we add one new feature and evaluate the result, stopping when the desired criteria is reached. This is known as *sequential forward selection*. Alternative schemes are discussed in Sect. 14.3.4.

For pattern classification, a popular criteria for choosing the optimum subset of features is *minimum redundancy maximum relevance* (mRMR): Consider a set of $N$ objects. Each object $O_i, i \in \{1, 2, \ldots, N\}$, is characterized by a class variable

$C$ and a set of feature variables $\{F_1, F_2, \ldots, F_K\}$. Let $\rho(X, Y)$ denote the degree that two variables $X$ and $Y$ are related. Then $\rho(F_k, C)$ represents how useful is the $k$th feature $F_k$ in determining the classification $C$ of the objects ("relevancy") and $\rho(F_k, F_l)$ represents how useful is the addition of the feature $F_k$ to the feature $F_l$ in determining the classification $C$ ("redundancy"). Let $S$ denote a given subset of features. Then

$$D(S) = \frac{1}{|S|}\sum_{F_l \in S} \rho(F_l, C) \quad \text{and} \quad R(S) = \frac{1}{|S|^2}\sum_{F_k, F_l \in S} \rho(F_k, F_l),$$

denotes the average relevancy and redundancy of the features $F_l \in S$. In mRMR, the optimum subset $S^*$ is given by

$$S^* = \max_S \Big( D(S) - R(S) \Big),$$

where [31] recommends using the mutual information for $\rho(X, Y)$:

$$\rho(X, Y) = \int\int p(X, Y) \log_2 \frac{p(X, Y)}{p(X)p(Y)} dX dY.$$

In this equation, $p(X)$ and $p(X, Y)$ denote, respectively, the marginal and joint pdf of the variables $X$ and $Y$. For further details on MI and its calculation, see Sect. 6.4.

## 4.7 Subspace Methods*

Subspace methods [7, 9] are widely used to produce a low-dimensional common representational format. The low-dimensional format is formed by creating a small number of new features which are specific combinations of the original high-dimensional input.

Table 4.7 lists some of the principal subspace techniques which are commonly used for dimension-reducing purposes.

### 4.7.1 Principal Component Analysis (PCA)

Principal component analysis (PCA) [7, 9, 20] is an *unsupervised* technique which builds a low-dimensional common representational format given a set of $N$ input vectors $\mathbf{y}_n, n \in \{1, 2, \ldots, N\}$, where the $\mathbf{y}_n$ are of $M$ dimensions.

The aim of PCA is to find a $L$-dimensional ($L < M$) *linear* projection that best represents the $\mathbf{y}_n, n \in \{1, 2, \ldots, N\}$, in a least squares sense. If we define a reduced $L$-dimensional space using a set of orthonormal axes $\mathbf{u}_l, l \in \{1, 2, \ldots, L\}$, then

**Table 4.7** Subspace techniques

| Technique | Description |
|---|---|
| PCA | Principal component analysis. Linear transformation chosen so projected components have maximum variance |
| LDA | Linear discriminant analysis. Linear transformation chosen for $K \geq 2$ classes so projected components for each class are maximally separated from projected components of other classes |
| NNMF | Non-negative matrix factorization. Finds factors with non-negative elements |
| CCA | Canonical correlation analysis. Assumes two separate sensors sets of measurements (feature vectors). For each set of measurements find linear transformation so the projected components of each class are maximally correlated |

$$\boldsymbol{\theta}_n = \mathbf{U}^T(\mathbf{y}_n - \boldsymbol{\mu}),$$

is a $L$-dimensional representation of $\mathbf{y}_n$, where $\boldsymbol{\theta}_n = \left(\theta_n^{(1)}, \theta_n^{(2)}, \ldots, \theta_n^{(L)}\right)^T$ and $\mathbf{U} = (\mathbf{u}_1, \mathbf{u}_2 \ldots, \mathbf{u}_L)$. *Note*: Given $\boldsymbol{\theta}_n$, the optimal reconstruction of $\mathbf{y}_n$ is $\widetilde{\mathbf{y}}_n = \mathbf{U}\boldsymbol{\theta}_n + \boldsymbol{\mu} = \mathbf{U}\mathbf{U}^T(\mathbf{y}_n - \boldsymbol{\mu}) + \boldsymbol{\mu}$.

Mathematically, the orthonormal axes $\mathbf{u}_l, l \in \{1, 2, \ldots, L\}$, are given by the $L$ dominant eigenvectors of the sample covariance matrix $\boldsymbol{\Sigma}$, i.e.

$$\boldsymbol{\Sigma}\mathbf{u}_l = \lambda_l \mathbf{u}_l,$$

where

$$\boldsymbol{\mu} = \frac{1}{N}\sum_n \mathbf{y}_n \quad \text{and} \quad \boldsymbol{\Sigma} = \frac{1}{N}\sum_n (\mathbf{y}_n - \boldsymbol{\mu})(\mathbf{y}_n - \boldsymbol{\mu})^T.$$

Figure 4.7a shows the PCA of a set of 2D input measurements $\mathbf{y}_n, n \in \{1, 2, \ldots, N\}$. The following example illustrates the use of PCA to form a low-dimensional common representational format.

**Example 4.20** *PCA for Vehicle Sound Signature Application.* All moving vehicles make some kind of noise in which vehicles of the same kind and working in similar conditions generate similar noises. This fact is used to detect and classify vehicles moving in a given surveillance region: Given $N$ vehicles, each vehicle is classified by its power spectrum in the range $\sim [5, 6500]$ Hz. If each spectrum is sampled every $\sim$5 Hz, the corresponding common representational format has $M \sim 1200$ dimensions. By using PCA we may reduce the number of dimensions from $M \sim 1200$ to a more manageable $L \sim 30$.

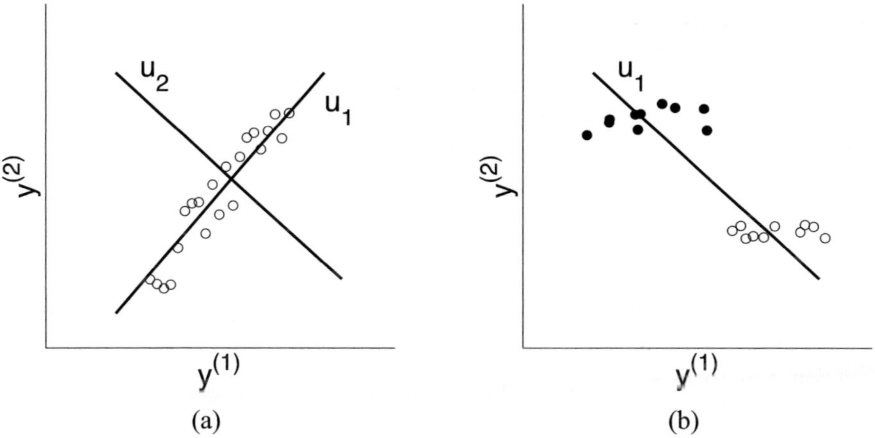


**Fig. 4.7** Shows 2D input measurements $\mathbf{y}_n = (y_i^{(1)}, y_i^{(2)}), n \in \{1, 2, \ldots, N\}$. **a** PCA representation consists of vectors $\mathbf{u}_1$ and $\mathbf{u}_2$, where the dominant vector is $\mathbf{u}_1$. **b** Shows the input measurements with their class labels. Measurements which belong to class $c_1$ and $c_2$ are denoted, respectively, by an open and a closed circle. For the two-class problem, the LDA representation is a single vector labeled $\mathbf{u}_1$ which joins $\boldsymbol{\mu}_1$ and $\boldsymbol{\mu}_2$

**Table 4.8** Principal component analysis (PCA) and its variants

| Technique | Description |
|---|---|
| PCA | Basic PCA algorithm. Maximizes the variance of all linear projections |
| Weighted PCA | Each input measurement $\mathbf{y}_i$ is given a different weight before PCA is performed |
| Robust PCA | Performs PCA using robust techniques. See Chap. 12 |
| Kernel PCA | Generalizes the PCA to non-linear transformations |

In PCA it is common practice to select the $L$ principal components with the largest eigenvalues. However, [38] suggests selecting the *most discriminant* principal components. For this purpose they use the LDA algorithm (see Sect. 4.7.2).

An important class of PCA algorithms are the robust PCA algorithms. These are used when the input data is contaminated with outliers. See Sect. 12.5.

In Table 4.8 we list some PCA variants which are in common use.

### 4.7.2 Linear Discriminant Analysis (LDA)

The method of *linear discriminant analysis* (LDA) is a *supervised* technique in which we suppose each input measurement $\mathbf{y}_n$ is associated with a given class $C = c_k, k \in \{1, 2, \ldots, K\}$. The aim of LDA is to find a $L$-dimensional subspace $\mathbf{U} =$

$(\mathbf{u}_1, \mathbf{u}_2, \ldots, \mathbf{u}_L)$ in which the different classes are maximally separated (Fig. 4.7b). If $\boldsymbol{\theta}_n = \big(\theta_n^{(1)}, \theta_n^{(2)}, \ldots, \theta_n^{(L)}\big)^T$ denotes the $L$-dimensional LDA representation of $\mathbf{y}_n$, then

$$\boldsymbol{\theta}_n = \mathbf{U}^T \mathbf{y}_n, \tag{4.7}$$

where $\mathbf{u}_l$ is a solution of the eigenvector equation:

$$\mathbf{H}\mathbf{u}_l = \lambda_l \mathbf{u}_l. \tag{4.8}$$

In Eq. (4.8), $\mathbf{H} = \boldsymbol{\Sigma}_W^{-1}\boldsymbol{\Sigma}_B$, where covariance matrices $\boldsymbol{\Sigma}_B$ and $\boldsymbol{\Sigma}_W$ measure, respectively, scatter of $\mathbf{y}_n$ "between-classes" and "within-classes". Mathematically, the covariance matrices are:

$$\boldsymbol{\Sigma}_B = \frac{1}{N}\sum_k n_k(\boldsymbol{\mu}_k - \boldsymbol{\mu}_G)(\boldsymbol{\mu}_k - \boldsymbol{\mu}_G)^T \quad \text{and} \quad \boldsymbol{\Sigma}_W = \frac{1}{N}\sum_k n_k \boldsymbol{\Sigma}_k, \tag{4.9}$$

where $n_k$ is the number of input measurements which belong to class $c_k$, $\boldsymbol{\mu}_k$ and $\boldsymbol{\Sigma}_k$ are, respectively, the mean vector and covariance matrix for the $k$th class, and $\boldsymbol{\mu}_G = \frac{1}{N}\sum_k n_k \boldsymbol{\mu}_k$ is the global mean vector.

The decomposition of $\mathbf{H}$ in Eq. (4.8) is only possible if the matrix is non-singular. This requires the number of training samples, $N$, to be larger than $L$, i.e. to be larger than $K-1$. Even if $N > L$, the LDA has a tendency to "overfit" when $N \lesssim 10L$ (see Example 10.8) [17, 42]. One way to prevent overfitting is to *regularize* the solution by performing the eigenvalue decomposition of a regularized matrix, $\widetilde{\mathbf{H}}$, as illustrated by the following example.

**Example 4.21** *Principal Discriminant Analysis (PDA) for Small Samples* [42]. PDA is based on the eigenvalue of the regularized matrix $\widetilde{\mathbf{H}}$:

$$\widetilde{\mathbf{H}} = (1-\epsilon)\boldsymbol{\Sigma}_W^{-1}\boldsymbol{\Sigma}_B + \epsilon\boldsymbol{\Sigma}_G,$$

where $\boldsymbol{\Sigma}_G = E\big((\mathbf{x} - \boldsymbol{\mu}_G)(\mathbf{x} - \boldsymbol{\mu}_G)^T\big) = \boldsymbol{\Sigma}_W + \boldsymbol{\Sigma}_B$ is the global covariance matrix and $\epsilon \in [0, 1]$ is a regularization parameter. For $\epsilon > 0$ the LDA solution is regularized by incorporating information on $\boldsymbol{\Sigma}_G$. *Note*. The PDA also increases the number of non-zero eigenvalues beyond the upper limit set by LDA.

In Eq. (4.9) the rank of $\boldsymbol{\Sigma}_B$ is bounded from above to $K-1$. This, in turn, implies $L \leq (K-1)$. In many data fusion applications we require a richer, i.e. a higher dimensional, common representational format. We may obtain such a representation by using the recursive LDA algorithm [44].

**Example 4.22** *Recursive LDA Algorithm** [44]. In the recursive LDA the feature vectors are obtained recursively one at a time as follows. The first feature vector $\mathbf{u}_1$ is obtained as in the conventional LDA algorithm:

$$\mathbf{\Sigma}_W^{-1}\mathbf{\Sigma}_B\mathbf{u} = \lambda\mathbf{u}_1 .$$

Before the second feature vector $\mathbf{u}_2$ is computed, the information represented by the first feature vector $\mathbf{u}_1$ is discarded from all the input vectors $\mathbf{y}_n$:

$$\mathbf{y}_n^{(2)} = \mathbf{y}_n - (\mathbf{u}_1^T\mathbf{y}_n)\mathbf{u}_1 .$$

The second feature vector $\mathbf{u}_2$ is then calculated by using the conventional LDA algorithm on $\mathbf{y}_n^{(2)}$:

$$(\mathbf{\Sigma}_W^{(2)}) - 1\mathbf{\Sigma}_B^{(2)}\mathbf{u} = \lambda\mathbf{u}_1 ,$$

where $\mathbf{\Sigma}_B^{(2)}$ and $\mathbf{\Sigma}_W^{(2)}$ denote the "between-classes" and "within-classes" covariance matrices of $\mathbf{y}_n^{(2)}$. Mathematically, we may rewrite this as:

$$(\mathbf{W}_2^T\mathbf{W}_2)^{-1}\mathbf{W}_2^T\mathbf{B}_2\mathbf{u}_2 = \lambda\mathbf{u}_2 ,$$

where $\mathbf{W}_2 = (\mathbf{\Sigma}_W^{(2)}\mathbf{u}_1^T)^T$ and $\mathbf{B}_2 = (\mathbf{\Sigma}_B^{(2)}\mathbf{0})^T$.

Similarly, the $k$th feature vector $\mathbf{u}_k$ is the solution with the largest eigenvalue of the eigenvector equation:

$$(\mathbf{W}_k^T\mathbf{W}_k)^{-1}\mathbf{W}_k^T\mathbf{B}_k\mathbf{u}_k = \lambda\mathbf{u}_k ,$$

where $\mathbf{W}_k = (\mathbf{\Sigma}_W^{(k)}\mathbf{u}_1^T \ldots \mathbf{u}_{k-1}^T)^T$, $\mathbf{B}_k = (\mathbf{\Sigma}_B^{(k)}\mathbf{0}\ldots\mathbf{0})^T$ and the within-class scatter matrix $\mathbf{\Sigma}_W^{(k)}$ and the between-class scatter matrix $\mathbf{\Sigma}_B^{(k)}$ are calculated from the pre-processed samples $\mathbf{y}_n^{(k)}$ in which all the information represented by previous feature vectors $\mathbf{u}_1, \mathbf{u}_2, \ldots, \mathbf{u}_{k-1}$ are eliminated using the following equation:

$$\mathbf{y}_n^{(k)} = \mathbf{y}_n^{(k-1)} - (\mathbf{u}_{k-1}^T\mathbf{y}_n^{(k-1)})\mathbf{u}_{k-1} , \qquad i = 1, 2, \ldots, N .$$

Outliers and atypical observations might have an undue influence on the results obtained, since the LDA is based on non-robust estimates of the population parameters. If outliers are thought to be present, the population parameters should be calculated using robust techniques (see Sect. 12.5).

In Table 4.9 we list some of the LDA variants which are in common use.

**Table 4.9** LDA and its variants

| Name | $\mathbf{H}$ |
|---|---|
| Original LDA | $\boldsymbol{\Sigma}_W^{-1}\boldsymbol{\Sigma}_B$ |
| PCA+LDA | $(\mathbf{V}^T\boldsymbol{\Sigma}_W\mathbf{V})^{-1}(\mathbf{V}^T\boldsymbol{\Sigma}_B\mathbf{V})$, where $\mathbf{V}$ denotes the principal component matrix. Variant of PCA+LDA is to replace PCA with kernel PCA |
| PDA | $(\boldsymbol{\Sigma}+\lambda\boldsymbol{\Omega})^{-1}\boldsymbol{\Sigma}_B$, where $\boldsymbol{\Omega}$ is regularization matrix. See Example 4.21 |
| Generalized LDA | $\boldsymbol{\Sigma}_G^{+}\boldsymbol{\Sigma}_B$, where $\boldsymbol{\Sigma}_G^{+}$ denotes the pseudo-inverse of $\boldsymbol{\Sigma}_G$ [18] |
| Recursive LDA | See Example 4.22 |

## *4.7.3 Canonical Correlation Analysis (CCA)**

Canonical Correlation Analysis (CCA) [40] is an unsupervised technique which builds a low-dimensional *coordinated* common representational format given *two* input datasets $\mathbf{X}=[\mathbf{x}_1,\mathbf{x}_2,\ldots,\mathbf{x}_N]\in\mathbb{R}^{p\times N}$ and $\mathbf{Y}=[\mathbf{y}_1,\mathbf{y}_2,\ldots,\mathbf{y}_N]\in\mathbb{R}^{q\times N}$, where $\mathbf{x}_n=(x_n^{(1)},x_n^{(2)},\ldots,x_n^{(p)})^T$ and $\mathbf{y}=(y_n^{(1)},y_n^{(2)},\ldots,y_n^{(q)})^T$.

Assuming $\mathbf{X}$ and $\mathbf{Y}$ are *zero-mean* matrices, then the aim of CCA is to jointly find $K$ pairs of linear projection directions (called "canonical" vectors) $\mathbf{U}=[\mathbf{u}_1,\mathbf{u}_2,\ldots,\mathbf{u}_K]\in\mathbb{R}^{p\times K}$ and $\mathbf{V}=[\mathbf{v}_1,\mathbf{v}_2,\ldots,\mathbf{v}_K]\in\mathbb{R}^{q\times K}$ so that $\mathbf{U}^T\mathbf{X}$ and $\mathbf{V}^T\mathbf{Y}$ are maximally correlated.

Taking a pair of canonical vectors $\mathbf{u}$, $\mathbf{v}$, then CCA maximizes the correlation coefficient $\rho$ between $\mathbf{X}^T\mathbf{u}$ and $\mathbf{Y}^T\mathbf{v}$:

$$\rho(\mathbf{X}^T\mathbf{u},\mathbf{Y}^T\mathbf{v})=\frac{\mathbf{u}^T\mathbf{X}\mathbf{Y}^T\mathbf{v}}{\sqrt{(\mathbf{u}^T\mathbf{X}\mathbf{X}^T\mathbf{u})\times(\mathbf{v}^T\mathbf{Y}\mathbf{Y}^T\mathbf{v})}}. \tag{4.10}$$

The solution of Eq. (4.10) is given by the $K$ largest eigenvalues of the generalized eigenvalue problem:

$$\begin{pmatrix}0 & \boldsymbol{\Sigma}_{xy}\\ \boldsymbol{\Sigma}_{yx} & 0\end{pmatrix}\begin{pmatrix}\mathbf{u}\\ \mathbf{v}\end{pmatrix}=\lambda\begin{pmatrix}\boldsymbol{\Sigma}_{xx} & 0\\ 0 & \boldsymbol{\Sigma}_{yy}\end{pmatrix}\begin{pmatrix}\mathbf{u}\\ \mathbf{v}\end{pmatrix}, \tag{4.11}$$

where $\boldsymbol{\Sigma}_{xx}=\mathbf{X}\mathbf{X}^T$, $\boldsymbol{\Sigma}_{yy}=\mathbf{Y}\mathbf{Y}^T$, $\boldsymbol{\Sigma}_{xy}=\mathbf{X}\mathbf{Y}^T$ denote, respectively, the covariances of $\mathbf{X}$, $\mathbf{Y}$ and the cross-covariance. The correlation $\rho(\mathbf{u}_k^T\mathbf{X},\mathbf{v}_k^T\mathbf{Y})$ is equal to the $k$th leading generalized eigenvalue.

When the feature dimensionality is high, especially when $p>N$ or $q>N$, the corresponding covariance matrix $\boldsymbol{\Sigma}_{xx}$ or $\boldsymbol{\Sigma}_{yy}$ is singular. To remedy this problem the covariance matrices can be regularized by adding a small non-negative value to the diagonal of the covariance matrices:

$$\boldsymbol{\Sigma}_{xx}^{\text{reg}}=\boldsymbol{\Sigma}_{xx}+r_x\mathbf{I},\quad\text{and}\quad\boldsymbol{\Sigma}_{yy}^{\text{reg}}=\boldsymbol{\Sigma}_{yy}+r_y\mathbf{I},$$

where $r_x$ and $r_y$ are non-negative regularization coefficients.

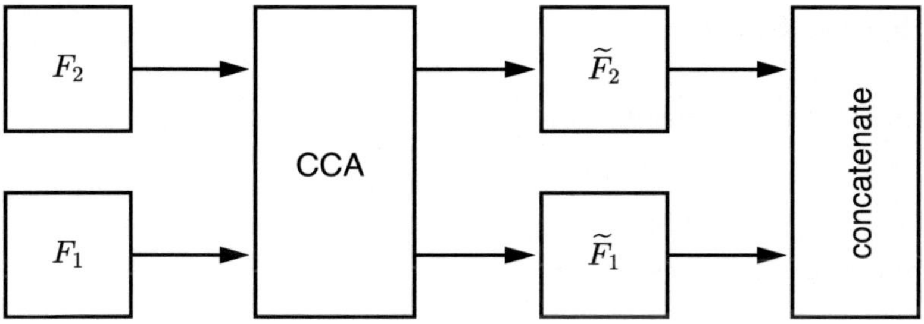


**Fig. 4.8** Shows generation of the new feature vectors $\widetilde{F}_1$ and $\widetilde{F}_2$ using CCA. The feature vectors $\widetilde{F}_1$ and $\widetilde{F}_2$ are fused together by concatenation

The following example illustrates the use of CCA for deep-learning feature fusion.

> **Example 4.23** *Deep CCA: Feature Fusion for Remote Sensing Classification* [5]. We use a pre-trained VGG convolutional neural network (CNN) (see Sect. 5.5.2) to extract features from the very high resolution (VHR) input images. For this purpose the authors use the first two *fully-connected* layers $FC_1$ and $FC_2$. These layers generate feature vectors $F_1$ and $F_2$ of lengths $M_1$ and $M_2$. The feature vectors are fed into a CCA algorithm which outputs two new vectors $\widetilde{F}_1$ and $\widetilde{F}_2$. Both of the new vectors are of length $\widetilde{M}$ and are fused together by concatenation. See Fig. 4.8.

CCA is affected by noise and outliers. It is common practice therefore to pre-process the data using an appropriate smoothing filter, e. g. Hempel filter (see Example 12.6). Alternatively, a robust CCA algorithm may be used (see Sect. 12.5).

If the inputs **X** and **Y** are asynchronous then we need to allow for *temporal* changes. One way of doing this is to use *canonical time warping* (CTW): this extends CCA by incorporating time warping and is discussed in Sect. 7.4.

In Table 4.10 we list some of the CCA variants which are in common use [15].

**Table 4.10** CCA and its variants

| Name | Description |
|---|---|
| CCA | Original canonical correlation analysis developed by Hotelling in 1936 |
| Kernel CCA | Use nonlinear kernels. Allows for nonlinear correlation between $X$ and $Y$ |
| Robust CCA | Uses robust statistics to make CCA robust against outliers. See Sect. 12.5 |
| Canonical time warp | Extends CCA by incorporating time warping. See Sect. 7.4 |
| Multi-view CCA | Generalize CCA to more than two views |
| Deep CCA | Use deep learning to extract nonlinear features. Then use CCA to calculate canonical vectors. See Example 4.23 |

## 4.8 Sparse Coding*

Another popular technique for producing a low-dimensional representation is the method of *sparse coding* [37]. Given a "dictionary" or set of $L$ overcomplete $M$-dimensional basis vectors $\mathbf{u}_l = (u_l^{(1)} u_l^{(2)} \ldots u_l^{(M)})^T, l \in \{1, 2, \ldots, L\}$, the idea of sparse coding is to represent a $M$-dimensional input vector $\mathbf{y} = (y^{(1)} y^{(2)} \ldots y^{(M)})^T$ as a *sparse* linear combination of $\mathbf{u}_l$. That is,

$$\mathbf{y} \approx \sum_l \alpha_l \mathbf{u}_l,$$

where $\boldsymbol{\alpha} = (\alpha_1 \alpha_2 \ldots, \alpha_L)$ is a sparse vector, i.e. most of the coefficients $\alpha_l$ are zero. The optimum combination is the representation with the fewest number of non-zero coefficients. It is given by

$$\boldsymbol{\alpha}^* = \arg\min_{\boldsymbol{\alpha}} \|\boldsymbol{\alpha}\|_0 \quad \text{subject to} \quad \left\|\mathbf{y} - \sum_l \alpha_l \mathbf{u}_l\right\|_2^2 \leq \epsilon,$$

where $\|\boldsymbol{\alpha}\|_0$ denotes the $L_0$ norm of $\boldsymbol{\alpha}$ and counts the number of non-zero entries in $\boldsymbol{\alpha}$. However, solving this equations is numerically unstable and NP-hard. For this reason, we more usually define the coefficients $\boldsymbol{\alpha}^*$ as the solution of the following $L_1$-norm optimization problem:

$$\boldsymbol{\alpha}^* = \arg\min_{\boldsymbol{\alpha}} \|\boldsymbol{\alpha}\|_1 \quad \text{subject to} \quad \left\|\mathbf{y} - \sum_l \alpha_l \mathbf{u}_l\right\|_2^2 \leq \epsilon,$$

which we rewrite as

$$\boldsymbol{\alpha}^* = \arg\min_{\boldsymbol{\alpha}} \left(\lambda \|\boldsymbol{\alpha}\|_1 + \left\|\mathbf{y} - \sum_l \alpha_l \mathbf{u}_l\right\|_2^2\right), \tag{4.12}$$

where $\lambda$ is a regularization term.

Given a finite set of training vectors $\mathbf{y}_n, n \in \{1, 2, \ldots, N\}$, a standard method for constructing the dictionary $\mathbf{u}_l, l \in \{1, 2, \ldots, L\}$, is the K-SVD algorithm [37]. This algorithm works iteratively: in each iteration we first sparse code the $\mathbf{y}_n$ using the current dictionary and then update the $\mathbf{u}_l$ to better fit the $\mathbf{y}_n$.

Sparse representation is a powerful method for producing low-dimensional representations. The following example illustrates its use in forensic sketch synthesis.

**Example 4.24** *Sparse Face Sketch Synthesis (FSS)** [13]. In law enforcement, it is often necessary to identify suspects using a forensic sketch drawn by a forensic expert according to eyewitness description. Directly matching forensic sketches to photographs in a database of criminals is challenging mainly due to the large discrepancy in texture. To reduce the texture mismatch a FSS procedure is used to form a common representational format. In general, FSS uses one of the following formats:

**Synthetic sketch**. Each photograph, or image $I$ in the database is transformed into a corresponding *synthetic* sketch $S$. In this case, the common representational format is *sketch-based.*

**Synthetic photograph**. The hand-drawn sketch $S$ is transformed into a *synthetic* photograph or image $I$. In this case, the common representational format is *photograph-based.*

We illustrate FSS by generating a synthetic sketch $S$ for a "test" photograph $T$: We use a training dataset $\mathcal{D}$ with $M$ face sketch-photograph pairs. Divide each photograph and sketch in $\mathcal{D}$ into $N$ overlapping $p \times p$ patches and represent each patch by a $p^2$ vector. Let $\{\mathbf{a}_1, \mathbf{a}_2, \ldots\}$ and $\{\mathbf{b}_1, \mathbf{b}_2, \ldots\}$ denote, respectively, the training photograph and sketch patches. We use the same procedure on $T$: dividing it into $N$ patches $\mathbf{t}_n, n \in \{1, 2, \ldots, N\}$. Formally we represent $T$ as:

$$T \leftrightarrow \{\mathbf{t}_1, \mathbf{t}_2, \ldots, \mathbf{t}_N\}.$$

For each patch $\mathbf{t}_n$ we find the $K$ patches $\{\mathbf{a}_{(1)}, \mathbf{a}_{(2)}, \ldots, \mathbf{a}_{(K)}\}$ which are most similar to $\mathbf{t}_n$, where the $\mathbf{a}_{(k)}, k \in \{1, 2, \ldots, K\}$, are selected one from each training photograph within a search region centered around the location of $\mathbf{t}_n$. We then approximate each patch $\mathbf{t}_n$ by a linear combination of $\mathbf{a}_{(k)}$:

$$\mathbf{t}_n \sim \sum_k w_{nk} \mathbf{a}_{(k)} = \mathbf{w}_n \cdot \mathbf{A}_n,$$

where $\mathbf{A}_n = \{\mathbf{a}_{(1)}, \mathbf{a}_{(2)}, \ldots, \mathbf{a}_{(K)}\}$. The corresponding *synthetic* patch $\mathbf{s}_n$ is:

$$\mathbf{s}_n = \sum_k w_{nk} \mathbf{b}_{(k)} = \mathbf{w}_n \mathbf{B}_n,$$

where $\mathbf{B}_n = \{\mathbf{b}_{(1)}, \mathbf{b}_{(2)}, \ldots, \mathbf{b}_{(K)}\}$ and the synthetic sketch $S$ is given by

$$S \leftrightarrow \{\mathbf{s}_1, \mathbf{s}_2, \ldots, \mathbf{s}_N\}.$$

In sparse FSS the optimum weights $\mathbf{w}_n^*$ are given by:

$$\mathbf{w}_n^* = \arg\min_w \left( \lambda \sum_k w_{nk} + \left\| \mathbf{t}_n - \sum_k w_{nk} \mathbf{a}_{(k)} \right\|_2^2 \right),$$

where $\lambda$ is a regularization parameter.

## 4.9 Software

**MATLAB.** MATLAB COMPUTER VISION and STATISTICS AND MACHINE LEARNING toolboxes includes m-files on: Handcrafted features (LBP, HOG, SIFT) and subspace models (PCA, LDA, CCA). SPARSE REPRESENTATION toolbox (Author: Li) has m-files on sparse coding.

**PYTHON.** SCIKIT-LEARN and SCIKIT-IMAGE includes python modules for handcrafted features (LBP, HOG, SIFT) and subspace models (PCA, LDA, CCA).

## 4.10 Further Reading

The question of what constitutes an appropriate common representational format has been intensely investigated for mobile robots [10]. For target tracking the choice of representational format has been discussed in [4]. The subject of invariant color spaces has been intensely investigated in recent years. A modern reference to the subject is: [12]. An excellent up-to-date review of feature learning (both traditional and deep learning) is [46].

A modern review of dimension reduction is [3].

## Problems

**4.1** Develop the algorithm for Bayesian triangulation (Example 4.2).

**4.2** Show the conversion of range-bearing measurements to Cartesian coordinates using $x = r\cos(\theta)$, $y = r\sin(\theta)$ gives biased estimates. Use a simple example to show that the conversion $x_{DB} = r\cos(\theta)/\lambda$, $y_{DB} = r\sin(\theta)/\lambda$ are less biased.

**4.3** Describe the main characteristics of the PCA, LDA, NMF and CCA subspaces.

**4.4** Describe a nearest neighbor face recognition scheme using (a) PCA (b) LDA.

## References

1. Ahadit AB, Jatoth RK (2022) A novel multi-feature fusion deep neural network using HOG and VGG-Face for facial expression classification. Mach Vis Appl 33:Article no 55
2. Arun KS, Govindan VK, Kumar SDM (2020) Enhanced bag of visual words representations for content based image retrieval: a comparative study. Art Intell Rev 53:1615–1653
3. Ayesha S, Hanif MK, Talib R (2020) Overview and comparative study of dimensionality reduction techniques for high dimensional data. Inf Fusion 59:44–58
4. Blackman SS, Popoli RF (1999) Design and analysis of modern tracking systems. Published by Artech House, Norwood, MA

5. Chaib S, Liu H, Gu Y, Yao H (2017) Deep feature fusion for VHR remote sensing scene classification. IEEE Trans Geo Rem Sens 55:4775–4784
6. Chen F-C, Jahanshahi MR, Wu R-T (2017) A texture-based video processing methodology using Bayesian data fusion for autonomous crack detection on metallic surfaces. Comput-Aided Civ Infrastruct Eng 32:271–287
7. Cunninghamd JP, Ghahramani Z (2015) Linear dimensionality reduction: survey, insights and generalizations. J Mach Learn Res 16:2859–2900
8. Dalal N, Triggs B (2005) Histograms of oriented gradients for human detection. Int Conf Comp Vis Patt Recogn (CVPR 2005)
9. Torre F (2012) A least-squares framework for component analysis. IEEE Trans Pattern Anal Mach Intell 34:1041–1055
10. Durrant-Whyte HF (1988) Sensor models and multisensor integration. Int J Robot Res 7:97–113
11. Fletcher FK, Kershaw DJ (2002) Performance analysis of unbiased and classical measurement conversion techniques. IEEE Trans Aerosp Electron Syst 38:1441–1444
12. Gevers T, Gijsenij A, van de Weijer J, Geusebroek J-M (2012) Color in computer vision: fundamentals and applications. Wiley
13. Gao X, Wang N, Tao D, Li X (2012) Face sketch-photo synthesis and retrieval using sparse representation. IEEE Trans Circuits Syst Video Tech 22:1213–1226
14. Goovaerts P (1999) Geostatistics in soil science: state-of-the-art and perspectives. Geoderma 89:1–45
15. Guo C, Wu D (2019) Canonical correlation analysis (CCA) based multi-view classification: an overview. arXiv:1907.01693
16. Hartley R, Zisserman A (2004) Multiple view geometry in computer vision, 2nd edn. Cambridge University Press, Cambridge
17. Howland P, Wang J, Park H (2006) Solving the small sample size problem in face recognition using generalized discriminant analysis. Patt Recogn 39:277–287
18. Ji S, Ye J (2008) A unified framework for generalized linear discriminant analysis. In: IEEE conference on computer vision and pattern recognition (CVPR)
19. Joshi N, Cohen MF (2010) Seeing Mt. Rainer: lucky imaging for multi-image denoising, sharpening and haze removal. In: IEEE international conference on computational photography (ICCP), pp 1–8
20. Jolliffe IT (1986) Principal component analysis. Published by Springer, Berlin
21. Lafiosca P, Ceccaroni M (2022) Rectifying homographies for stereo vision: analytical solution for minimal distortion. Proc Comp Conf 2:484–503
22. Liao J, Wang Y, Yin J, Zhang S, Zhu D (2018) Segmentation of rice seedlings using the YCrCB color space and an improved Otsu method. Agronomy 8:269
23. Lin Y, Yu Q, Medioni G (2011) Efficient detection and tracking of moving objects in geo-coordinates. Mach Vis Appl 22:505–520
24. Lowe D (2004) Distinctive image features from scale-invariant keypoints. Int J Comp Vis 20:91–110
25. Marinescu M (2024) Explaining and connecting Kriging with Gaussian process regression. arXiv:2408.02331
26. Mihailescu R-C, Davidsson P, Eklund U et al (2018) A survey and taxonomy on intelligent surveillance from a system perspective. Know Eng Rev 33:1–25
27. Nawreen F, Halder KK, Tahtali M et al (2023) Stablization of turbulence-degraded video using patch-based reference frame. Opt Contin 2:2484–2499
28. Nguyen K, Fookes C, Ross A et al (2018) Iris recognition with off-the-shelf CNN features: a deep learning perspective. IEEE Access 6:18848–18855
29. Nowinski WL (2017) Human brain atlasing: past, present and future. Neuroradiol J 30:504–519
30. Nunn WR (1979) Position finding with prior knowledge of covariance parameters. IEEE Trans Aero Elect Sys 15:204–208
31. Peng H, Long F, Ding C (2005) Feature selection based on mutual information: criteria of max-dependency, max-relevance and min-redundancy. IEEE Trans Pattern Anal Mach Intell 27:1226–1238

32. Razavian AS, Azizpour H, Sullivan J, Carlsson S (2014) CNN features off-the-shelf: an astounding baseline for recognition. In: IEEE conference on computer vision and pattern recognition workshops, pp 512–519
33. Schabenberger O, Gotway CA (2005) Statistical methods for spatial data analysis. Published by Chapman and Hall
34. Sanchez-Brea LM, Bernabeu E (2002) On the standard deviation in charge-coupled device cameras: a variogram-based technique for non-uniform images. J Electron Imaging 11:121–126
35. Shukla JS, Rastogi K, Patel H, Jain G, Sharma S (2022) Bag of visual words methodology in remote sensing—a review. Proceedings of the international e-conference on intelligent systems and signal processing
36. St-Charles P-L, Bilodeau G-A (2014) Improving background subtrction using local binary similarity patterns. In: IEEE winter conference on applications of computer vision, pp 509–515
37. Thiagarajan JJ, Ramamurthy KN, Turaga P, Spanias A (2022) Image understanding using sparse representations. In: Springer, synthesis lectures on image, video and multimedia processing
38. Thomaz CE, Giraldi GA (2010) A new ranking method for principal component analysis and its application to face image analysis. Image Vis Comput 28:902–913
39. Topi M (2003) The local binary pattern approach to texture analysis—extensions and application. PhD thesis, University of Oulu
40. Uurito V, Monteiro JM, Kandola J et al (2017) A tutorial on canonical correlation methods. ACM Comput Surv 50:Article 95
41. Valera M, Velastin SA (2005) Intelligent distributed surveillance systems: a review. IEE Proc Vis Image Signal Process 152:192–204
42. Wang M, Perera A, Gutierrez-Osuna R (2004) Principal discriminant analysis for small-sample-size problems: application to chemical sensing. Proceedings of the 3rd IEEE Conference Sensors
43. Wilson D, Alshaabi T, Oort C, van Zhang X, Nelson J (2022) Object tracking and geo-localization from street images. Remote Sens 14:2575
44. Xiang C, Fan XA, Lee TH (2006) Face recognition using recursive fisher linear discriminant. IEEE Trans Image Proc 15:2097–2105
45. Xu K, Li H, Cao W et al (2020) Recognition of weeds in wheat fields based on the fusion of RGB images and depth images. IEEE Access 8:110362
46. Zhao F, Zhang C, Geng B (2024) Deep multimodal data fusion. ACM Comput Surv 56:Article no. 216
47. Zhao L, Peng Q, Huang B (2015) Shape matching algorithm based on shape contexts. IET Comput Vis 9:681–690
48. Zhao H, Lai Z, Leung H et al (2024) Feature learning and understanding: algorithms and applications. Published by Springer, Berlin
49. Zomet A, Levin A, Peleg S, Weiss Y (2006) Seamless image stitching by minimizing false edges. IEEE Trans Image Process 15:969–977

# Chapter 5
# Deep Learning

**Abstract** In this chapter we present a detailed introduction to deep learning and more specifically its use in data fusion. In the first half of the chapter we concentrate on the convolutional neural network (CNN) discussing: training the CNN, CNN architectures (VGGNet, GoogleNet, ResNet, DenseNet, SiameseNet), recent developments (Image fusion CNN) and hybrid networks (SIFT-CNN). In the second half of the chapter we turn to unsupervised deep learning concentrating on autoencoders (AE) discussing AE architectures (convolutional, denoising, denseBlock) and recent developments (denseFuse). Altogether the chapter includes more than 15 examples illustrating CNN's and AE's in different computer vision applications.

## 5.1 Introduction

The subject of this chapter is deep learning and more specifically its use in data fusion. We start with a brief description of the artificial neural network (ANN) since this forms the basis of all subsequent work on deep learning. We discuss the limitations of the ANN and how these eventually led to the development of the convolutional neural network (CNN). Historically the CNN played an important role in the development of deep learning (see Example 5.1) and it remains today the most widely used deep learning architecture. The CNN is a supervised network which requires training on a large training dataset. We discuss the CNN in detail and include discussions on different CNN architectures, recent developments and its use in data fusion. We then turn to *unsupervised* deep learning, specifically, deep auto-encoders. We follow the same approach used for the CNN and discuss the auto-encoder in detail, including discussions on recent developments and the use of auto-encoders in data fusion.

**Example 5.1** *ILSVRC and AlexNet* [18]. Historically the deep learning revolution started when Krizhevsky et al. [18] won the 2012 ImageNet Large Scale Visual Recognition Competition (ILSVRC) [31] by a large margin using a convolutional neural network (CNN) known as AlexNet. AlexNet has 5 con-



H. B. Mitchell, *Data Fusion*, https://doi.org/10.1007/978-3-662-71023-4_5

volutional layers followed by 3 fully-connected-layers (see Sect. 5.5.1). In subsequent years the winning entries of the ILSVCR have all involved deep learning. These competitions were based on ImageNet: a dataset of over 15 million labeled high-resolution images belonging to approximately 22000 categories. The images were collected from the world wide web and labeled by human labelers. The 2012 ILSVRC used a subset of ImageNet with approximately 1000 images in each of the $K = 1000$ categories, or classes, $c_k, k \in \{1, 2, \ldots, K\}$. In total, for the competition there are 1.2 million images available for training, 50000 images for validation and 150000 images for testing.

## 5.2 Artificial Neural Network (ANN)

The first attempts at building *artificial neural networks* (ANN) were motivated by the desire to create a computational model of the animal and human brain. The basic building block of an ANN is the perceptron or neuron. A neuron is a simple processing block with $M$ inputs $x(i), i \in \{1, 2, \ldots, M\}$, and one output $y$ (see Fig. 5.1). After weighting by a factor $w(i)$, the inputs are summed together. A bias $b$ is added to the sum which is then passed through a non-linear *activation* function $f$. This is the output, $y$, of the neuron:

$$y = f\Big(\sum_{i=1}^{M} w(i)x(i) + b\Big). \tag{5.1}$$

Early artificial neural *networks* were constructed by simply placing the neurons in multiple layers and connecting each neuron of one layer to all the neurons of the previous layer. In this case the output of the $n$th neuron in layer $l$ is

$$y(n) = f\Big(\sum_{m} W(m, n)x(m) + b(n)\Big), \tag{5.2}$$

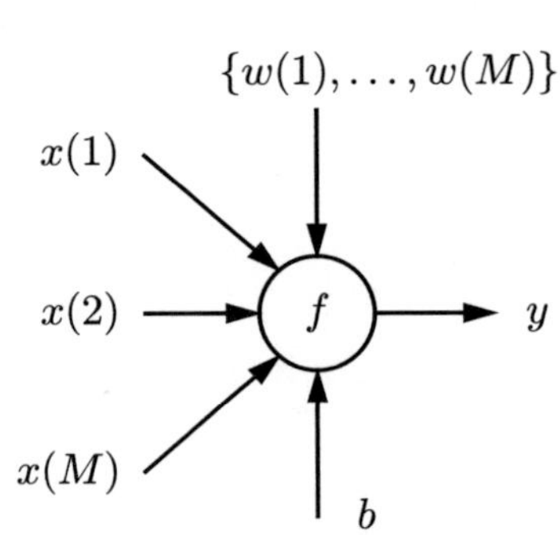


**Fig. 5.1** Shows the neuron with inputs and weights $\big(x(i), w(i)\big), i \in \{1, 2, \ldots, M\}$, bias $b$ and output $y$

where $W(m, n)$ denotes the weight connecting the $m$th neuron in layer $l-1$ to the $n$th neuron in layer $l$ and $b(n)$ is the added bias for the $n$th neuron in layer $l$.

Apart from the layers $l = \{1, 2, \ldots, L\}$, some authors introduce an additional *input sensory layer* which we label $l = 0$. No processing takes place in this layer: its only purpose is to distribute the input signal $\mathbf{x} = \big(x(1), x(2), \ldots, x(M)\big)^T$ to the $l = 1$ layer.

The main limitation of the above ANN is the large number of weights which are used. If $M_l$ denotes the number of neurons in the $l$th layer, then a ANN with $L$ layers has $\sum_{l=1}^{L-1} M_l M_{l+1}$ weights. As we increase $L$, so the number of weights increases exponentially and so does the number of training samples required to train the weights. To prevent this exponential increase in the number of weights, alternative schemes were developed for connecting the neurons.

We now consider two of these schemes ("sparse" and "convolutional") along with the original "fully-connected" scheme.

**Fully-connected.** Each neuron in layer $l$ is connected to all neurons in the previous layer $l-1$ (see Fig. 5.2a). Thus if layers $l-1$ and $l$ are fully-connected there are altogether $M \times N$ independent weights $W(m, n)$ and $N$ independent bias values $b(n)$.

**Sparse.** Each neuron in layer $l$ is connected to $K' = (2K + 1)$ neurons in the previous layer $(l-1)$, where $K' < M$. If layers $l-1$ and $l$ are sparsely connected, there are altogether $K' \times N$ independent weights and $N$ independent bias values $b(n)$. Given a neuron $n$ in layer $l$, we often choose the neurons in layer $l-1$ to

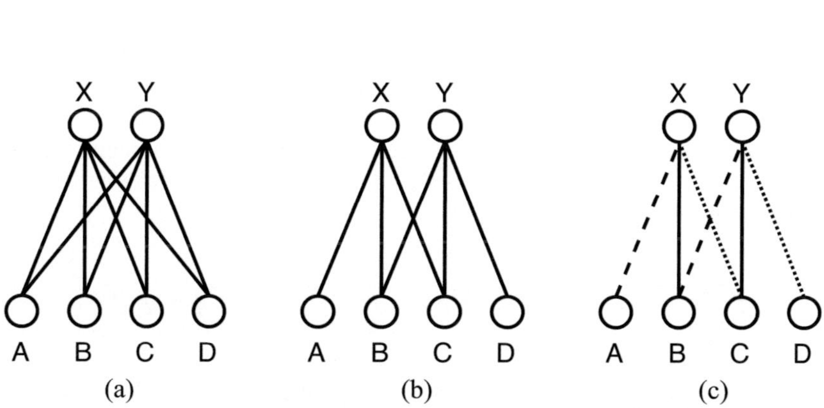


**Fig. 5.2** Shows layer $l$ with two neurons X, Y and layer $l-1$ with four neurons A, B, C, D. The layers are connected in three different ways. **a Fully-connected**: There are $2 \times 4 = 8$ weights: $\{W(A, X), W(A, Y), W(B, X), \ldots, W(D, Y)\}$. **b Sparsely-connected**: Each neuron in layer $l$ is connected to its nearest $K' = 3$ neurons in layer $l-1$. Thus $X$ is connected to $\{A, B, C\}$ and $Y$ is connected to $\{B, C, D\}$. Altogether there are $2 \times 3 = 6$ weights: $\{W(A, X), W(B, X), W(C, X), W(B, Y), W(C, Y), W(D, Y)\}$. **c Convolutionally-connected**: Each neuron in layer $l$ is sparsely and convolutionally connected to its $K' = 3$ neurons in layer $l-1$. Altogether there are 3 weights: $V(-1)$, $V(0)$ and $V(+1)$, where $W(B, X) = W(C, Y) = V(-1)$; $W(A, X) = W(B, Y) = V(0)$ and $W(C, X) = W(D, Y) = V(+1)$

be the $K'$ nearest neighbors of $n$ (see Fig. 5.2b). In this case, for each $n$, the set of (sparse) weights are:

$$W_{\text{sparse}}(m,n) = \begin{cases} \neq 0 & \text{if} \quad m \text{ is among the } K' \text{ nearest neighbors of } n, \\ 0 & \text{otherwise.} \end{cases} \tag{5.3}$$

**Convolutional.** Each neuron $n$ in layer $l$ is sparsely connected to the $K' = (2K+1)$ nearest neurons in the previous layer $l-1$ and *additionally*, the same set of weights and bias values are used for each $n$ (see Fig. 5.2c). If the layers $l$ and $l-1$ are convolutionally-connected, then for each $n$, we have $K' = (2K+1)$ non-zero (convolutional) weights $V(k)$, $k \in \{-K, -K+1, \ldots, K\}$, where

$$W_{\text{conv}}(m,n) = \begin{cases} V(k) & \text{if} \quad m \text{ is the } (k+K+1)\text{th nearest neighbor of } n, \\ 0 & \text{otherwise.} \end{cases} \tag{5.4}$$

As we shall see, the fully-connected and convolutional layers are both used in the CNN, although recently, sparse layers have also been used mainly in place of the fully-connected layers.

### Convolutional Layer: A Sliding-Window Interpretation

The convolutional layer itself may be interpreted as a sliding-window filter. This interpretation [26, 34] is simple and intuitive and we shall often use it in the sections which follow. Mathematically, we may understand the action of the convolutional layer as follows. Given two layers $l-1$ and $l$ with equally-spaced neurons, we use the convolutional weights $V(k)$, $k \in \{-K, -K+1 \ldots, +K\}$, to calculate the output $y(n)$ of the $n$th neuron in layer $l$:

$$y(n) = f\Big(\sum_{m=1}^{M} W_{\text{conv}}(m,n)x(m) + b\Big) = f\Big(\sum_{k=-K}^{K} V(k)x\big(\phi(k;n)\big) + b\Big), \tag{5.5}$$

where $\phi(k;n)$ denotes the neuron in layer $l-1$ which is the $(k+K+1)$th nearest neighbor of the neuron $n$ in layer $l$ and $b$ denotes the one independent bias value.[1]

If the two layers have the same number of neurons, i.e. $M = N$, and are spatially aligned with each other, then $\phi(k;n) = (n+k)$. In this case, Eq. (5.5) becomes

$$y(n) = f\Big(\sum_{k=-K}^{K} V(k)x(n+k) + b\Big). \tag{5.6}$$

[1] In Fig. 5.2, $\phi(-1,n) = \text{B}$, $\phi(0,n) = \text{A}$ and $\phi(1,n) = \text{C}$ if $n = \text{X}$ and $\phi(-1,n) = \text{C}$, $\phi(0,n) = \text{B}$ and $\phi(1,n) = \text{D}$ if $n = \text{Y}$.

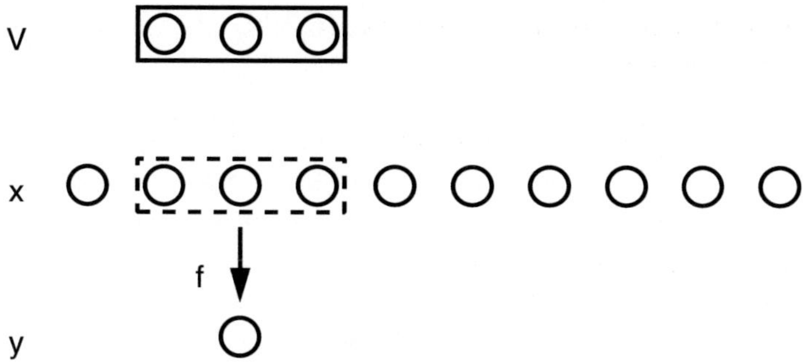


**Fig. 5.3** Shows the calculation of $y(n)$ by (1) convolving **x** with the kernel (sliding-window) **V**, (2) Adding the bias $b$ and (3) Passing the result through the activation function $f$

where, by definition, $\sum_{k=-K}^{K} V(k)x(n+k)$ is the convolution[2] between $\mathbf{x} = \big(x(1), x(2), \ldots, x(M)\big)^T$ and the filter (sliding window) $\mathbf{V} = \big(V(-K), V(-K+1) \ldots, V(+K)\big)^T$.

Inspection of Eq. (5.6) shows that the output of the $n$th neuron, $y(n), n \in \{1, 2, \ldots, N\}$, may be calculated in the following two stages. See Fig. 5.3.

- **Sliding-window**. Move a sliding window **V** of length $(2K+1)$ over the vector **x** (i.e. over the neurons in layer $(l-1)$). At each position of the sliding window add a bias $b$ to the weighted sum of $x$.
- **Activation**. Pass the weighted sum plus bias through the activation function $f(\cdot)$.

Further examination of Eq. (5.6), shows that the summation $\sum_k V(k)x(n+k)$ is only defined when $(n+k) \in \{1, 2, \ldots, N\}$. This in turn means that $y(n)$ can only be calculated for $n \in \{K+1, 2, \ldots, N-K\}$. To ensure that we obtain a full complement of $y(n), n \in \{1, 2, \ldots, N\}$, values, we may *artificially* extend the input vector **x** by "padding" **x** with $K$ zeros at each end. Let $\mathbf{x}_{\mathrm{PAD}}$ denote the padded vector:

$$\mathbf{x}_{\mathrm{PAD}} = \Big(\overbrace{0\,0\,0\,\ldots\,0}^{K \text{ zeros}}\; x(1)\; x(2), \ldots\; x(M)\; \overbrace{0\,0\,\ldots\,0}^{K \text{ zeros}}\Big)^T. \tag{5.7}$$

Then convolving $\mathbf{x}_{\mathrm{PAD}}$ with **V** gives $y(n)$ for all $n, n \in \{1, 2, \ldots, N\}$. The technique of zero-padding is widely used in CNNs, and more generally in deep learning. However, alternative padding schemes are sometimes used.

We now consider the CNN in detail.

[2] Strictly speaking $\sum_k V(k)x(n+k)$ is the cross-correlation between **x** and **V**. However, in computer vision we often refer to it as convolution.

## 5.3 Convolutional Neural Network (CNN)

Since its appearance in 2012, the CNN has been widely used in pattern recognition and image classification. The CNN may be understood as sequence of signal processing modules, whose architecture is a mixture of different modules and includes convolution (sliding-window filter), activation functions, normalization functions, down-sampling (pooling) operators and full-connected layers [26]. Each module is called a "layer" and is given a label $l, l \in \{1, \ldots, L\}$.

If we represent each layer $l$ by a function $g_l$, then formally we write the CNN as a composition of the $L$ functions $g_l$:

$$\mathbf{y}_L = g_L\Big(\cdots g_2\big(g_1(\mathbf{x}_1; \mathbf{W}_1); \mathbf{W}_2\big) \cdots \mathbf{W}_L\Big), \tag{5.8}$$

where each function $g_l$ takes as input a vector $\mathbf{x}_l$ and a set of parameters $\mathbf{W}_l$ and outputs a vector $\mathbf{y}_l$, where $\mathbf{y}_l = g_l(\mathbf{x}_l; \mathbf{W}_l)$.

The layers themselves may be usefully divided into "early" layers and "late" layers. Given an input image $I$, the function of the early layers is to extract a hierarchy of features from $I$ while the function of the late layers is to make a prediction regarding the classification of $I$.

To use the CNN, it is first trained on a large dataset of labeled images $\mathcal{D}$ which contain the objects of interest. During training, the CNN learns to associate the extracted features with the correct classification labels through an optimization process known as *backpropagation*. Once the CNN has been trained, it is used to classify a new, previously unseen test image $I_{\text{TEST}}$, by passing the image through the network and selecting the label with the highest predicted probability.

To make the above discussion more concrete we now consider a basic generic CNN and its training.

## 5.4 A Generic Basic CNN

Figure 5.4 shows a generic CNN consisting of two blocks of layers. The first block contains the "early" layers and its purpose is to extract a hierarchy of features from the input signal. The second block contains the "late" layers and its purpose is to classify the input signal.

#### Convolutional (Sliding-Window) Layer

This is the core building block of a CNN. It performs a convolution operation on the input signal $\mathbf{x}$ with a set of learnable filters. The filters are small $K' \times L' \times D$ matrices that slide over $\mathbf{x}$, computing a dot product between the filter and a small region of the input at each step. The output of the convolutional layer is a set of feature maps that represent different features in $\mathbf{x}$. The feature maps are stacked

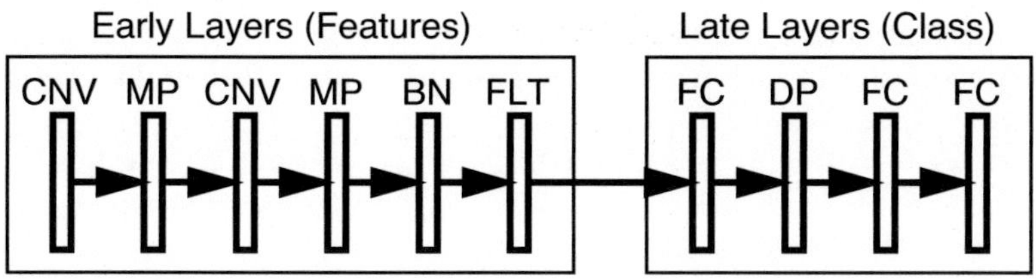


**Fig. 5.4** Shows the early and late layers in a generic CNN. The early layers include convolutional layers CNV (plus zero-padding and ReLU activation); max pooling layers MP; a batch normalization layer BN and a flatten layer FLT. The late layers include a dropout layer DP and fully-connected layers FC

together to form the output signal $\mathbf{y}$. The feature maps may also be sub-sampled. The sub-sampling is defined by a stride factor $s$ along each spatial dimension of the feature maps.

Suppose the size of $\mathbf{x}$ is $M \times N \times D$. If $\mathbf{x}$ is zero-padded and the stride $s = 1$, then each feature map is of size $M \times N$ and the $h$th feature map is given by

$$F_h(m,n) = \sum_{k=-K}^{K} \sum_{l=-L}^{L} \sum_{d=1}^{D} x(m+k, n+l, d) V_h(k,l,d) + b_h, \tag{5.9}$$

where $K' = (2K+1)$, $L' = (2L+1)$, $m \in \{1, 2, \ldots, M\}$ and $n \in \{1, 2, \ldots, N\}$. $V_h(k,l,d)$ denotes the weights in the $h$th sliding-window filter $\mathbf{V}_h$ and $b_h$ is the bias. The feature maps $F_h$, $h \in \{1, 2, \ldots, H\}$, are stacked together to give the output signal output signal $\mathbf{y}$, where the element $y(m,n,h)$ is given by:

$$y(m,n,h) = F_h(m,n). \tag{5.10}$$

The size of $\mathbf{y}$ is $M \times N \times H$, where $H$ is the total number of filters used in the convolutional layer.

### Zero-Padding Layer

The zero-padding layer pads the input signal $\mathbf{x}$ with zeros. Most times, this layer is included within the convolutional layer. Nowadays, zero-padding is the most popular padding scheme. Nevertheless, alternative padding schemes [23] are sometimes used e.g. in small object detection.

### Max-Pooling Layer

This layer is used to reduce the spatial dimensions of the input signal $\mathbf{x}$. Initially $\mathbf{x}$ (size $M \times N \times D$) is decomposed into $D$ feature maps of size $M \times N$. Then taking each feature map in turn, we slide a small $K \times L$ window over the feature map taking the maximum value in each window. The result is a "max" feature map $F_d(m, n)$:

$$F_d(m,n) = \max\big(x(m,n,d), x(m+1,n,d), \ldots, x(m+K, n+L, d)\big), \tag{5.11}$$

where $m \in \{1, 2, \ldots, M\}$, $n \in \{1, 2, \ldots, N\}$ and $d \in \{1, 2, \ldots, D\}$. The $F_d(m, n)$ are down-sampled by a factor $s$ in each spatial dimension and then stacked to give the output signal $\mathbf{y}$ of size $(M/s \times N/s \times D)$. The element $y(i, j, d)$ in $\mathbf{y}$ is given by:

$$y(i, j, d) = F_d((i-1) \times s + 1, (j-1) \times s + 1). \tag{5.12}$$

In general the results of pooling are to substantially reduce the size of the network and to make it robust to small translations in the input scene.

Although max pooling is the most popular pooling method, in some applications alternative pooling methods [38] are used instead.

### ReLU Activation Layer

The activation layer applies a non-linear activation function to the output of the previous layer. This introduces non-linearity into the network, allowing it to learn more complex features. The most widely used activation function is ReLU:

$$ReLU(x) = \begin{cases} x & \text{if } x \geq 0, \\ 0 & \text{otherwise.} \end{cases} \tag{5.13}$$

Most times this layer is included within the convolutional layer.

Apart from ReLU, different activation functions including modified ReLU functions, are sometimes used. See [6] for a review of modern activation functions.

### Batch Normalization Layer

The batch normalization layer normalizes the output of the previous layer by subtracting the mean and dividing by the standard deviation of the batch. This helps to reduce the internal covariate shift and improves the convergence of the network.

### Flatten Layer

The flatten layer simply rewrites the input to the layer as a single 1D vector.

**Table 5.1** Basic CNN architecture

| # | Input | Output | Type | Kernel | Remarks |
|---|---|---|---|---|---|
| 1 | $226 \times 226 \times 3$ | $224 \times 224 \times 32$ | CNV1 | $3 \times 3 \times 3(32)$ | Stride $s = 1+$ zero-pad $+$ ReLU |
| 2 | $224 \times 224 \times 32$ | $112 \times 112 \times 32$ | MP1 | $2 \times 2$ | Stride $s = 2$ |
| 3 | $112 \times 112 \times 32$ | $112 \times 112 \times 32$ | CNV2 | $3 \times 3 \times 32(32)$ | Stride $s = 1+$ zero-pad $+$ ReLU |
| 4 | $112 \times 112 \times 32$ | $28 \times 28 \times 32$ | MP2 | $2 \times 2$ | Stride $s = 4$ |
| 5 | $28 \times 28 \times 32$ | $28 \times 28 \times 32$ | BN | – | Batch normalization |
| 6 | $28 \times 28 \times 32$ | $25088 \times 1$ | FLT | – | Flattening |
| 7 | $25088 \times 1$ | $64 \times 1$ | FC1 | – | Includes ReLU activation |
| 8 | $64 \times 1$ | $64 \times 1$ | DP | – | Dropout 20% of weights |
| 9 | $64 \times 1$ | $64 \times 1$ | FC2 | – | Includes ReLU activation |
| 10 | $64 \times 1$ | $4 \times 1$ | FC3 | – | Includes softmax activation |

**Dropout Layer**

The dropout layer randomly drops out a fraction $P$ of the neurons in the previous layer during training. This helps to prevent over-fitting by forcing the network to learn more robust features.

**Fully-Connected Layer**

The fully-connected layer is a traditional neural network layer that connects every neuron in the previous layer to every neuron in the current layer. It is typically used at the end of the network to produce the final output.

## 5.4.1 *Analysis*

We now analyze the basic CNN shown in Fig. 5.4. The modules in the basic CNN are listed in Table 5.1 together with their parameters.[3] Input to the network is a $224 \times 244 \times 3$ color image $I$. In the "early" layers we extract a hierarchy of features from $I$ using the CNV1, MP1, CNV2, MP2 layers. The features are normalized by the BN layer and are stored in a $28 \times 28 \times 32$ image $\mathbf{y}_5$. In the flattening layer FLT this image is rewritten as a column vector $\mathbf{y}_6$ which is fed to the dropout DP and

[3] In Table 5.1 we explicitly write the kernels of the convolutional layers as the tuples $K' \times L' \times D$ followed by the number of kernels $(H)$. However, it is common practice to leave out the depth $D$. From now on, we shall therefore, unless indicated otherwise, write the kernels of the convolutional layers and their number as $K' \times L'(H)$.

fully-connected ("late") layers FC1, FC2, FC3. The output of the last layer, FC3, is a vector $\mathbf{p}$:

$$\mathbf{p} = (p(1), p(2), \ldots, p(K))^T, \tag{5.14}$$

where $p(k)$ is interpreted as the *a posteriori* probability that $I$ belongs to the $k$th class $c_k, k \in \{1, 2, \ldots, K\}$. The probabilities $p(k)$ are generated by passing the outputs of FC3, $y(k), k \in \{1, 2, \ldots, K\}$, through a *softmax* activation function:

$$p(k) = \frac{\exp\big(y(k)\big)}{\sum_{l=1}^{K} \exp\big(y(l)\big)}. \tag{5.15}$$

The architecture of the basic CNN is widely used in many applications with only minor variations. In the following example we use a "region"-CNN (R-CNN) to perform generic object detection [30].

**Example 5.2** *Fast R-CNN and Faster R-CNN: Two-Stage Object Detection using CNN* [9, 29, 30]. An early application of the CNN was object detection. Given an input image $I$, containing an unknown number of objects, we perform:

**Classification.** Classify each object. For this purpose we use $K$ object classes and one "background" class.

**Localization.** Specify the position and size of each object in the image. The method adopted for doing this is to specify a compact rectangular "bounding box" (bbox) for each object. The bounding boxes are specified by a four-tuple $(R, C, H, W)$, where $(R, C)$ and $(H, W)$ are, respectively, the top left corner row $(R)$ and column $(C)$ coordinates of the bounding box and its height $(H)$ and width $(W)$.

Figure 5.5 illustrates the architecture of the fast R-CNN algorithm. The algorithm works as follows.

The input image $I$ is first processed by a non deep-learning region proposal module (RPM) whose aim is to identify regions of interest (ROI) which have a high probability of containing an object. Simultaneously, $I$ is processed through the early layers of a CNN which outputs a $M \times N \times D$ feature map $F$. For each ROI (size $H \times W$) we extract the corresponding projected feature block which we convert to a fixed size (using a "region-of-interest" (ROI) pooling layer) and then send to the fully-connected (FC) layers for classification and positioning.

In faster R-CNN the RPM is integrated into the algorithm by using a specially designed CNN which acts on the feature map and outputs multiple candidate region. Faster R-CNN is a complete object detector which is trained end-to-end and operates in near real-time.

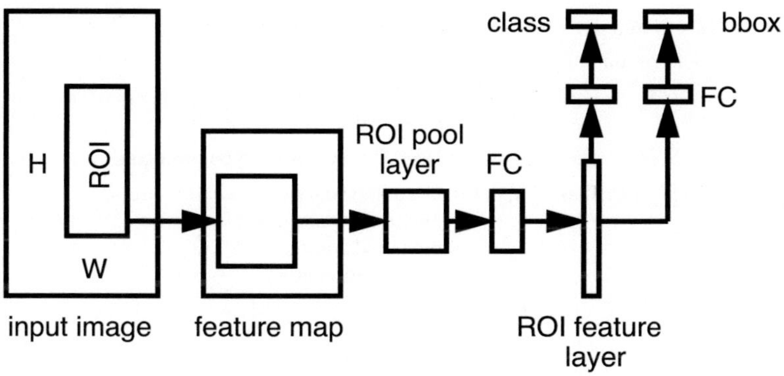


**Fig. 5.5** Shows the architecture of the fast R-CNN object detector. The original R-CNN used AlexNet. However, the later fast R-CNN and faster R-CNN algorithms also used GoogleNet and ResNet. See Sect. 5.5

Apart from the class of R-CNN object detectors alternative CNN-based detectors have been developed. Prominent among them are: YOLO ("You-only-look-once") and SSD ("Single shot detector") [28, 30].

## 5.4.2 *Training*

We now consider how to train the CNN, i.e. adjusting the weights and biases of the network so that it produces accurate predictions on the training data. This is achieved using an optimization algorithm that minimizes a loss function $\mathcal{L}$ which measures the difference between the predicted output of the network and the true output. The most commonly utilized optimization algorithm for training CNNs is stochastic gradient descent (SGD), which updates the weights and biases of the model in small steps based on the gradient of the loss function with respect to these parameters. The process of updating the weights and biases using the gradient is called *backpropagation*, and it involves propagating the error from the output layer back through the network to update the weights and biases of each layer.

**Loss Functions**

The loss function $\mathcal{L}$ uses two parameters to calculate the error, the first parameter is the estimated output of the CNN model and the second one is the actual output. Table 5.2 lists some common loss functions used in CNN learning.

**Table 5.2** CNN loss functions

| Name | Formula | Description |
|---|---|---|
| Cross-entropy | $-\sum_i y_i \log(p_i)$ | Used for classification, $p_i$ is predicted output, $y_i$ is true label |
| Euclidean | $\sum(x_i - y_i)^2$ | Used for regression-type problems |
| Hinge Loss | $\sum_i \max(0, m - \widetilde{y}_i p_i)$ | Used for binary classification problems. Aim is to maximize margin $m$ between the two classes; $p_i$ is predicted output; $\widetilde{y}_i = 2y_i - 1$ and $y_i \in \{0, 1\}$ is true classification label |

## Overfitting

The learning rate controls the size of the updates made to the weights and biases at each iteration of the optimization algorithm. The learning rate should be carefully chosen: If the rate is too low the optimization takes too long to converge while if rate is too high the algorithm may fail to converge.

A major problem in training the CNN is *overfitting*. This occurs when the network performs well on the training data but poorly on validation or test data. Regularization techniques are used to help prevent overfitting. They work by adding a penalty term to the loss function. Table 5.3 lists some common regularization techniques.

Another technique used to prevent overfitting is to increase the size of the training dataset. This is known as *data augmentation* and is used to generate additional training data from an existing training dataset [8, 35]. Some common data augmentation techniques are listed in Table 5.4.

## Transfer Learning

A useful technique for training a CNN is to adapt an already trained CNN to a new task. This is known as *transfer learning*. For example, suppose we wish to adapt a

**Table 5.3** Regularization techniques in CNN learning

| Name | Description |
|---|---|
| $L1$ regularize | This technique adds penalty to $\mathcal{L}$ which is proportional to absolute value of weights in CNN. Equation for $L1$ regularized loss function is $\mathcal{L}_1 = \mathcal{L} + \lambda \sum_{l=1}^{L} \lvert \mathbf{W}_l \rvert$, where $\mathcal{L}$ is original non-regularized loss function, $\mathbf{W}_l . l \in \{1, 2, \ldots, L\}$, are weights in layer $l$ in network, $\lambda$ is regularization strength |
| $L2$ regularize | Same as $L1$ regularization except penalty is proportional to square of weights |
| Dropout | Technique randomly drops some of neurons during training. This prevents network relying too much on any one feature. Dropout is implemented by randomly setting output of each neuron to zero with certain probability $P$ during training. During testing, dropout is turned off and output is scaled by $(1 - P)$ to compensate for effect of dropout [33] |
| Early stopping | Technique stops training process when performance on training dataset starts to deviate from performance on validation dataset |

**Table 5.4** Data augmentation techniques

| Name | Description |
|---|---|
| Flipping | Create new image by image mirroring around random horizontal or vertical line |
| Cropping | Create new image by randomly cropping training image |
| Affine | Create new image by applying random affine transform to training image |
| Color jitter | Create new image by randomly adjusting color, intensity of training image. Color and intensity found by principal component analysis of training dataset |
| Cutout | Create new image by random masking square region in training image |
| Mixup | Create new image $\tilde{x} = \lambda x_1 + (1 - \lambda)x_2$ and new one-hot label $\tilde{y} = \lambda y_1 + (1 - \lambda)y_2$ from training images $x_1, x_2$, one-hot labels $y_1, y_2$, where $\lambda$ is random # $\in [0, 1]$ |
| RICAP | Create new image by random image crop & patch. Crop 4 training images. Patch them to form new image |

pre-trained CNN to a new set of classification labels. In this case, all layers of the new model (CNN), apart from the last few layers, are fixed and are set equal to the pre-trained CNN. In the new CNN, the last few layers are randomly initialized, where the number of neurons in the last layer is set equal to the number of class labels in the new dataset. See Example 5.4.

We conclude this discussion on the convolutional neural network (CNN) with an example illustrating deep learning multi-sensor data fusion using multiple CNN's.

**Example 5.3** *Deep Learning Multi-Sensor Fusion for Waste Image Classification* [2]. The aim of waste recycling is recover valuable materials from the waste and thereby minimize the environmental impact of extracting these materials. An important application of waste recycling is to use waste wood as a substitute for virgin wood. To work efficiently, waste wood recycling requires effective sorting and classification of recycled wood. Traditional sorting methods are labor intensive with limited accuracy and high error rates. To overcome these limitations [2] suggest multi-sensor deep learning data fusion classification.

For this application, $K = 4$ image sensors (RGB, hyperspectral near-infrared, thermography, and terahertz) were used. Each sensor recorded an image $I_k, k \in \{1, 2, \ldots, K\}$, of bulky waste. For each sensor a simple CNN (similar to the basic CNN of Sect. 5.4) was separately trained on an appropriate training dataset. Given a test sample $O$, images are captured with the $K$ image sensors. Each image $I_k$ is separately classified as wood or no-wood. Suppose $S_k$ denotes the corresponding classification score for the $k$th sensor. Then the $S_k$ are concatenated together and the final classification score $\widetilde{S}$ and decision $\widetilde{D}$ are found using an artificial neural network consisting of $L$ fully-connected layers (see Fig. 5.6).

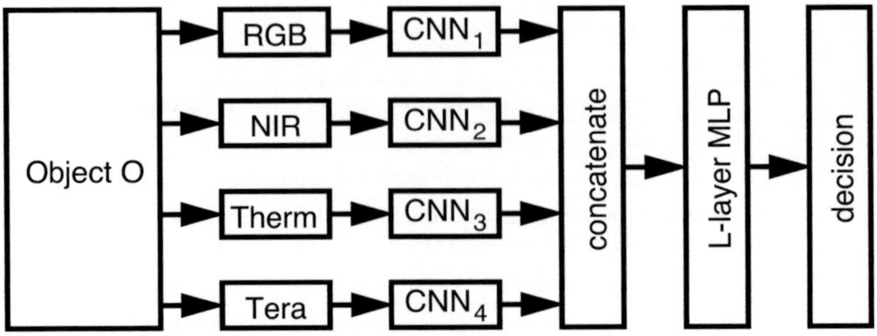


**Fig. 5.6** Shows architecture of multisensor deep learning classifier. Four multimodal cameras (RGB, near infra- red, thermography and terahertz) capture images $I_k, k \in \{1, 2, \ldots, 4\}$, of the bulky waste object $O$. Each image $I_k$ is separately classified as wood or no-wood, using the $k$th network $\text{CNN}_k$. The corresponding classification scores $S_k$ are concatenated together after which they are fed into a multi-layer perceptron with $L$ fully-connected layers. The multi-layer perceptron outputs the final classification score $\widetilde{S}$ and (binary) decision $\widetilde{D}$ regarding the object $O$

## 5.5 CNN Architectures

In this section we briefly describe a few of the more common CNN architectures.

### *5.5.1 AlexNet*

Although not widely used nowadays it is of historical significance: it is the original CNN used by Krizhevsky et al. in the 2012 ILVRS competition and is listed in Table 5.5. AlexNet has five convolutional layers followed by three fully-connected

**Table 5.5** AlexNet input/output analysis

| # | Input | Output | Type | Kernel | Remarks |
|---|---|---|---|---|---|
| 1 | $224 \times 224 \times 3$ | $55 \times 55 \times 96$ | CNV1 | $11 \times 11 \times 3(96)$ | Stride 4+ RelU |
| 2 | $55 \times 55 \times 96$ | $27 \times 27 \times 96$ | MP1 | $3 \times 3$ | Stride 2 |
| 3 | $27 \times 27 \times 96$ | $27 \times 27 \times 96$ | LRN1 | – | |
| 4 | $27 \times 27 \times 96$ | $27 \times 27 \times 256$ | CNV2 | $5 \times 5 \times 96(256)$ | Zero-pad + ReLU |
| 5 | $27 \times 27 \times 256$ | $13 \times 13 \times 256$ | MP2 | $3 \times 3$ | Stride 2 |
| 6 | $13 \times 13 \times 256$ | $13 \times 13 \times 256$ | LRN2 | – | |
| 7 | $13 \times 13 \times 256$ | $13 \times 13 \times 384$ | CNV3 | $3 \times 3 \times 256(384)$ | Zero-pad + ReLU |
| 8 | $13 \times 13 \times 384$ | $13 \times 13 \times 384$ | CNV4 | $3 \times 3 \times 384(384)$ | Zero-pad + ReLU |
| 9 | $13 \times 13 \times 384$ | $13 \times 13 \times 256$ | CNV5 | $3 \times 3 \times 384(256)$ | Zero-pad + ReLU |
| 10 | $13 \times 13 \times 256$ | $6 \times 6 \times 256$ | MP3 | $3 \times 3$ | Stride 2 |
| 11 | $6 \times 6 \times 256$ | $9216 \times 1$ | FL | – | |
| 12 | $9216 \times 1$ | $4096 \times 1$ | FC1 | – | ReLU |
| 13 | $4096 \times 1$ | $4096 \times 1$ | FC2 | – | ReLU |
| 14 | $4096 \times 1$ | $1000 \times 1$ | FC3 | – | Softmax |

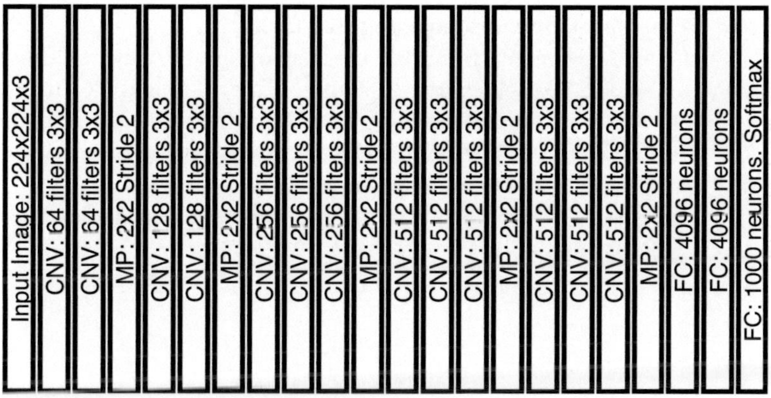


**Fig. 5.7** Shows the architecture of VGGNet-16. It consists of 13 convolutional layers (with $3 \times 3$ sliding-window kernels), five $2 \times 2$ max pooling layers (with stride of 2) and 3 fully-connected layers. Each convolutional layer includes zero-padding and ReLU activation function

layers. The first two convolutional layers use $11 \times 11$ and $5 \times 5$ kernels. The remaining convolutional layers use $3 \times 3$ kernels. Batch normalization is not used. Instead a local response normalization (LRN) is used.

### 5.5.2 *VGGNet*

After CNN was found to be very effective for image recognition (cf. Example 5.1), Simonyan and Zisserman introduced an simple and efficient design principle for CNN. This design was called Visual Geometry Group (VGG) in which convolutional layers were restricted to size $3 \times 3$ and max pooling layers were all of size $2 \times 2$.

Altogether there are six VGGNet configurations of which the most successful are VGGNet-16 and VGGNet-19. All VGGNets were trained on the ILSVRC training set. The architecture of VGGNet-16 is shown in Fig. 5.7.

The VGGNet-16 and VGGNet-19 networks are widely used in different computer vision and machine learning applications, although their computational cost is excessive with approximately 140 million parameters.

The following example illustrates using VGGNet in shadow detection.

**Example 5.4** *Shadow Detection Using VGGNet-16 Network* [7, 16]. Traditionally, shadow detection algorithms exploit the fact that the brightness of shadows generally decrease while their chromaticity value remains fixed. In this example we describe a new algorithm based on deep learning. Figure 5.8 shows a flowchart of the algorithm. In the first step a foreground region is detected through background subtraction (see Examples 4.14 or 6.7). Then a $M \times M$ window is extracted from the foreground region. After resizing the window to $224 \times 224 \times 3$ it is used as input to a pre-trained VGGNet-16. The number of neurons in the last layer is reduced from 1000 to 2, and the network is re-trained by *transfer learning* on a experimental shadow training database $\mathcal{D}_{\text{shadow}}$. The output from the network is a binary decision map $B_{\text{shadow}}$, where $B_{\text{shadow}}(i, j) = 1$ if pixel $(i, j)$ is classified as a shadow; otherwise $B_{\text{shadow}}(i, j) = 0$.

### 5.5.3 GoogleNet

The winner of the ILSVRC competition held in 2014 was GoogleNet. This CNN was trained on the ILSVRC training set and was designed for high accuracy with low computational cost. To this end, GoogleNet uses a parallel network of filters of different sizes. The basic building block of the network is the "inception module" (Fig. 5.9). The inception module consists of $1 \times 1$, $3 \times 3$ and $5 \times 5$ convolutional layers and a $3 \times 3$ max pooling layer all of which are processed in parallel. The corresponding feature maps are then combined by concatenation. Since this results in a very high-dimensional feature, the output of the inception module is reduced in size by pre-filtering with $1 \times 1$ filters.

The architecture of GoogleNet with the inception modules is shown in Fig. 5.10.

### 5.5.4 ResNet

The winner of the ILSVRC competition held in 2015 was ResNet. This CNN was trained on the ILSVRC training set. The objective in ResNet was to design a network with improved performance by designing an ultra-deep neural network.

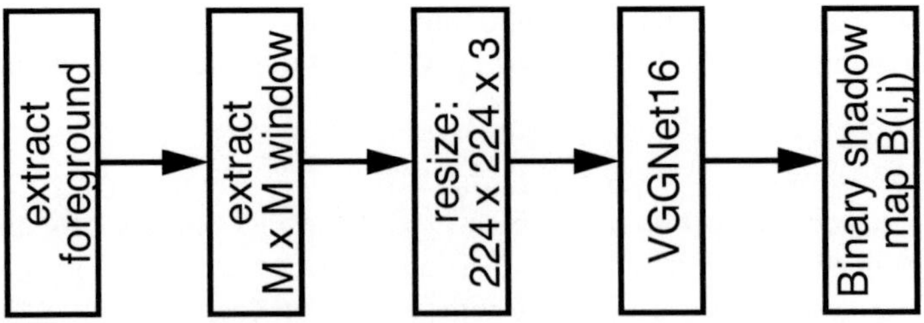


**Fig. 5.8** Shows the CNN-based shadow detection algorithm. It consists of 5 steps: (1) Foreground extraction. (2) Extract $M \times M$ window $W$ from foreground, (3) Resize $W$ to $224 \times 224 \times 3$ (4) VGGNet16. (5) Detect shadow region

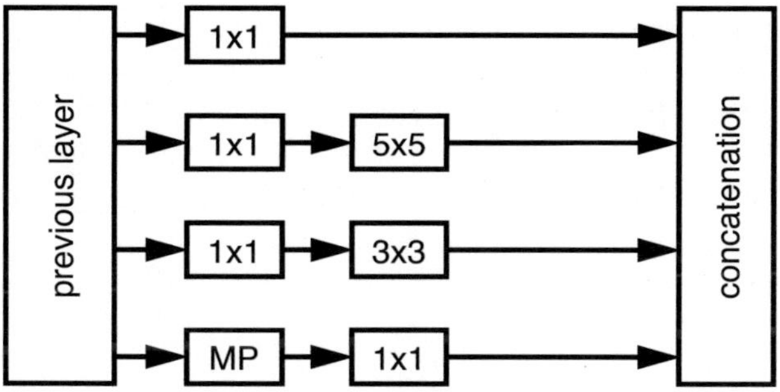


**Fig. 5.9** Shows the structure of the inception module

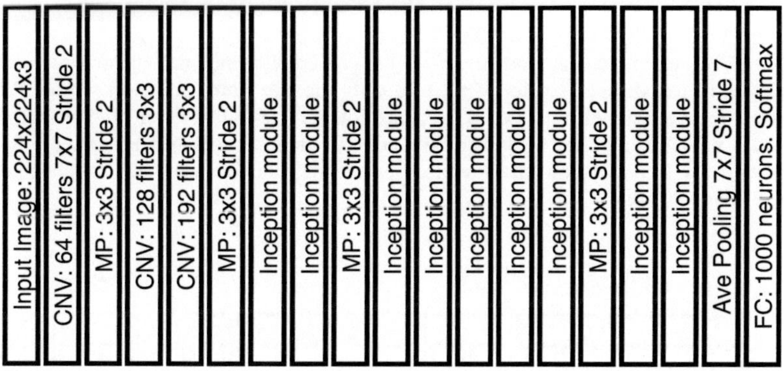


**Fig. 5.10** Shows the architecture of GoogleNet. It consists of traditional convolutional and max-pooling layers followed by 9 inception modules. In place of fully-connected layers it uses an average pooling layer

Theoretically, deep neural networks (DNN) outperform shallow ones as the former can extract more complicated and sufficient features. However, with the increase in the number of layers, the gradient of the loss function becomes vanishingly small as they propagate back through the network, and this in turn makes it difficult to train a DNN. ResNet addresses this issue by introducing a "residual block" constructed by using a skip, or shortcut, connection. Three different types of residual blocks are shown in Fig. 5.11. The architecture of ResNet-34 with 13 residual blocks is shown in Fig. 5.12. Resnet-34 consists of convolutional layers with filters of size $3 \times 3$. Only two pooling layers are used: one at the beginning and the other at the end of the network. Identity connections are used between every two convolutional layers.

**Example 5.5** *A Comparative Analysis of ResNet Architectures* [22]. The use of skip connections means that theoretically the ResNet architecture can be designed to any depth. Nagpal et al. [22] performed a comparative analysis of

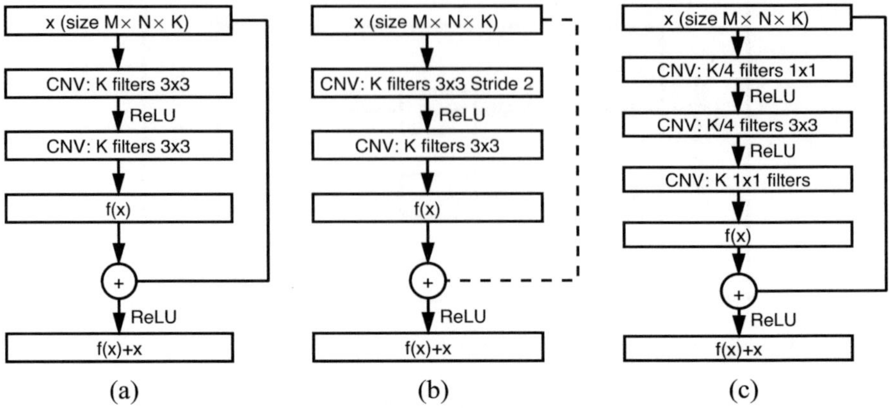


**Fig. 5.11** Shows three residual module architectures. **a** Identity module. The original vector **x** is summed with its transformed version $f(\mathbf{x})$. **b** Identity module with $1 \times 1$ convolution. When the dimensions of **x** is different from $f(\mathbf{x})$, skip connection includes a $1 \times 1$ convolution to change the dimension of **x**. This is indicated by dashed skip line. **c** Bottleneck module. The module which starts and ends with $1 \times 1$ convolution which reduces number of parameters and improves the nonlinear response of the network

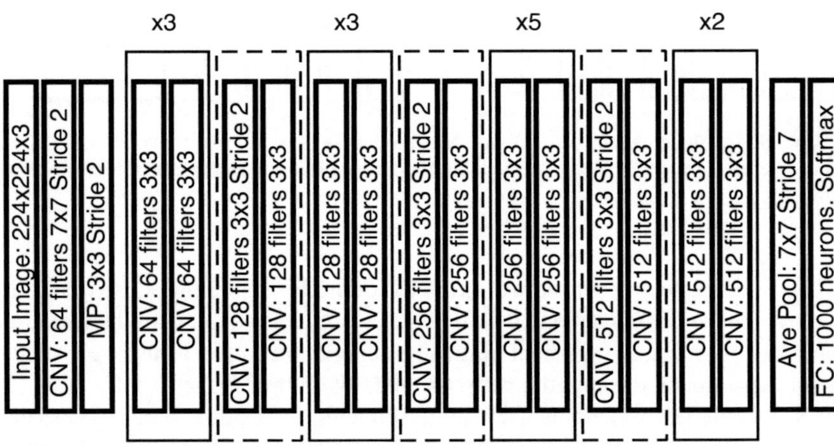


**Fig. 5.12** Shows the architecture of ResNet-34. It consists of traditional convolutional and max-pooling layers followed by 13 residual modules. Solid lines indicate identity mappings while dashed lines indicate identity mappings with $1 \times 1$ convolution

ResNet architectures by training ResNet architectures with different depths on the same dataset $\mathcal{D}$. Table 5.6 shows the performance of the different architectures. We see, as expected, a steady improvement in performance as the depth of the network increases.

**Table 5.6** Comparison between ResNet architectures

| Name | Accuracy (%) | Loss (%) |
|---|---|---|
| ResNet-34 | 82.5 | 81.11 |
| ResNet-50 | 91.39 | 34.73 |
| ResNet-101 | 94.89 | 19.39 |
| ResNet-152 | 95.29 | 18.5 |

The reader should keep in mind that although in general, deeper networks have better performance, this is not always true as the following example shows.

**Example 5.6** *Low-Resolution Images: Comparative Analysis of AlexNet, VGGNet, GoogleNet and ResNet* [13]. The results obtained in the ILSVRC competitions show that classification using GoogleNet and ResNet give substantially better results than AlexNet and VGGNet. These results are, however, only true for high resolution images. For low-resolution images, [13] obtained superior classification results with "classical" AlexNet and VGGNet networks

### 5.5.5 DenseNet

DenseNet extends the idea of ResNet by defining a "dense block". This is a generalization of the residual module. It consists of $L$ layers in which each layer $l, l \in \{1, 2, \ldots, L\}$, takes as input the concatenation of all the preceding feature maps $\mathbf{x}_1, \mathbf{x}_2, \ldots, \mathbf{x}_{l-1}$. Thus each layer $l$ produces an output which is function of all previous feature maps:

$$\mathbf{x}_l = f_l([\mathbf{x}_1, \mathbf{x}_2, \ldots, \mathbf{x}_{l-1}]),$$

where $f_l(x)$ is composed of three operations in sequence: batch normalization, ReLU and $3 \times 3$ convolution (see Fig. 5.13).

A denseNet may be formed from a *sequence* of dense blocks in which the dense blocks are separated by a transition layer composed of batch normalization, a $1 \times 1$ convolution, followed by a $2 \times 2$ average pooling with a stride of 2. DenseNet was the winner of the ILSVRC competition held in 2016.

### 5.5.6 Convolutional Siamese Networks

Convolutional Siamese networks (CSN) are a class of neural network architecture that contain several parallel branches with identical CNN's, i.e. each CNN has the

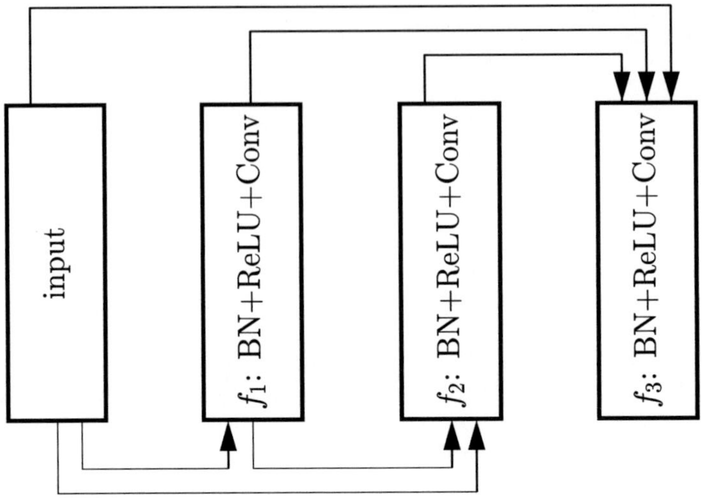


**Fig. 5.13** Shows the structure of the dense block composed of three functions $f_l$

same configuration with the same parameters and weights. During the training phase, parameter updating is mirrored across all branches.

The following example illustrates a siamese CNN image fusion network.

**Example 5.7** *IFCNN: A siamese image fusion framework* [40]. We consider the fusion of $K(K \geq 2)$ spatially aligned images $I_k, k \in \{1, 2, \ldots, K\}$, using a siamese image fusion network (IFCNN). Salient features are separately extracted from each image $I_k$ using convolutional layers CNV1 and CNV2 (Fig. 5.14). The features are fused together using an appropriate element-by-element fusion rule. Possible fusion rules include:

**Max Rule**. $\widetilde{F}(m,n) = \max\left(F_1''(m,n), F_2''(m,n), \cdots, F_K''(m,n)\right)$,
**Min rule**. $\widetilde{F}(m,n) = \min\left(F_1''(m,n), F_2''(m,n), \cdots, F_K''(m,n)\right)$,
**Mean rule**. $\widetilde{F}(m,n) = \frac{1}{K}\left(F_1''(m,n) + F_2''(mn,)+, \cdots, +F_K''(m,n)\right)$,

where $F_k''(m,n)$ denotes the $(m,n)$th feature generated CNV2 from the image $I_k$ and $\widetilde{F}(m,n)$ is the corresponding fused feature.

Convolutional layers CNV3 and CNV4 use the fused features $\widetilde{F}(m,n)$ to generate the fused image $\widetilde{I}$. To train the IFCNN a database $\mathcal{D}$ of *spatially-aligned* sets of images are used. IFCNN has been successfully used for fusing multi-modal medical images, multi-exposure images and infrared-visual images.

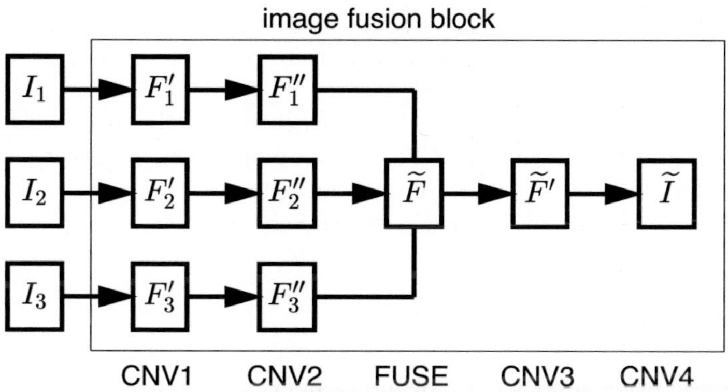


**Fig. 5.14** Shows a general image fusion framework consisting of a siamese CNN with three parallel branches (layers CNV1 and CNV2) followed an element-by-element fusion block and a conventional CNN (layers CNV3 and CNV4)

The following example illustrates a novel combination of IFCNN and ResNet.

**Example 5.8** *Wind blade Inspection Using IFCNN and ResNet* [37]. Over the years the blades used wind turbines have become progressively larger which has led to an increase in the proportion of blade failures. For inspection of wind turbine blades, Wu et al. [37] used both images of the blade as well as vibration data obtained using a vibration sensor. The vibration data was transformed into a 2D time-frequency (wavelet) map or image. The time-frequency map $I_1$ and the image of the blade $I_2$ were fused together into the image $\widetilde{I}$ using a IFCNN. The fused image $\widetilde{I}$ was then classified using ResNet-34.

## 5.6 CNN Features

When the CNN is trained on very large labeled datasets, it learns to extract a hierarchy of features from the input data. Studies [11, 27] have shown the features learnt on ImageNet may then be used as a *generic* set of features or common representational format (see Example 4.13). We define the features as the output of a given layer $l^*$ in the CNN.

We identify three ways we may use these features:

**"Off-the-shelf" Features.** The features are used *directly* in a image processing algorithm. These algorithms may include object classification, scene recognition, image retrieval, image segmentation and clustering.

**Hybrid Features.** The CNN features are combined with a set of handcrafted features to form a new common representational format. The new format is then used in a variety of image processing algorithms.

**Transfer learning.** The features are used when the CNN is *re-trained* on a *new* and different training dataset $\mathcal{D}'$. In retraining the CNN we *fix* the weights and biases of all layers $l, l \leq l^*$. By only training the layers $l > l^*$ we reduce both time and number of train samples required. See Sect. 5.4.2.

### "Off-the-Shelf" Features

The following example illustrates the use of the CNN features in a Bag-of-Visual-Words (BOVW) classifier (cf. Example 4.17).

**Example 5.9** *"Off-the-Shelf" Bag-of-Visual-Words (BOVW) for Melanoma Detection* [24]. The general idea of BOVW is to represent an image as a *histogram* of invariant features which are known as "visual words".

In this application, we create a set, or dictionary, of "visual words" as follows. Let $\mathcal{D}$ denote a set of malignant and benign images $I_i, i \in \{1, 2, \ldots, N\}$. For each image $I_i$ we move a $w \times w$ sliding window over the image and extract $M$ $w \times w$ sub-images $\widetilde{I}_i(m), m \in \{1, 2, \ldots, M\}$. Each sub-image is subsequently processed by a CNN and a feature vector $\widetilde{\mathbf{F}}_i(m)$ is extracted from the flattening layer. The feature vectors from *all* the sub-images and *all* the train images, $\widetilde{\mathbf{F}}_i(m), i \in \{1, 2, \ldots, N\}, m \in \{1, 2, \ldots, M\}$, are then clustered into $K$ groups using a $K$-means clustering algorithm (see Sect. 8.4). The representative vector of each cluster is regarded as an invariant feature, or "visual word" $\mathbf{y}_k, k \in \{1, 2, \ldots, K\}$, in a BOVW dictionary.

Given the BOVW dictionary, we encode each feature vector $\widetilde{\mathbf{F}}_{\mathrm{TEST}}(m), m \in \{1, 2, \ldots, M\}$, extracted from the test image $I_{\mathrm{TEST}}$ into the *nearest* visual word. We thus obtain a global representation of $I_{\mathrm{TEST}}$ by forming a histogram $\mathcal{H}_{\mathrm{TEST}}$ of the visual words, i.e. by counting the number of times each visual word is used to encode the feature vectors extracted from $I_{\mathrm{TEST}}$. By following a similar procedure we also find the BOVW representation of each training image $I_i, i \in \{1, 2, \ldots, N\}$ in $\mathcal{D}$.

Then to classify $I_{\mathrm{TEST}}$ we simply find the histogram $\mathcal{H}_{i^*}$ which is closest to $\mathcal{H}_{\mathrm{TEST}}$:

$$i^* = \arg\min_i dist(\mathcal{H}_{\mathrm{TEST}}, \mathcal{H}_i),$$

where $dist(\cdot, \cdot)$ is an appropriate histogram distance metric (see Sect. 9.3). We classify $I_{\mathrm{TEST}}$ as malignant or benign according to whether $I_{i^*}$ is malignant or benign:

$$I_{\mathrm{TEST}} = \begin{cases} \text{Malignant} & \text{if } I_{i^*} \text{ is malignant,} \\ \text{Benign} & \text{otherwise.} \end{cases}$$

For another example of "Off-the-Shelf" CNN features see Example 4.13.

### Hybrid Features

Despite the success of hand-crafted features in computer vision, nowadays it is more common to use deep learning features. These features are trained end-to-end and are extracted from deep CNN's. By combining deep learning features with hand-crafted features we may further improve their performance. In the following example we combine CNN features with SIFT to give local rotation invariance.

**Example 5.10** *Hybrid Features. SIFT-CNN* [36]. Given a gray input image $I$, a SIFT descriptor (Sect. 4.4.3) is computed for every pixel. The result is a "SIFT-image" in which every pixel is represented by a 128-dimensional histogram. A CNN is then trained on a training set $\mathcal{D}$ of SIFT-images. By training on SIFT images, the network implicitly acquires a local rotation invariance property. This is a desired property for problems in which regions within the image can be rotated without affecting the overall classification result.

For another example of hybrid features see Example 4.16.

## 5.7 Auto-Encoder

We now turn to auto-encoders (AE). AE is a neural network that aims to find a low-dimensional (latent) representation of the input signal [3, 5, 19]. It does this using a special form of neural network which is trained using an *unsupervised* learning technique. The auto-encoder comprises two main components:

**Encoder**. The encoder is a neural network that learns the saliant features of the input data. As a result the encoder compresses and transfers the input data into a meaningful low-dimensional latent common representational format.

**Decoder**. The decoder is another neural network that takes the low-dimensional latent representation from the encoder and *reconstructs* the original input signal as closely as possible. The goal is to minimize the reconstruction error between the input signal and the reconstructed output.

Figure 5.15a shows a basic AE. It is composed of 3 *fully-connected* neural network layers. The input layer and the hidden layer form the encoder and the output layer forms the decoder.

Mathematically the basic AE works as follows: The encoder takes the input vector $\mathbf{x} = \big(x(1), x(2), \ldots, x(M)\big)^T$ vector and maps it to a latent representation $\mathbf{y} = \big(y(1), y(2), \ldots, y(K)\big)^T$ using a deterministic mapping:

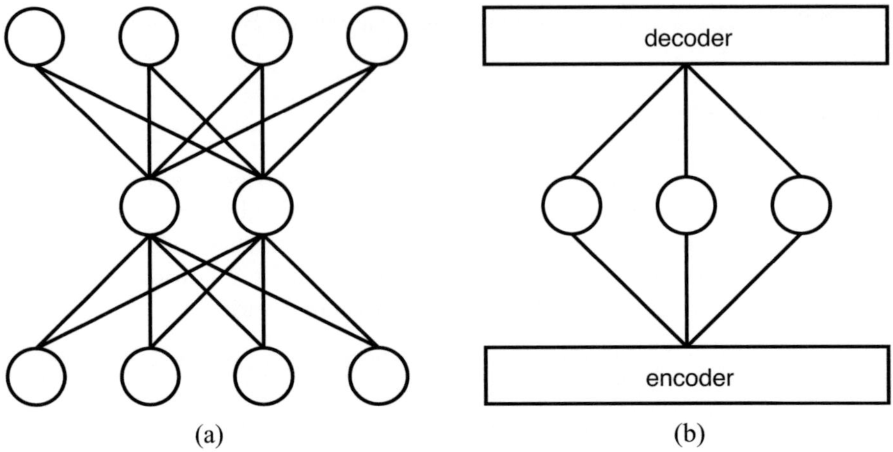


**Fig. 5.15** Shows the architecture of an autoencoder (AE). **a** Shows a simple AE with three layers. **b** Shows a convolutional AE with a single common fully-connected layer

$$y(k) = f_E\Big(\sum_m W(m,k)x(m) + b(k)\Big), \tag{5.16}$$

where $W$ is a $M \times K$ weight matrix, $\mathbf{b} = \big(b(1), b(2), \ldots, b(K)\big)^T$ is a bias vector and $f_E$ is the encoder activation function. The latent representation $\mathbf{y}$ is mapped back into a reconstructed vector $\widetilde{\mathbf{x}} = \big(\widetilde{x}(1), \widetilde{x}(2), \ldots, \widetilde{x}(M)\big)^T$ using the deterministic mapping:

$$\widetilde{x}(m) = f_D\Big(\sum_k \widetilde{W}(k,m)y(k) + \widetilde{b}(m)\Big), \tag{5.17}$$

where $\widetilde{W}$, $\widetilde{\mathbf{b}}$ and $f_D$ are the corresponding weight matrix, bias vector and activation function for the decoder. The parameters $\boldsymbol{\theta} = (W, \widetilde{W}, \mathbf{b}, \widetilde{\mathbf{b}}, f_E, f_D)$ of the AE are optimized to minimize an appropriate loss function $\mathcal{L}$:

$$\boldsymbol{\theta}^* = \arg\min_{\boldsymbol{\theta}} \sum_i \mathcal{L}(\mathbf{x}_i, \widetilde{\mathbf{x}}_i); , \tag{5.18}$$

where the summation is taken over all signals $\mathbf{x}_i$, $i \in \{1, 2, \ldots, N\}$, in $\mathcal{D}$.

### Autoencoder and PCA

An important property of AE is its equivalence to PCA (cf. Sect. 4.7.1): PCA and the basic AE are mathematically equivalent if (1) $f_E$ and $f_D$ are linear functions; (2) $\frac{1}{N}\sum_{i=1}^N \mathcal{L}(\mathbf{x}_i, \widetilde{\mathbf{x}}_i)$ is the mean square error and (3) $\mathbf{x}_i$ are normalized (zero-mean and unit covariance).

**Table 5.7** Autoencoder variants

| Technique | Description |
|---|---|
| Basic | Basic AE uses three fully-connected layers |
| Robust | Designed to enhance the AE's resilience when dealing with noisy or corrupt input data |
| Denoising | Designed to reconstruct clean data from noisy input by introducing noise during training |
| Convolutional | Uses CNN's in the AE |
| DenseBlock | Uses denseblock in the AE |

## 5.8 Autoencoder Variants

Since the introduction of the basic AE, a great number of different AE variants have been developed. A few of the more common variants are listed in Table 5.7.

### Convolutional Autoencoder

In the convolutional AE we generalize the basic AE by adding a CNN to both the encoder and decoder (Fig. 5.15b). In the encoder we add convolutional and pooling layers while in the decoder we add convolutional and upsampling layers.

In general, the convolutional AE is much more powerful than the basic AE. It is widely used for performing dimension reduction and feature extraction as illustrated by the following examples.

**Example 5.11** *Dimensionality reduction using convolutional AE* [32]. Shinde et al. [32] use a convolutional AE to construct a low-dimensional latent representation of brittle fracture images. Brittle fracture images are difficult to analyze and conventional dimension reducing techniques (cf. Sect. 4.7) do not work. The AE used in [32] is shown in Fig. 5.16. Input to the AE is a $62 \times 62 \times 3$ input image. The encoder has two convolutional layers CNV1 and CNV2, each followed by a max-pooling layer. After MP2, a flattening layer FLT forms the data into a 1D column vector (length $15616 \times 1$) which is passed to a fully-connected layer FC1. The output from FC1 is a feature vector **F** of size 5. The decoder part of the AE takes the feature vector **F** which is passed through the fully-connected layer FC2 which restores it to a 1D vector of length $15616 \times 1$. The output from FC2 is then reshaped into a $16 \times 16 \times 61$ image. This is followed by two upsampling layers (UP1 and UP2) and one convolution layer CNV3. The output from CNV3 is a reconstructured image of the original image size: $62 \times 62 \times 3$.

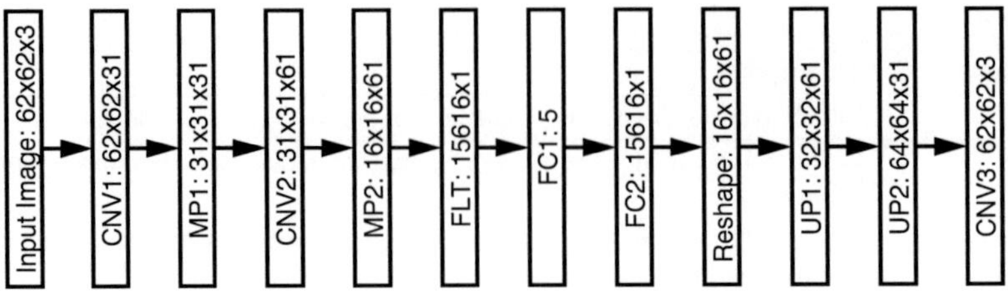


**Fig. 5.16** Shows the architecture of the convolutional autoencoder used by Shinde et al.

**Example 5.12** *Convolutional AE for Finger-Vein Verification* [15]. Automatic personal verification is widely used in an increasing number of applications such as: border control, access control, attendance systems and financial security. For this purpose, we often use biometric techniques which rely on physical or physiological characteristics. Among the different biometrics which may be used is the finger-vein image. Compared to other biometrics, the finger-vein image has several advantages: (1) *Fast*. The finger-vein images are small for fast processing. (2) *No-contact*. The images are not affected by skin conditions. (3) *Biometric*. Finger-vein patterns can be identified on living body. (4) *Robust*. The finger-vein pattern is an internal feature and is not easy to replicate.

Hou and Yan [15] used convolutional AE with a support vector machine (SVM) for state-of-the-art finger-vein verification. The AE is used to automatically learn a small set of salient features from a database of finger-vein images without any prior knowledge. A SVM is then used to classify (verify) each finger-vein test image.

### Denoising Auto-Encoder

Denoising auto-encoders are a variant of AE that is specifically designed to learn robust representations of data in the presence of noise. They are trained to reconstruct the original, non-noisy version of the input data from a noisy version of the data.

To train a denoising AE we minimize the difference between the original "clean" input data and the reconstructed data. In this process, the denoising AE identifies salient features in the input data while disregarding noise and irrelevant aspects.

Given a clean input signal $\mathbf{x}$, let $\mathbf{x}'$ denote the corresponding noisy signal. If $\widetilde{\mathbf{x}}'$ denotes the reconstructed noisy signal, then the denoising loss function is

$$\mathcal{L}(\mathbf{x}, \widetilde{\mathbf{x}}') = \|\mathbf{x} - \widetilde{\mathbf{x}}'\|^2 .$$

Note: Strictly speaking training the denoising AE is *not* unsupervised learning since the input $\mathbf{x}'$ is not equal to $\mathbf{x}$.

### DenseBlock Autoencoder

One drawback to the convolutional AE is that only the results of the last layer in the encoder are used as features. This means that information produced by the middle layers is not directly used as features. This problem may be solved by using a dense block in the encoder [20].

The reasoning behind this suggestion is that in a dense block, the output of each layer is used as input for the succeeding layers. This in turn means that the results of each layer are used directly to construct the feature map.

## 5.9 Auto-Encoder Applications

In this section we illustrate the power of AE with several computer vision applications which employ AE.

### Anomaly Detection

Convolutional AE's are used to identify any object deviating from the normal pattern. By training an AE to reconstruct a normal pattern, with a small reconstruction error, we are then able to detect any *unknown* non-normal object by its large reconstruction error.

**Example 5.13** *Anomaly Detection for Agricultural Vehicles Using AE* [21]. The safe operation of autonomous agricultural vehicles requires the detection of *all* objects that could pose a risk of collision. Such a list of objects would include humans, animals, other vehicles, holes, standing pools of water, buildings, tools, equipment and all other objects left in the field. The very wide and diverse nature of this list precludes the use of the current vision-based object detection algorithms (cf. Example 5.2). Instead [21] suggested using a convolutional AE as an *anomaly* detector. In the convolutional AE, both encoder and decoder have six convolutional layers. The decoder corresponds to the encoder using 6 convolutional layers. The encoder and decoder share one fully-connected layer with 16 channels. The AE was trained on a large set images drawn from a normal dataset.

### Speckle Noise Removal

We illustrate the use of denoising AE to remove speckle noise in ultrasound images.

**Example 5.14** *Speckle Noise Removal in Breast and Lung Ultrasound Images* [25]. Ultrasound (US) imaging is widely used in diagnosis and monitoring of breast and lung cancer. However the presence of speckle noise in ultrasound images severely limits the resolution for of the images which in turn impairs the correct diagnostic interpretation of the image. In [25] the authors successfully used a denoising CNN-AE to remove speckle noise from ultrasound images. The result was fewer misclassifications of breast cancer in comparison to the use of the original US images after filtering with a traditional local filter.

### Infrared-Visible Light Fusion

We illustrate infrared-visible light fusion using a dense-block AE.

**Example 5.15** *DenseFuse: Infrared and Visible Image Fusion* [20]. Li and Wu [20] presented a novel deep learning approach to fuse infrared ($I_{IR}$) and visible ($I_{VI}$) images which have been pre-registered (see Chap. 6). The network has three parts: encoder, fusion layer and decoder.

The encoder contains a convolutional layer followed by a 3-layer dense block. Both the convolutional layer and the dense block use $3 \times 3$ filter kernels. The decoder has four convolutional layers with $3 \times 3$ kernel filters.

In the training phase, we initially discard the fusion layer and train the encoder-decoder pair to reconstruct input images with minimal distortion. The AE weights (i.e. encoder and decoder parameters) are then fixed.

In the fusion layer different fusion strategies may be used. The simplest is element-by-element addition (see Example 5.7).

The fused features are then reconstructured by the decoder to produce the final fused image. DenseFuse has demonstrated good generalization ability and has been successfully used in a variety of image fusion applications (Fig. 5.17).

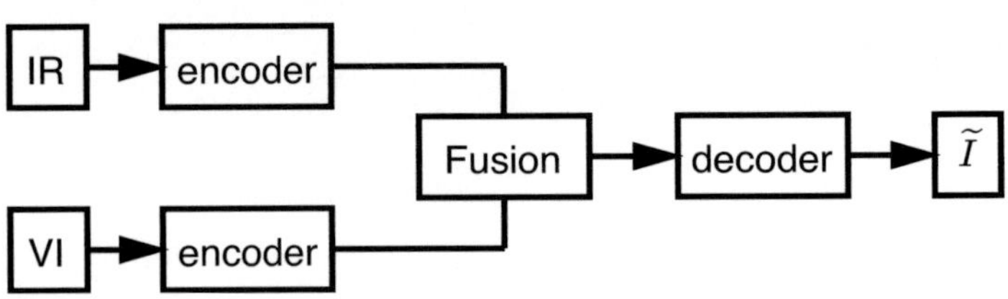


**Fig. 5.17** Shows infrared-visible fusion network

## 5.10 Software

**MATLAB.** MATLAB DEEP LEARNING toolbox includes m-files for both CNN (including pre-trained models) and AE. The toolbox includes m-files for object detection (including fast R-CNN and faster R-CNN).

**PYTHON.** TensorFlow, PyTorch and Keras are three python libraries for deep learning.

## 5.11 Further Reading

Excellent modern overview of CNN in computer vision [41]. Review of deep learning for visual understanding [10]. Modern reviews of deep learning in various areas include: [4] (meteorology), [12] (optics), [17] (fruit detection), [14] (agriculture), [39] (image fusion). A thorough survey of autoencoders in machine learning is [1].

## References


1. Berahmand K, Daneshfar F, Salehi ES et al (2024) Autoencoders and their applications in machine learning: as survey. Art Intel Rev 57:Article no 28
2. Bihler M, Roming L, Jiang Y et al (2023) Multi-sensor data fusion using deep learning for bulky waste image classification. Proc SPIE 12623
3. Charte D, Charte F, Garcia S et al (2018) A practical tutorial on autoencoders for nonlinear feature fusion: taxonomy, models, software and guidelines. Inform Fusion 44:78–96
4. Chase RJ, Harrison DR, Lackman GM et al (2023) A machine learning tutorial for operational meteorology. Part II: Neural networks and deep learning. Weather Forecast 38:1271–1293
5. Chen S, Guo W (2023) Auto-encoders in deep learning - a review with new perspectives. Mathematics 11:1777
6. Dubey SR, Singh SK, Chaudhuri BB (2022) Activation functions in deep learning: a comprehensive survey and benchmark. Neurocomputing 503:92–108
7. Fan H, Han M, Li J (2019) Image shadow removal using end-to-end using end-to-end deep convolutional neural networks. Appl Sci 2019:1009
8. Garcea F, Serra A, Lamberti F et al (2023) Data augmentation for medical imaging: a systematic literature review. Comp Biol Med 152:106391
9. Girshick R (2015) Fast R-CNN. Proceedings of IEEE international conference on computer vision, pp 1440–1448
10. Guo Y, Liu Y, Oerlemans A et al (2016) Deep learning for visual understanding: a review. Neurocomputing 187:27–48
11. Guerin J, Gibaru O, Thiery S et al (2018) CNN features are also great at unsupervised classification. In: The 8th International Conference on Computer Science and Application Engineering, pp 83–95
12. Hadad B, Froim S, Yosef E et al (2023) Deep learning in optics - a tutorial. J Opt 25:123501
13. Herrmann C, Willersinn D, Beyerer J (2017) Residual versus inception versus classical networks for low-resolution face recognition. Lect Notes Comp Sci 10270
14. Kamilaris A, Prenafeta-Bold FX (2018) Deep learning in agriculture: a survey. Comput Electron Agric 147:70–90

15. Hou B, Yan R (2020) Convolutional autoencoder model for finger-vein verification. IEEE Trans Instrum Meas 69:2067–2074
16. Kim DS, Arsalan M, Park KR (2018) Convolutional neural network-based shadow detection in images using visible light camera sensor. Sensors 2018:960
17. Koirala A, Walsh KB, Wang Z et al (2019) Deep learning—method overview and review of use for fruit detection and yield estimation. Comput Electron Agric 162:219–234
18. Krizhevsky A, Sutskever I, Hinton GE (2012) ImageNet classification with deep convolution neural networks. In: Advances in neural information processing systems, pp 1097–1105
19. Li P, Pei Y, Li J (2023) A comprehensive survey on design and application of autoencoder in deep learning. App Soft Learn 128:110176
20. Li H, Wu X-J (2019) DenseFuse: a fusion approach to infrared and visible images. IEEE Trans Image Process 28:2614–2623
21. Mujkic E, Philipsen MP, Moeslund TB et al (2022) Anomaly detection for agricultural vehicles using autoencoders. Sensors 22:3608
22. Nagpal P, Bhinge SA, Shitole A (2022) A comparative analysis of ResNet architectures. In: IEEE international conference on smart generation computing, communication and networking (SMART GENCOM)
23. Ning C, Gan H, Shen M et al (2023) Learning-based padding: from connectivity on data borders to data padding. Eng Appl Artif Intell 121:106048
24. Okur E, Turkan M (2024) Weighted bag of visual words with enhanced deep features for melanoma detection. Expert Syst Appl 237:121531
25. Oliveira-Saraiva D, Mendes J, Leote J et al (2023) Make it less complex: autoencoder for speckle noise removal—application to breast and lung ultrasound. J Imaging 9:9100217
26. Ponti MA, Ribeiro SF, Nazare TS et al (2017) Everything you wanted to know about deep learning for computer vision but were afraid to ask. In: 2017 30th SIBGRAPI conference on graphics, patterns and images tutorials, pp 17–41
27. Razavian AS, Azizpour H, Sullivan J et al (2014) CNN features off-the-shelf: an astounding baseline for recognition. In: IEEE conference on computer vision and pattern recognition workshops
28. Reddy S, Pillay N, Singh N (2024) Comparative evaluation of convolutional neural network object detection algorithms for vehicle detection. J Imaging 10:162
29. Ren S, He K, Girshick R et al (2017) Faster R-CNN: towards real-time object detection with region proposal networks. IEEE Trans Pattern Anal Mach Intell 39:1137–1149
30. Ren J, Wang Y (2022) Overview of object detection algorithms using convolutional neural networks. J Comput Commun 10:115–132
31. Russakovsky O, Deng J, Su H et al (2015) Imagenet large scale visual recognition challenge. Int J Comput Vis 115:211–252
32. Shinde K, Itier V, Mennesson J et al (2023) Dimensionality reduction through convolutional autoencoders for fracture patterns prediction. Appl Math Model 114:94–113
33. Srivastava N, Hinton G, Krizhevsky A et al (2014) Dropout: a simple way to prevent neural networks from overfitting. J Mach Learn Res 15:1929–1958
34. Stankovic L, Mandie D (2023) IEEE Trans Syst, Man, Cybern 53:3614–3628
35. Takahashi R, Matsubara T, Uehara K (2020) Data augmentation using random image cropping and patching for deep CNNs. IEEE Trans Circuits Syst Video Technol 30:2917–2931
36. Tsourounis D, Kastaniotis D, Theoharatos C et al (2022) SIFT-CNN: when convolutional neural networks meet dense sift descriptors for image and sequence classification. J Imaging 8:8100256
37. Wu S, Wang G, Jiang N et al (2024) Fault diagnosis of wind turbine blades based on image fusion and ResNet. J Phys Conf 2718:012059
38. Zafar A, Aamir M, Nawi NM et al (2022) A comparison of pooling methods for convolutional neural networks. Appl Sci 12:8643
39. Zhang H, Xu H, Tian X et al (2021) Image fusion meets deep learning: a survey and perspective. Inf Fusion 76:323–336

40. Zhang Y, Liu Y, Sun P et al (2020) IFCNN: a general image fusion framework based on convolutional neural network. Inf Fusion 54:99–118
41. Zhao X, Wang L, Zhang Y et al (2024) A review of convolutional neural networks in computer vision. Artif Intell Rev 57:99

# Chapter 6
# Spatial Alignment

**Abstract** In this chapter we discuss the spatial alignment of all spatial positions to a common coordinate system and forms the first stage in building a common representational format. To keep the discussion focused we limit ourselves to image registration defined as spatial alignment of two-dimensional images. The images themselves may belong to the same sensor type (monomodal registration) or different sensor types (multimodal registration). Image registration methods include: area matching (using mutual information), sparse parametric matching (using both handcrafted and deep-learning keypoints) and dense non-parametric matching (using both Lucas-Kanade and Horn-Schunck optical flow, SIFT flow and deep-learning flow). Finally we consider the problem of multiple image registration and bundle adjustment.

## 6.1 Introduction

The subject of this chapter is *spatial alignment*. This is the conversion of local spatial positions to a common coordinate system and forms the first stage in the formation of a common representational format. To keep our discussion focused we shall limit ourselves to two-dimensional $(x, y)$ image sensors. In this case the process of spatial alignment is more commonly referred to as *image registration*.

In many multi-sensor data fusion applications spatial alignment is the primary fusion algorithm. In Table 6.1 we list some of these applications together with the classification of the type of fusion algorithm involved.

## 6.2 Image Registration

Let $I_1$ and $I_2$ denote two input images, where $I_m(i, j)$, $m \in \{1, 2\}$, denotes the gray-level of $I_m$ at the pixel $(i, j)$.[1] The registration process is defined as finding the transformation $\mathcal{T}$ which "optimally" maps coordinates in $I_1$ to coordinates in $I_2$. Thus,

[1] In this chapter we shall assume, unless indicated otherwise, that $I_1$ and $I_2$ are gray-scale images.



H. B. Mitchell, *Data Fusion*, https://doi.org/10.1007/978-3-662-71023-4_6

**Table 6.1** Applications in which spatial alignment is the primary fusion algorithm

| Class | | Application |
|---|---|---|
| Early-DaO | Example 4.1 | A Standardized Brain Atlas |
| | Example 4.4 | A Distributed Surveillance System |
| | Example 4.10 | Myocardial Imaging |
| | Example 6.3 | Spatial Alignment in Cardiac MR Images |
| | Example 6.4 | Fingerprint warping using thin-plate spline (TPS) |
| | Example 6.10 | Global Image Registration Using RANSAC |
| | Example 6.15 | Converting a Smiling Face to a Neutral Face |
| | Example 6.17 | Optical flow for Driver Fatigue Detection |
| | Example 6.19 | SIFT Flow for Kinship Verification |
| | Example 6.21 | Mosaic Fingerprint Image |

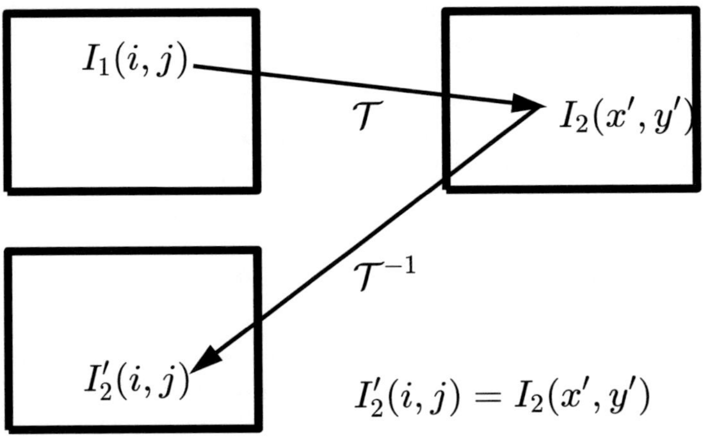


**Fig. 6.1** Shows input images $I_1$ and $I_2$ and corresponding warped image $I_2'$. The transformation $\mathcal{T}$ maps the pixel $(i, j)$ in $I_1$ to $(x', y')$ in $I_2$, where $(x', y') = \mathcal{T}(i, j)$. The inverse transformation $\mathcal{T}^{-1}$ maps $(x', y')$ in $I_2$ to $(i, j)$ in $I_2'$, where $(i, j) = \mathcal{T}^{-1}(x', y')$

given a pixel $(i, j)$ in $I_1$, the corresponding location in $I_2$ is $(x', y') = \mathcal{T}(i, j)$. Using the inverse transformation $\mathcal{T}^{-1}$, we may then define the corresponding "warped" image $I_2'$, where $I_2'(i, j) = I_2(x', y')$ (see Fig. 6.1).

Often a transformed point $(x', y')$ does not fall on a pixel in $I_2$. In this case, strictly speaking, we have no gray-level for the point $(x', y')$. To handle such situations we resample/interpolate $I_2$ so that a gray-level is defined for all points $(x', y')$. This we do by using an resampling/interpolation algorithm.

The following example illustrates the use of medical image registration in modern medicine.

**Example 6.1** *Medical Image Registration.* For disease diagnosis and treatment design, a surgeon will often rely on images acquired at different times or by different sensors at the same time, for an optimal description of anatomical structures and metabolic function of the patient. It is therefore essential to transform the different images $I_m, m \in \{1, 2, \ldots, M\}$, so they align to the same coordinate frame or common representational format. The corresponding points on the transformed images $\mathcal{T}(I_m)$ are supposed to represent the same anatomical position after registration.

Image registration algorithms may be broadly divided into three groups:

**Area Matching**. Given two input images $I_1$ and $I_2$ it works by directly matching $I_1$ and the warped image $I'_2$. The optimal transformation $\mathcal{T}^*$ is found by searching for a transformation which gives the best match.

**Sparse parametric matching**. Works by separately extracting a *sparse* set of features in input images $I_1$ and $I_2$. The optimal transformation $\mathcal{T}^*$ is defined as the parametric transformation which best aligns the two sets of features.

**Dense non-parametric matching**. Works by separately extracting a *dense* set of features in input images $I_1$ and $I_2$. The optimal transformation $\mathcal{T}$ is defined as a *pixel-by-pixel* transformation which best aligns the two sets of features.

We now consider the transformation $\mathcal{T}$ and the resample/interpolation process.

### 6.2.1 Transformation $\mathcal{T}$

The (pairwise) transformation $\mathcal{T} : (x', y') = \mathcal{T}(x, y)$ is a mathematical relationship that maps a spatial location $(x, y)$ in one image $I_1$ to a new location, $(x', y')$, in another image $I_2$. It arises in many image analysis problems, whether we wish to remove optical distortions introduced by a camera or a particular viewing perspective, or to register an image with a given map or template, or simply to align two input images. The choice of transformation is always a compromise between a smooth distortion and a distortion which achieves a good match. One way to ensure smoothness is to assume a parametric form of low-order for the transformation [10] such as that given in Table 6.2. In many applications, the transformation $\mathcal{T}$ is chosen on the grounds of mathematical convenience. However, in some cases, we may use physical arguments in order to *derive* the transformation $\mathcal{T}$.

**Example 6.2** *Transformation for Retinal Images* [6]. Let $\mathcal{T}$ denote the pairwise transformation between a pair of retinal images. If we model the retina as a quadratic surface and assume a rigid weak-perspective transformation between the two images, then the corresponding transformation is

**Table 6.2** Spatial transformations $\mathcal{T} : (x', y') = \mathcal{T}(x, y)$

| Name | Formula |
|---|---|
| Translation | $x' = x + \alpha_1,\ y' = y + \alpha_2$ |
| Similarity | $x' = \alpha_1 x + \alpha_2 y + \alpha_3,\ y' = -\alpha_2 x + \alpha_1 y + \alpha_4$ |
| Affine | $x' = \alpha_1 x + \alpha_2 y + \alpha_3,\ y' = \alpha_4 x + \alpha_5 y + \alpha_6$ |
| Planar perspective | $x' = (\alpha_1 x + \alpha_2 y + \alpha_3)/(\alpha_7 x + \alpha_8 y + \alpha_9)$,<br>$y' = (\alpha_4 x + \alpha_5 y + \alpha_6)/(\alpha_7 x + \alpha_8 y + \alpha_9)$ |
| Polynomial | $x' = \sum \alpha_{ij} x^i y^j,\ y' = \sum \beta_{ij} x^i y^j$ |

$$x' = \alpha_1 x^2 + \alpha_2 xy + \alpha_3 y^2 + \alpha_4 x + \alpha_5 y + \alpha_6 \ ,$$
$$y' = \alpha_7 x^2 + \alpha_8 xy + \alpha_9 y^2 + \alpha_{10} x + \alpha_{11} y + \alpha_{12} \ .$$

The assumptions made in deriving this transformation may be justified on physical grounds [6]. Alternatively we may "derive" the transformation by performing a Taylor expansion of the general polynomial transformation in $x$ and $y$ and only retaining terms upto, and including, second order.

In many applications, the images undergo local deformations. In this case, we cannot describe the alignment of two images using a single low-order transformation. Instead, we often use a "non-rigid" transformation, which we write as a sum of a low-order global transformation $\mathcal{T}_G(x, y)$ and a local transformation $T_L(x, y)$:

$$(x', y') = \mathcal{T}(x, y) = \mathcal{T}_G(x, y) + \mathcal{T}_L(x, y) \ , \tag{6.1}$$

where the parameters of $\mathcal{T}_L$ change with $(x, y)$. The following example illustrates the use of such a transformation.

**Example 6.3** *Spatial Alignment in Cardiac MR Images* [32]. Magnetic imaging (MR) of the cardiovascular system is playing an increasingly important role in the diagnosis and treatment of cardiovascular diseases. During a cardiac cycle the heart changes size and shape. To perform spatial alignment of two MR images taken at different times during the cardiac cycle we require a transformation $\mathcal{T}(x, y)$ which includes both a global transformation and a local transformation, where the local transformation $\mathcal{T}_L(x, y)$ corrects for changes in the size and shape of the heart during a cardiac cycle.

The thin-plate spline (TPS) function is widely used to model non-rigid spatial transformations $\mathcal{T}_L$. This is illustrated in the following example.

**Example 6.4** *Fingerprint Warping Using a Thin-plate spline (TPS)* [35]. The performance of a fingerprint matching system is affected by the nonlinear deformations introduced in the fingerprint during image acquisition. This nonlinear deformation may be represented using a TPS [49]:

$$x' = \alpha_1 + \alpha_2 x + \alpha_3 y + \sum_i \beta_i r_i^2 \ln r_i^2 ,$$

$$y' = \alpha_4 + \alpha_5 x + \alpha_6 y + \sum_i \gamma_i r_i^2 \ln r_i^2 ,$$

where $(x_i, y_i), i \in \{1, 2, \ldots, N\}$, is a set of known anchor points and $r_i^2 = (x - x_i)^2 + (y - y_i)^2 + d^2$. Apart from $d$, the TPS model has six parameters, $\alpha_1, \alpha_2, \ldots, \alpha_6$, corresponding to a global affine transformation and $2N$ parameters $(\beta_i, \gamma_i), i \in \{1, 2, \ldots, N\}$, corresponding to a local deformation, and which satisfy the following constraints:

$$\sum_i \beta_i = 0 = \sum_i \gamma_i , \quad \sum_i x_i \beta_i = 0 = \sum_i x_i \gamma_i , \quad \sum_i y_i \beta_i = 0 = \sum_i y_i \gamma_i .$$

## 6.2.2 *Resample/Interpolation*

Sometimes the transformed coordinates $(x', y') = \mathcal{T}(i, j)$ do not fall on an image pixel. In this case we resample/interpolate the image so that a gray-level is defined for all points $(x', y')$. The resample/interpolation process plays a crucial role in all registration schemes because it uniquely defines the behaviour of the transformation $\mathcal{T}$ when the samples do not fall on a pixel. In image interpolation we reconstruct a 2D continuous image $I(x, y)$ from its discrete pixel values $I(i, j)$. This is often modeled as a convolution of the discrete image with a continuous 2D impulse response $h(x, y)$:

$$I(x, y) = \sum_i \sum_j I(i, j) h(x - i, y - j) . \tag{6.2}$$

It is common practice to use symmetrical and separable interpolation kernels:

$$h(x, y) = h_x(x) h_y(y) , \tag{6.3}$$

where $h_x(x) = h_x(-x)$ and $h_y(y) = h_y(-y)$. In Table 6.3 we list several popular kernel functions $h(x)$ used for image interpolation [20, 45].

Apart from the fixed coefficient kernels listed in Table 6.3 we may also use the spatial Kriging estimators listed in Table 4.4.

**Table 6.3** Kernel Functions $h(x)$ for image interpolation

| Name | $h(x)$ |
|---|---|
| Nearest neighbour | 1 if $0 \leq \lvert x \rvert < \frac{1}{2}$; otherwise 0 |
| Linear | $1 - \lvert x \rvert$ if $0 \leq \lvert x \rvert < 1$; otherwise 0 |
| Quadratic | $\begin{cases} -2\lvert x\rvert^2 + \frac{1}{4} & \text{if } 0 \leq \lvert x \rvert < \frac{1}{2}, \\ \lvert x\rvert^2 - \frac{5}{2}\lvert x\rvert + \frac{3}{8} & \text{if } \frac{1}{2} \leq \lvert x \rvert < \frac{3}{2}, \\ 0 & \text{otherwise} \end{cases}$ |
| Cubic ($N = 4$) | $\begin{cases} (a+2)\lvert x\rvert^3 - (a+3)\lvert x\rvert^2 + 1 & \text{if } 0 \leq \lvert x \rvert < 1, \\ a\lvert x\rvert^3 - 5a\lvert x\rvert^2 + 8a\lvert x\rvert - 4a & \text{if } 1 \leq \lvert x \rvert < 2, \\ 0 & \text{otherwise}, \end{cases}$ where $a$ is limited to values $a = -1/2, -2/3, -3/4, -1, -4/3$ |

The formulas in the table assume $x$ and $y$ are given in units of the sampling interval

**Table 6.4** Image registration similarity measures

| Name | Formula |
|---|---|
| Sum Absolute Diff. | $SAD = \sum_{(i,j)} \lvert I_1(i,j) - I_2(i,j)\rvert$ |
| Normalized correlation coefficient (NCC) | $NCC = \sum_{(i,j)} \prod_m \big(I_m(i,j) - \mu_m\big)/\big(\sum_{(i,j)} \prod_m (I_m(i,j) - \mu_m)\big)^{1/2}$, where $\mu_m$ is mean gray-level of $I_m$ |
| Mutual Information (MI) | See Eq. (6.6) |
| Regional MI (RMI) | See Eq. (6.9) |

### 6.2.3 *Uncertainty of $\mathcal{T}$* *

The accuracy of the transformation $\mathcal{T}$ plays an essential role in many data fusion applications: In many cases the registration accuracy determines whether, and to what extent, we may fuse the individual sensors.

A simple but effective algorithm for measuring the transformation accuracy is bootstrap resampling (see Sect. 15.4.3). The is illustrated in the following example.

**Example 6.5** *Bootstrap Resampling for Image Registration Uncertainty* [19]. Given two input images $X$ and $Y$ we create $B$ bootstrapped sets of pixels $X^{(b)}, Y^{(b)}, b \in \{1, 2, \ldots, B\}$. For each $b$, we perform registration on $(X^{(b)}, Y^{(b)})$. Suppose $\hat{\boldsymbol{\theta}}^{(b)}$ denotes corresponding estimated transformation parameters. If $\bar{\boldsymbol{\theta}} = \sum_b \hat{\boldsymbol{\theta}}^{(b)}/B$, then the uncertainty in $\boldsymbol{\theta}$ is given by the covariance matrix:

$$\boldsymbol{\Sigma} = \frac{1}{B}\sum_b (\hat{\boldsymbol{\theta}}^{(b)} - \bar{\boldsymbol{\theta}})(\hat{\boldsymbol{\theta}} - \bar{\boldsymbol{\theta}})^T .$$

## 6.3 Area Matching

The basic concept behind this approach is to find a global transformation $\mathcal{T}$ : $(x', y') = \mathcal{T}(x, y)$, which optimally matches the image $I_1$ with the warped image $I'_2$, where $I'_2(x, y) = I_2(x', y')$.

We measure the match between $I_1$ and $I'_2$ using a similarity operator $SIM$. If $I_1$ and $I_2$ were captured by sensors of the same type ("monomodal"), then sum of absolute differences or normalized correlation coefficient may be used (see Table 6.4). However, when the sensors are of different types ("multimodal") we require a similarity measure which is robust against changes which arise from using different sensors. For this purpose mutual information (MI) and its variants has been found to be an effective criterion that can be applied automatically and reliably without any pre-processing on a large variety of applications.

The basic idea behind MI is to measure the amount of information that is shared between $I_1$ and $I_2$. The success of MI image registration [40] lies in its inherent simplicity. It makes few assumptions regarding the relationship that exists between $I_1$ and $I_2$ instead it only assumes a *statistical* dependence. The idea is that although different sensors may produce very different signals, since they are imaging the same underlying scene, there will exist some inherent MI between the signals. This is illustrated in the following thought experiment.

**Example 6.6** *Medical Image Registration: A Thought Experiment.* Consider a pair of medical images from different modalities (e.g. Magnetic Resonance and Computerized Tomography) that are in perfect alignment. At many corresponding spatial locations in the two images, the associated gray-levels will exhibit a consistent mapping. In other words, there will exist some kind of global relation, or mapping, between the gray-levels of one image with the gray-levels of the second image. Whatever the mapping is (one-to-one, one-to-many, many-to-one or many-to-many) the statistical dependency between the images is strong and the mutual information (MI) measure has a high value.

Suppose now that we move out of registration by gradually deforming the test image. The mapping between the gray-levels of the two images will degrade, i.e. will become less consistent, and the value of the MI will decrease. In the extreme case, the two images are completely independent of one another, and the MI will take a value of zero which signifies that it is not possible to predict any part of one image from the information stored in the other.

Using MI as the similarity measure, the corresponding optimal transformation is:

$$T^* = \arg\max_{T \in \Omega} \left( MI\left(I_1, I'_2\right) \right) , \tag{6.4}$$

where $\Omega$ is a chosen set of transformations. Solutions to Eq. (6.4) may be found through exhaustive search over all possible transformations.

## 6.4 Mutual Information (MI)

The mutual information (MI) between two random variables $u$ and $v$ is given by

$$MI(u, v) = H(u) + H(v) - H(u, v) , \tag{6.5}$$

where $H(u)$ and $H(v)$ are the marginal entropies of the two variables $u$ and $v$ and $H(u, v)$ is their joint entropy. By rewriting Eq. (6.5) as

$$MI(u, v) = H(u) - H(u|v) = H(v) - H(v|u) ,$$

we may interpret MI as representing the reduction in the uncertainty of $u$ when $v$ is known, or equivalently, as representing the reduction in the uncertainty in $v$ when $u$ is known.

In the continuous domain, the (differential) entropy of a random variable $u$ and the joint (differential) entropy of two random variables $u$ and $v$ are defined as

$$\begin{aligned} H(u) &= -\int_{-\infty}^{\infty} p_U(u) \log_2 p_U(u) du , \\ H(u, v) &= -\int_{-\infty}^{\infty} p_{UV}(u, v) \log_2 p_{UV}(u, v) du dv , \end{aligned}$$

where $p_U(u)$ and $p_{UV}(u, v)$ are, respectively, the marginal and joint pdf's of the random variables $u$ and $v$. In this case the corresponding formula for MI is

$$\begin{aligned} MI(u, v) &= -\int_{-\infty}^{\infty} p_U(u) \log_2\big(p_U(u)\big) du - \int_{-\infty}^{\infty} p_V(v) \log_2\big(p_V(v)\big) dv \\ &\quad + \int_{-\infty}^{\infty} p_{UV}(u, v) \log_2\big(p_{UV}(u, v)\big) du dv , \\ &= \int\int p(u, v) \log_2 \frac{p_{UV}(u, v)}{p_U(u) p_V(v)} du dv . \end{aligned} \tag{6.6}$$

In the case of two input images $I_1$ and $I_2$, $u$ and $v$ represent, respectively, the gray-levels $I_1(i, j)$ and $I_2(i, j)$ of the same pixel $(i, j)$.

Apart from its use in image registration, MI is widely used as a robust similarity measure.

### 6.4.1 Histogram Estimation

Given $N$ simultaneous measurements $(u_i, v_i), i \in \{1, 2, \ldots, N\}$, the most straight-forward, and widely used, approach to calculate the mutual information $MI(u, v)$ is to estimate the marginal pdf's $p_U(u)$ and $p_V(v)$ and the joint pdf $p_{UV}(u, v)$

**Table 6.5** Methods for calculating optimum number bins in a regular histogram

| Name | Formula |
|---|---|
| Sturges | $K_{\mathrm{STURGES}} = 1 + \log_2(N)$, where $N$ is the number of data points |
| Scott | $K_{\mathrm{SCOTT}} = RN^{1/3}/(3.49s)$, where $R$ is the range of the data and $s$ is the sample standard deviation of the input data |
| Knuth | $K_{\mathrm{KNUTH}} = \arg\max_K \big(N \ln K + \ln \Gamma(K/2) - K \ln \Gamma(N + (K/2)) - \sum_{k=1}^{K} \ln \Gamma(n_k + (1/2))\big)$, where $n_k$ is the number of data points which fall in the $k$th bin and $\Gamma$ is the gamma function [18] |
| Birge and Rosenhole | See Example 9.3 |

using a histogram technique. We divide the $(u, v)$ space into $K$ rows and $L$ columns. The intersection of the rows and columns define a set of rectangular bins $B_{k,l}, k \in \{1, 2, \ldots, K\}, l \in \{1, 2, \ldots, L\}$. If $n_{kl}$ is the number of simultaneous measurements which fall in the $(k, l)$ bin $B_{kl}$, then MI is defined as

$$MI(u, v) \approx \log_2 N + \frac{1}{N} \sum_{k,l} n_{kl} \log_2 \frac{n_{kl}}{N_k^r + N_l^c} , \tag{6.7}$$

where $N_k^r = \sum_{l=1}^{L} n_{kl}$ and $N_l^c = \sum_{k=1}^{K} n_{kl}$ are, respectively, the number of measurement pairs which lie in the $k$th row and $l$th column.

In Eq. (6.7) the number of histogram bins, $K$ and $L$, is crucial: If $K$ and $L$ are too small the estimated pdf's are accurate but do not follow the shape of the true density curves. On the other hand, if $K$ and $L$ are too large the estimated pdf's follow the shape of the true density curves but are inaccurate. Table 6.5 lists several formulas which are used to estimate the optimum number of bins in a regular histogram [42].

Instead of using histograms with fixed bin widths, we may improve the MI estimate by adaptively partitioning the $(u, v)$ space. E.g. we may construct a hierarchy of partitions which recursively divide $(u, v)$ space into smaller and smaller bins. The idea is that regions in $(u, v)$ space, where pairs of measurements $(u_i, v_i)$ are uniformly distributed, cease to contribute further to the estimated MI under further partitioning. For these regions, there is therefore no point in subdividing them further. The result is that dense regions containing many simultaneous measurements are covered with small bins, while regions containing few, or no, simultaneous measurements are covered with large bins.

### 6.4.2 *Kernel Density Estimation*

Instead of using discrete bins we may use continuous bins, or *kernels*, to calculate the pdf's $p_U(u)$, $p_V(v)$ and $p_{UV}(u, v)$. This is known as kernel, or Parzen-window density estimation [38, 41] and is a generalization of histogram binning. The estimated pdf $p_U(u)$ is given by

**Table 6.6** Methods for calculating optimum one-dimensional bandwidth $H$

| Name | Description |
|---|---|
| Rule-of-Thumb | Suppose input data $y_i, i \in \{1, 2, \ldots, N\}$, is generated by parametric density function, e.g. Gaussian. Then $H = 1.06\hat{\sigma}N^{-1/5}$ where $\sigma$ is sample standard deviation. Robust versions of $H$ are: $H = 1.06 \min(\sigma, Q/1.34)N^{-1/5}$ and $H = 1.06SN^{-1/5}$; $Q$ is sample interquartile distance and $S$ is median absolute deviation (see Sect. 12.2) |
| Cross-Validation (CV) | Use CV to directly minimize MISE or AMISE (see Sect. 14.5). CV variants include: least square, biased, smoothed CV [17] |
| Plug-in | Minimize AMISE using second (*pilot*) bandwidth $L$. In *solve-equation plug-in* method we write $L$ as a function of kernel bandwidth $H$ [17] |
| Cross-entropy | See [5] |
| Boosting | See Sect. 15.9 |

MISE (Mean Integrated Square Error) $= \int \big(p(y) - \hat{p}_H(y)\big)^2 dy$; $\hat{p}_H(y)$ is kernel approximation to $p(y)$. Asymptotic MISE (AMISE) represents large number approximation of MISE

$$p_U(u) = \frac{1}{NH}\sum_i K\left(\frac{u - u_i}{H}\right), \tag{6.8}$$

where the input data consists of $N$ measurements $u_i, i \in \{1, 2, \ldots, N\}$, and $\int K(u)du = 1$. In general the density estimate $p(u)$ is more sensitive to the choice of bandwidth $H$ and less sensitive to the choice of kernel $K(u)$. For this reason we often let $K(u) = \mathcal{N}(0, 1)$, where $\mathcal{N}(0, 1)$ denotes a zero-mean Gaussian with unit standard deviation. Table 6.6 lists several schemes which are commonly used to calculate the optimal bandwidth $H$.

Apart from its use in MI calculations, accurate pdf estimation plays an important role in many data fusion algorithms as illustrated in the following example.

**Example 6.7** *Background Subtraction in Video Surveillance* [8]. In video surveillance, stationary cameras are typically used to monitor activity at a given site. Since the cameras are stationary, moving objects can be detected by comparing each incoming image with a representation of the background scene. This process is called *background subtraction*. Let $x_1, x_2, \ldots, x_N$ be $N$ consecutive gray-level values for a given pixel. Suppose the pixel has a gray-level $x_t$ at time $t$, then probability that $x_t$ belongs to the background is:

$$p(x_t|x_1, x_2, \ldots, x_N, I) = \frac{1}{NH}\sum_i K\left(\frac{x_t - x_i}{H}\right),$$

where $\int K(u)du = 1$. The pixel is considered to be in the foreground if $p(x_t|x_1, x_2, \ldots, x_N, I) < T$, where $T$ is a given threshold. We estimate the bandwidth $H$ using:

$$H = \frac{1}{\sqrt{2} \times 0.68} \mathrm{med}(|x_1 - x_2|, |x_2 - x_3|, \ldots, |x_{N-1} - x_N|) \ .$$

The motivation for using this equation is that although the intensity of a given pixel may vary in time, differences between consecutive intensity values are generally small, and only a few pairs are expected to exhibit a jump in intensity. By using the median the estimate is not affected by a few jumps in intensity.

### *6.4.3 Regional Mutual Information (RMI)**

A drawback to the use of mutual information (MI) is that it does not take into account local relationships which may exist in the input data. The following example explains a simple way to incorporate these relationships into MI.

**Example 6.8** *Neighbourhood Mutual Information (NMI)* [36]. In NMI we simply replace the random variables $u$ and $v$ in Eq. (6.6) with 2D random vectors $\mathbf{u} = \left(u^{(1)}, u^{(2)}\right)^T$ and $\mathbf{v} = \left(v^{(1)}, v^{(2)}\right)^T$. In the case of two input images $I_1$ and $I_2$, $u^{(1)}$ and $v^{(1)}$ represent, respectively, the gray-levels $I_1(i, j)$ and $I_2(i, j)$ at a pixel $(i, j)$ and $u^{(2)}$ and $v^{(2)}$ represent, respectively, the mean gray-levels $\overline{I}_1(i, j)$ and $\overline{I}_2(i, j)$ at the same pixel $(i, j)$.

To incorporate more spatial relationships in the MI we require random vectors **u** and **v** with more dimensions. In practice, the number of dimensions is limited, since the number of data points required for an accurate estimate of the marginal and joint pdf's increases exponentially with the number of dimensions. One way to overcome this problem (known as the "curse of dimensionality") is described in the following example.

**Example 6.9** *Regional Mutual Information (RMI)** [36]. In RMI we include all the pixels which are in a $(2r + 1) \times (2r + 1)$ local neighborhood of $(i, j)$. By assuming the distribution of the points is Gaussian, we can *analytically* calculate the RMI as follows:

$$RMI = \ln\left(\Delta \det \times (\boldsymbol{\Sigma}_A)^{1/2}\right) + \ln\left(\Delta \times \det(\boldsymbol{\Sigma}_B)^{1/2}\right) - \ln\left(\Delta \times \det(\boldsymbol{\Sigma}_J)^{1/2}\right) , \tag{6.9}$$

where $\boldsymbol{\Sigma}_A$ denotes the covariance matrix of gray-scale image $A$ and its $D = (2r + 1)^2$ local neighbors, $\boldsymbol{\Sigma}_B$ is the covariance matrix of gray-scale image $B$

and its $D$ local neighbors, $\mathbf{\Sigma}_J$ is the joint covariance matrix of images $A$ and $B$ and their $2D$ local neighbors and $\Delta = (2\pi)^{D/2}$ (Fig. 6.2).

The main steps in calculating the RMI are given in Algorithm 1.

**Algorithm 1:** Regional Mutual Information

**Input** : Input Images $A(m,n)$ and $B(m,n)$ of size $M \times N$
**Output** : Regional mutual information $RMI$

1. Create a stack $\Phi$ of $2D = 2(2r+1)^2$ images
   $D$ images of $A$ and her corresponding neighbors and
   $D$ images of $B$ and her corresponding neighbors (see Fig. 6.2)
2. Calculate mean image $\overline{\Phi}(m,n) = \sum_d \Phi(m,n,d)/2D$
3. Remove $\overline{\Phi}$ from stack: $\widetilde{\Phi} = \Phi - \overline{\Phi}$
4. Calculate covariance $\mathbf{\Sigma}_J = \widetilde{\Phi}\widetilde{\Phi}^T/((M-2r)(N-2r))$
5. Calculate joint entropy $H(A,B) = \log((2\pi)^D \det(\mathbf{\Sigma}_J)^{1/2})$
6. Calculate marginal entropy $H(A) = \log((2\pi)^D \det(\mathbf{\Sigma}_A)^{1/2})$
   where $\mathbf{\Sigma}_A$ is $D \times D$ sub-matrix in top-left corner of $\mathbf{\Sigma}_J$
7. Calculate marginal entropy $H(B) = \log((2\pi)^D \det(\mathbf{\Sigma}_B)^{1/2})$
   where $\mathbf{\Sigma}_B$ is $D \times D$ sub-matrix in bottom-right corner of $\mathbf{\Sigma}_J$
8. Calculate regional mutual information $RMI = H(A) + H(B) - H(A,B)$

In the case of image registration using MI, special attention must be given to the process of resampling/interpolation, because the process introduces new gray-levels which may interfere with the computation of the MI. For example, in remote sensing applications, nearest neighbour interpolation is often found to perform better than linear and cubic convolution interpolation (Fig. 6.2).

## 6.5 Sparse Parametric Matching

In sparse-parametric matching we extract distinctive salient features from each input image $A$ and $B$. We then establish a global correspondence by matching individual features in $A$ to features in $B$. Using the corresponding pairs of features we estimate the optimal geometric transformation $\mathcal{T}^*$ between the images, where the corresponding transformation parameters are: $\alpha_k^*, k \in \{1, 2, \ldots, K\}$.

We now consider the main steps in a sparse parametric matching algorithm (see Fig. 6.3).

**Feature Detection.** Distinctive salient features are *separately* detected in $A$ and in $B$. The features are compact geometrical entities which may be points, curves or surfaces. More commonly, a keypoint detector/descriptor (see Table 6.7) is used for this purpose. For the sake of concreteness we shall assume a SIFT detec-

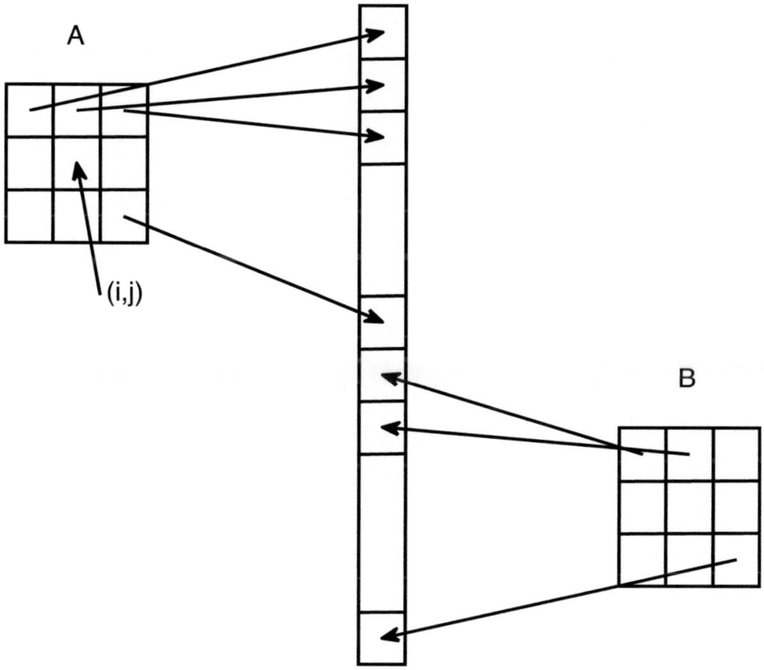


**Fig. 6.2** Shows the creation of a stack $\Phi$ of $2D = 2(2r+1)^2$ images for the pixel $(i, j)$. In the figure we assume $r = 1$, i.e. $D = (2r+1)^2 = 9$. The first $D$ images are derived from the image $A$ and the second $D$ images are derived from the image $B$

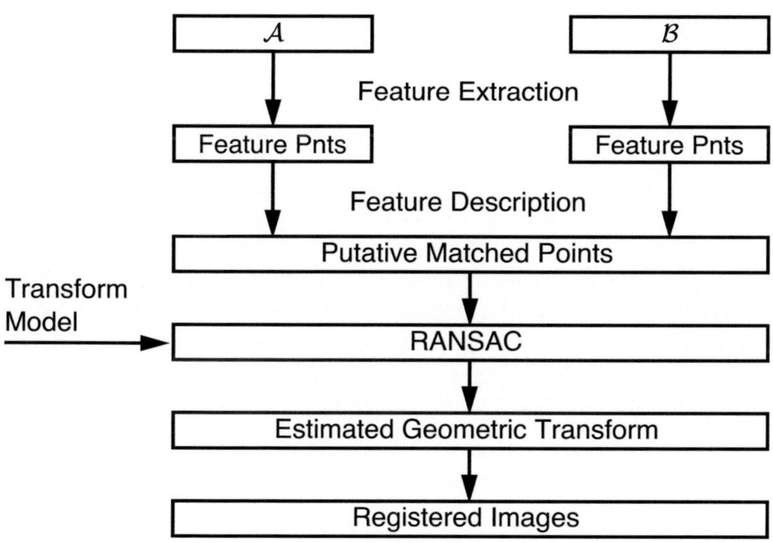


**Fig. 6.3** Shows the main steps in sparse parametric match algorithm

**Table 6.7** Handcrafted and deep learning keypoint detectors and descriptors [23]

| Detector | Scale | Rotation | Illumination | Descriptor | Scale | Rotation | Illumination |
|---|---|---|---|---|---|---|---|
| **Handcrafted** | | | | | | | |
| BRISK | ✓ | – | ✓ | BRISK | ✓ | ✓ | ✓ |
| ORB | – | ✓ | ✓ | ORB | – | ✓ | ✓ |
| – | | | | BRIEF | ✓ | ✓ | ✓ |
| SIFT | ✓ | ✓ | ✓ | SIFT | ✓ | ✓ | ✓ |
| SURF | ✓ | ✓ | ✓ | SURF | ✓ | ✓ | ✓ |
| **Deep-learning** | | | | | | | |
| – | | | | L2Net | ✓ | ✓ | ✓ |
| MatchNet | ✓ | ✓ | ✓ | MatchNet | ✓ | ✓ | ✓ |
| D2Net | ✓ | ✓ | ✓ | D2Net | ✓ | ✓ | ✓ |
| ALIKE | ✓ | ✓ | ✓ | ALIKE | ✓ | ✓ | ✓ |

tor/descriptor (see Sect. 4.4.3). This is the most successful and widely used handcrafted keypoint detector/descriptor. It works by separately identifying distinctive locations (keypoints) in each image that are invariant to changes in scale, rotation and illumination.

**Feature Descriptor**. For each keypoint we encode a descriptor, or feature vector, of the local patch centered on the keypoint. In SIFT the descriptor is a $L = 128$-long vector that is generated by dividing the local patch into $4 \times 4$ non-overlapping cells and conducting a histogram of gradient orientations with 8 bins in each cell. The SIFT descriptor is invariant to changes in scale, rotation and illumination.

**Putative Matching**. We create a global correspondence by matching SIFT keypoints detected in $A$ with a corresponding keypoints detected in $B$. A simple strategy for matching the keypoints is to use the *nearest neighbor* (NN) criterion i.e. minimum Euclidean distance between corresponding keypoint descriptors. Suppose the $i$th keypoint in $A$ has a descriptor $\mathcal{A}_i = (\mathcal{A}_i(1), \mathcal{A}_i(2), \ldots, \mathcal{A}_i(L))^T$. Then the NN match in $B$ is the $j^*(i)$ keypoint, whose descriptor is $\mathcal{B}_{j^*(i)}$, where

$$j^*(i) = \arg\min_j \|\mathcal{A}_i - \mathcal{B}_j\| \ , \tag{6.10}$$

$\|\mathcal{A}_i - \mathcal{B}_j\|$ is the Euclidean distance $\left(\sum_{l=1}^{L}(\mathcal{A}_i(l) - \mathcal{B}_j(l))^2\right)^{1/2}$ and the minimization is taken over all keypoints $\mathcal{B}_j$, $j \in \{1, 2, \ldots, N\}$. In practice the NN criterion generates too many false matches. To help identify false matches we may use the second best match (second minimum distance from $\mathcal{A}_i$) to validate a match: $j^*(i)$ is considered a valid match if

$$\|\mathcal{A}_i - \mathcal{B}_{j^*(i)}\| \leq T \times \|\mathcal{A}_i - \mathcal{B}_{r^*(i)}\| \ , \tag{6.11}$$

where

$$r^*(i) = \arg\min_{j, j \neq j^*(i)} \|\mathcal{A}_i - \mathcal{B}_j\| \ , \tag{6.12}$$

and $T$ is a given threshold. This is known as the NNDR (NN "distance ratio") criterion.

**Refined Putative Matching (RANSAC).** Although validated matching procedure reduces the number of false matches, the number of outliers may still be too high. To further reduce the number of false matches we refine the set of putative matches using geometrical constraints. One such method of doing this is RANSAC (Random Sample Consensus). This follows a hypothesize-and-verify paradigm and works as follows. We randomly select a set $S$ containing $|S|$ putative matches to compute a least square transformation $\mathcal{T}$.[2] For this transformation count the number of inliers (i.e. the number of putative keypoint pairs that are close to their predicted locations). The entire process is repeated a large number of times. The set with the largest number of inliers defines the optimal set $S^*$ and the corresponding optimal set of $P^*$ matching inlier pairs: $(x_{(p)}, y_{(p)}) \leftrightarrow (x'_{(p)}, y'_{(p)}), p \in \{1, 2, \ldots, P^*\}$.

**Transformation Estimation.** Given the optimal set $S^*$, use the set of $P^*$ matching inlier pairs $(x_{(p)}, y_{(p)}) \leftrightarrow (x'_{(p)}, y'_{(p)}), p \in \{1, 2, \ldots, P^*\}$, to compute a least square transformation $\mathcal{T}^*$ with parameters $\alpha_k^*, k \in \{1, 2, \ldots, K\}$.

The following example illustrates global image registration using RANSAC.

**Example 6.10** *Global Image Registration Using RANSAC.* Given two gray-scale input images $A$ and $B$, we detect $M$ keypoints in $A$ and $N$ keypoints in $B$. After RANSAC we have a set of $P^*$ inlier keypoint pairs: $(x_{(p)}, y_{(p)}) \leftrightarrow (x'_{(p)}, y'_{(p)}), p \in \{1, 2, \ldots, P^*\}$. To make the analysis concrete we consider an *affine* transformation:

$$x' = \alpha_1 x + \alpha_2 y + \alpha_3 \quad \text{and} \quad y' = \alpha_4 x + \alpha_5 y + \alpha_6 .$$

Substituting the $P^*$ inlier keypoint pairs into the affine transformation gives:

$$\begin{pmatrix} x'_{(1)} \\ y'_{(1)} \\ x'_{(2)} \\ \vdots \\ x'_{(P^*)} \\ y'_{(P^*)} \end{pmatrix} = \begin{pmatrix} x_{(1)} & y_{(1)} & 1 & 0 & 0 & 0 \\ 0 & 0 & 0 & x_{(1)} & y_{(1)} & 1 \\ x_{(2)} & y_{(2)} & 1 & 0 & 0 & 0 \\ \vdots & & & & & \\ x_{(P^*)} & y_{(P^*)} & 1 & 0 & 0 & 0 \\ 0 & 0 & 0 & x_{(P^*)} & y_{(P^*)} & 1 \end{pmatrix} \begin{pmatrix} \alpha_1 \\ \alpha_2 \\ \vdots \\ \alpha_K \end{pmatrix} .$$

---

[2] If the number of parameters $\alpha_k, k \in \{1, 2, \ldots, K\}$, in $\mathcal{T}$ is $K$, then we require $|S| \geq K/2$. For translation, similarity, affine and homography transformations, we thus require at least 1, 2, 3 and 4 putative matches. In practice, we often use a set whose size is slightly larger than the above minimal value.

or

$$\mathbf{x}' = X\boldsymbol{\alpha} .$$

The linear least squares solution of this equation is: $\boldsymbol{\alpha} = \left(X^T X\right)^{-1} X^T \mathbf{x}'$.

Pseudo-code for RANSAC is given in Algorithm 2. See also Sect. 12.6.

---

**Algorithm 2:** RANSAC

**Input**: Putative keypoint pairs $\mathcal{A}_{(p)} \leftrightarrow \mathcal{B}_{(p)}$, $p \in \{1, 2, \ldots, P\}$;
Maximum Number of Iterations $N$
**Output**: Optimized global transformation $\mathcal{T}^*$ and its parameters $\alpha_k^*$, $k \in \{1, 2, \ldots, K\}$

$i \leftarrow 1$
$P^* \leftarrow 0$
**while** $i \leq N$ **do**
  $S_i$ ←Randomly select a minimal set of keypoint pairs
  $\mathcal{T}_i$ ←Generate a hypothesis (transformation ) from $S_i$
  $m$ ← Count number of detected inliers (i. e. number of inlier keypoint pairs)
  **if** $m > P^*$ **then**
    $P^* \leftarrow m$
    $S^* \leftarrow S_i$
    $(x_{(p)}, y_{(p)}) \leftrightarrow (x'_{(p)}, y'_{(p)})$, $p \in \{1, 2, \ldots, P^*\}$ ← Set of putative keypoint pairs which are inliers
  **end**
**end**
$i \leftarrow i + 1$
Calculate $\mathcal{T}^*$ and parameters $\alpha_k^*$, $k \in \{1, 2, \ldots, K\}$, using $P^*$ in linear least square solution. See Example 6.10

---

## 6.5.1 KeyPoint Detector/Descriptors

It is clear that the keypoint detector/descriptors play a key role in sparse parametric matching. Of the detector/descriptors in use today, SIFT is probably the most widely used. Nevertheless alternative handcrafted detector/descriptors (which are often faster and/or use less memory) and deep learning detector/descriptors (which are often more powerful) are also used [23]. Some of these detector/descriptors and their properties are listed in Table 6.7.

The following example describes a deep learning keypoint descriptor which may be used as a *direct replacement* for the SIFT descriptor.

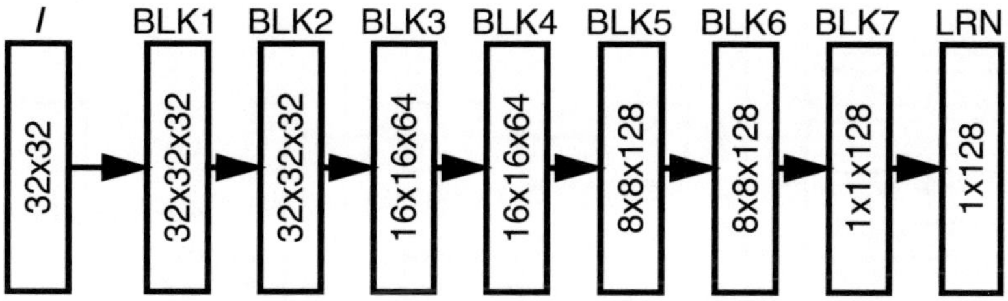


**Fig. 6.4** Shows the architecture of the L2-Net. Input to the network is a $32 \times 32$ image patch $I$. Final output is a vector of length 128

**Example 6.11** *L2-Net Keypoint Descriptors* [46]. Let $P$ denote a $32 \times 32$ patch in a given gray-scale input image $I$. Then the L2-Net outputs a 128-dimensional descriptor which may be used as a *direct replacement* of the SIFT descriptor, i.e. we match the L2-Net descriptors using Euclidean $L_2$-distance.

The L2-Net (see Fig. 6.4) consists of seven basic blocks and a local response normalization (LRN) layer. Except for BLK7, each block consists of a $3 \times 3$ convolutional layer (with zero padding), a batch normalization (BN) layer, and a ReLU activation layer. All blocks use a stride of 1 apart from BLK3 and BLK5 which used a stride of 2. BLK7 has no-padding and its kernel size and kernel number are $8 \times 8$ and 128 respectively.

An important development in deep learning keypoint detector/descriptors is the introduction of a *joint* detector/descriptor network [7].

**Example 6.12** *Joint Keypoint Detector/Descriptor Network* [21]. The multi-level joint detector/descriptor is an end-to-end deep learning network. The network takes a RGB input image $I$ and outputs a keypoint map and corresponding keypoint descriptions (see Fig. 6.5). The networks consists of three parts: VGGnet-like backbone, feature encoder, keypoint detection and description module. The feature encoder takes a RGB input image of size $H \times W \times 3$ and outputs four feature maps $F_k, k \in \{1, 2, 3, 4\}$, of sizes $H \times W \times 64$, $H/2 \times W/2 \times 64$, $H/4 \times W/4 \times 128$ and $H/8 \times W/8 \times 128$. The feature maps are restored to the original image resolution by up-sampling and bilinear interpolation. For each feature map, aliasing effects from up-sampling are reduced using a $3 \times 3$ convolution layer. The resulting full-resolution feature maps (denoted as $\widetilde{F}_k, k \in \{1, 2, 3, 4\}$) are aggregated by channel-wise concatenation and are fed into the keypoint detector (S) and descriptor module (D).

The following example describes a powerful lightweight deep learning keypoint detector/descriptor ALIKE.

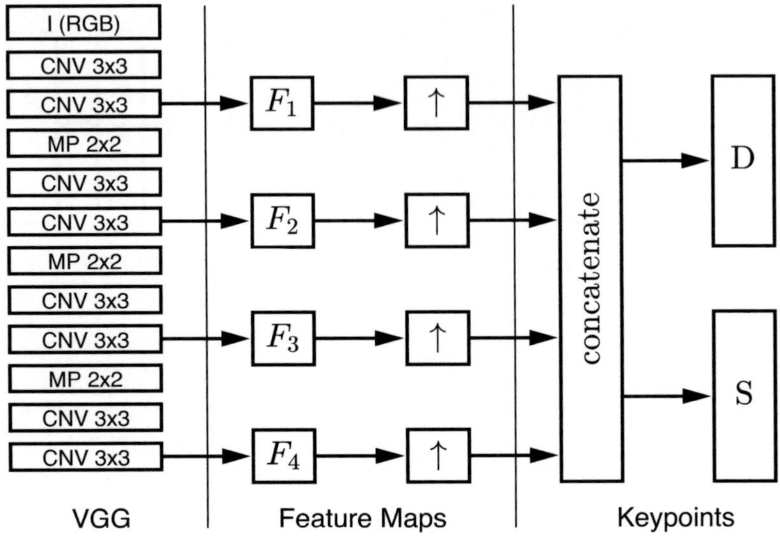


**Fig. 6.5** Shows the architecture of the joint keypoint detector/descriptor network. Input to the network is $H \times W$ image $I$ which outputs four feature maps $F_k$, $k \in \{1, 2, 3, 4\}$, of different sizes. The $F_k$ are restored to the original image size by up-sampling and bilinear interpolation. The full resolution feature maps are aggregated by concatenation and outputs the keypoint location ($S$) and descriptor ($D$)

**Example 6.13** *ALIKE: Accurate and Lightweight Keypoint Detection and Descriptor* [50]. ALIKE is an accurate and lightweight joint keypoint detector/descriptor. Its architecture is similar to the multi-stage joint detector/descriptor described in Example 6.12. In ALIKE the feature encoder consists of four blocks. The first block BLK1 is a two-layer $3 \times 3$ convolution with ReLU activation and the next three blocks BLK2, BLK3 and BLK4 contain a max-pooling layer and a $3 \times 3$ ResNet layer. Input to the feature encoder is a $H \times W \times 3$ input image and the output are feature maps $F_k$, $k \in \{1, 2, 3, 4\}$, of sizes: $H \times W \times 16$, $H/2 \times W/2 \times 32$, $H/8 \times W/8 \times 64$ and $H/32 \times W/32 \times 128$ (Fig. 6.6). The $F_k$ are aggregated using $1 \times 1$ convolution and are restored to the original image size by up-sampling and bilinear interpolation. Output of the aggregation stage is a $H \times W \times 65$ feature map, where the first 64 channels are $L_2$ normalized (forming the descriptor map $D$) and the last channel is sigmoid normalized (forming a keypoint score map $S$). Finally the keypoint locations are detected in $S$ and their corresponding descriptors taken from $D$.

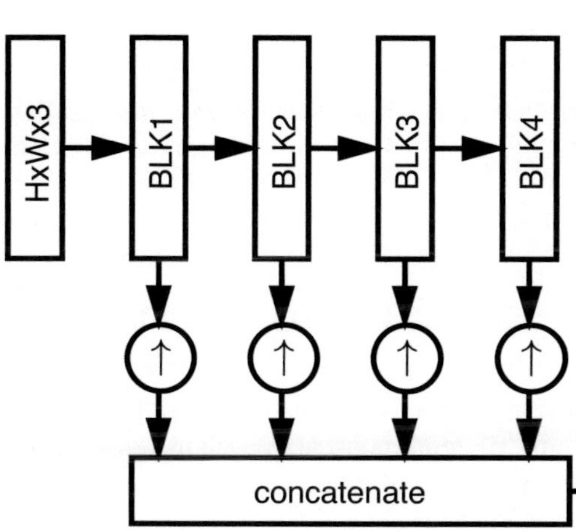


**Fig. 6.6** Shows the architecture of the ALIKEnet, where the upsampling of the feature maps is performed in the circular blocks labeled with an arrow

### 6.5.2 Putative Matching

Traditionally putative keypoint matching is carried out using the nearest neighbor criterion and the NNDR criterion. Additional criteria may however been used to eliminate false matches. A recent criterion which has proven very successful in eliminating false matches is *local patch matching* (LPM) [25].

**Example 6.14** *Local Patch Matching (LPM)* [25]. In LPM we convert each putative pair $(x_{(p)}, y_{(p)}) \leftrightarrow (x'_{(p)}, y'_{(p)})$ into a displacement vector $\mathbf{v}_{(p)}$, where the head and tail of $\mathbf{v}_{(p)}$ correspond to the spatial positions $(x_{(p)}, y_{(p)})$ and $(x'_{(p)}, y'_{(p)})$ in the two images $\mathcal{A}$ and $\mathcal{B}$. Then for each vector $\mathbf{v}_{(p)}$ we measure the similarity, $S_k(p)$, between $\mathbf{v}_{(p)}$ and its $k$th nearest neighbor (which we denote as $\mathbf{w}_{(k)}$). Let $SIM(|\mathbf{v}_{(p)}|, |\mathbf{w}_{(k)}|)$ denote the similarity between the lengths of $\mathbf{v}_{(p)}$ and $\mathbf{w}_{(k)}$ and let $SIM(\angle\mathbf{v}_{(p)}, \angle\mathbf{w}_{(k)})$ denote the similarity between the orientations of $\mathbf{v}_{(p)}$ and $\mathbf{w}_{(k)}$, then

$$S_k(p) = SIM(|\mathbf{v}_{(p)}|, |\mathbf{w}_{(k)}|) \times SIM(\angle\mathbf{v}_{(p)}, \angle\mathbf{w}_{(k)}) \ ,$$

The keypoint match $(x_{(p)}, y_{(p)}) \leftrightarrow (x'_{(p)}, y'_{(p)})$ is regarded as valid if $\sum_k S_k(p) > T$, where $T$ is a given threshold.

### 6.5.3 *Coherent Point Drift**

A completely different approach to matching keypoint *detections* is the Coherent Point Drift (CPD) algorithm. This is often used for registration of 3D point clouds and is described in Sect. 10.5.1.

## 6.6 Dense Non-parametric Matching (Optical Flow)

In dense non-parametric matching we compute a displacement for *every* pixel in the image. The most popular example of dense non-parametric matching is optical flow: defined as the pattern of apparent motion of objects in a visual scene caused by the relative motion between sensor and scene.

Formally, optical flow [9, 37, 47] is defined as a 2D field of velocities which is associated with the variation of brightness in the input image. The assumption is that the brightness of an image at any point on an object is invariant under motion. This is known as the *constant brightness assumption.*

Let $P$ denote some point in a given scene. Suppose $(x, y)$ is the position of $P$ in an image of the scene taken at time $t$ and $(u(x, y), v(x, y))$ is the projected velocity vector of this point in the image plane. Then after a time $\delta t$ the image of the scene will show the point $P$ in a new position $(x + u(x, y)\delta t, y + v(x, y)\delta t)$. According to the assumption of constant brightness, the brightness of the point $P$ in the two *gray-scale* input images should be the same, i.e.

$$I(x + u(x, y)\delta t, y + v(x, y)\delta t, t + \delta t) - I(x, y, t) = 0 \, , \tag{6.13}$$

where $I(x, y, t)$ is the brightness in the image taken at time $t$ at the position $(x, y)$. Solving Eq. (6.13) is a difficult optimization problem. It can, however, be made much more tractable by linearizing the equation. We do this by applying a first-order Taylor expansion to $I(x + u\delta t, y + v\delta t, t + \delta t)$:

$$I(x + u\delta t, y + v\delta t, t + \delta t) \approx I(x, y, t) + \Big(\frac{\partial I}{\partial x}u + \frac{\partial I}{\partial y}v + \frac{\partial I}{\partial t}\Big)\delta t \, .$$

The result is known as the *optical flow equation*

$$I_x(x, y)u(x, y) + I_y(x, y)v(x, y) + I_t(x, y) = 0 \, , \tag{6.14}$$

where $I_x = \partial I/\partial x$, $I_y = \partial I/\partial y$, $I_t = \partial I/\partial t$.

The solution of Eq. (6.14) are the velocity components $(u(x, y), v(x, y))$ which minimize $\sum_{x,y} R_{\text{data}}(x, y)$, where

$$R_{\text{data}}(x, y) = \big(I_x(x, y)u(x, y) + I_y(x, y)v(x, y) + I_t(x, y)\big)^2$$

Solving Eq. (6.14) is ill-posed since there is only one equation for each pixel but there are two unknowns (velocities $u$ and $v$). This is known as the *aperture problem* which states that only the optical flow component parallel to the image gradient can be computed. In order to determine both components we need impose additional constraints on Eq. (6.14).

The additional constraints are often written in the form of local or global spatial coherency.

#### Local Spatial Coherency: Lucas-Kanade Equations

Consider a small image patch $P$ containing $K$ pixels $\{(x_1, y_1), (x_2, y_2), \ldots, (x_K, y_K)\}$. Then the residual error for the image patch is

$$E = \sum_{(x,y)\in P} R_{\text{data}}(x, y) = \sum_{(x,y)\in P} \big(uI_x(x, y) + vI_y(x, y) + I_t(x, y)\big)^2 ,$$

Minimizing $E$ leads to the Lucas-Kanade equation which is identical to an ordinary least square solution:

$$\begin{pmatrix} u \\ v \end{pmatrix} = \begin{pmatrix} \sum I_x(x, y)I_x(x, y) & \sum I_x(x, y)I_y(x, y) \\ \sum I_x(x, y)I_y(x, y) & \sum I_y(x, y)I_y(x, y) \end{pmatrix}^{-1} \begin{pmatrix} -\sum I_x(x, y)I_t(x, y) \\ -\sum I_y(x, y)I_t(x, y) \end{pmatrix} , \tag{6.15}$$

where the summation $\sum$ is taken over all pixels $(x, y)$ in the patch $P$, i.e. $\sum \equiv \sum_{(x,y)\in P}$.

The Lucas-Kanade equation is found to yield a biased estimate of the flow [11]. The reason for this is the implicit assumption that the estimated derivatives $I_x$, $I_y$ and $I_t$ are noise-free.

#### Global Spatial Coherency: Horn-Schunck Equations

Consider a pixel $(x, y)$ and its velocity components $\big(u(x, y), v(x, y)\big)$. We define a smoothness measure $S(x, y)$ by summing the local differences in velocity:

$$\begin{aligned} R_{\text{smth}}(x, y) = \frac{1}{4}\Big(&\big(u(x + 1, y) - u(x, y)\big)^2 + \big(u(x, y + 1) - u(x, y)\big)^2 \\ &+ \big(v(x + 1, y) - v(x, y)\big)^2 + \big(v(x, y + 1) - v(x, y)\big)^2\Big) . \end{aligned}$$

We find the components $u(x, y)$ and $v(x, y)$ that minimize the error $E$:

$$E = \sum_{(x,y)} \big(R_{\text{data}}(x, y) + \lambda R_{\text{smth}}(x, y)\big) .$$

Differentiating $E$ with respect to $u(x, y)$ and $v(x, y)$ and setting to zero gives the Horn-Schunck Equations:

$$\begin{aligned}
\big(u(x, y) - \overline{u}(x, y)\big) + \lambda\big(I_x(x, y)u(x, y) + I_y(x, y)v(x, y) + I_t(x, y)\big)I_x(x, y) &= 0 \,, \\
\big(v(x, y) - \overline{v}(x, y)\big) + \lambda\big(I_x(x, y)u(x, y) + I_y(x, y)v(x, y) + I_t(x, y)\big)I_y(x, y) &= 0 \,,
\end{aligned} \tag{6.16}$$

where $\overline{u}(x, y)$ and $\overline{v}(x, y)$ are the local averages of $(u, v)$ around the pixel $(x, y)$. These equations may be solved iteratively:

$$\begin{aligned}
u(x, y)^{(k+1)} &= \overline{u}^{(k)}(x, y) - \alpha I_x(x, y) \,, \\
y(x, y)^{(k+1)} &= \overline{v}^{(k)}(x, y) - \alpha I_y(x, y) \,,
\end{aligned}$$

where $k$ is the iteration number and

$$\alpha = \frac{I_x(x, y)\overline{u}^{(k)}(x, y) + I_y(x, y)\overline{v}^{(k)}(x, y) + I(x, y)}{1 + \lambda(I_x(x, y)^2 + I_y(x, y)^2)} \,.$$

The following examples illustrate the use of optical flow in three data fusion applications.

**Example 6.15** *Converting a Smiling Face to a Neutral Face* [13]. The human face displays variety of expressions, e.g. smile, sorrow, surprise etc. These expressions lead to a significant change in the appearance of a facial image, which in turn, leads to a reduction in recognition rate of a face recognition system which in general have been trained on neutral faces and which represent the "common representational format". We may alleviate the reduction in performance by using optical flow to warp the test image to a neutral face.

**Example 6.16** *Radar-Camera Fusion Using Optical-Flow* [24]. Radar is a mainstream 3D sensor used in perception systems for driving assistance and autonomous driving. Apart from measuring 3D positions, radar estimates radial velocities of returned signals (using the doppler effect). To exploit this extra capability we require the full velocity. In [24] the radial doppler velocity $\mathbf{v}_R$ is fused with the 2D optical flow $(u, v)$ (provided by a RGB camera). By fusing $\mathbf{v}_R$ with $(u, v)$ we determine the full velocity $\mathbf{v}$ by constraining $\mathbf{v}$ to be consistent with $\mathbf{v}_R$ and $(u, v)$.

**Example 6.17** *Optical flow for Driver Fatigue Detection* [39]. A major cause of road accidents is lack of driver concentration due to fatigue. A strong indicator of driver fatigue is "percentage eye closure" $\mathcal{P}$, i.e. the percentage of the time the eye is closed. In [39] the movement of the upper eyelid is tracked using image velocities based on optical flow. Eye closures and eye openings are detected which are, in turn, used to estimate $\mathcal{P}$.

In practice, we do *not* estimate the optical flow field by directly applying Eqs. (6.15) and (6.16) to the images $I_1$ and $I_2$. Instead we introduce two modifications which we now describe. *Note*. For notational convenience we shall, in the remainder of this section, denote the input images $I(x, y, t)$ and $I(x, y, t + \delta t)$ as $A$ and $B$.

### Pre-processing

We pre-process the images $A$ and $B$ by smoothing them. Experiments [31] have shown this may substantially improve the accuracy of the optical flow field. For this purpose a 2D Hampel filter is recommended (see Sect. 12.2).

**Example 6.18** *Hampel Filter for Low-End Optical Flow Estimation* [31]. In this example we smooth the input images $A$ and $B$ using a 2D Hampel filter. Let $I(x, y)$ denote the value of an input image at the pixel $(x, y)$, then

$$\widetilde{I}(x, y) = \begin{cases} I(x, y) & \text{if } |I(x, y) - M(x, y)| \leq \alpha S(x, y)\ , \\ M(x, y) & \text{otherwise}\ , \end{cases}$$

where $\widetilde{I}(x, y)$ is the smoothed image value at pixel $(x, y)$, $M(x, y)$ and $S(x, y)$ are, respectively, the median and the median absolute deviation in a small window centered at $(x, y)$ and $\alpha$ is a suitable multiplier.

### Coarse-to-Fine

A basic assumption which underlies the use of the optical flow equation is that the Taylor expansion in Eq. (6.6) is valid, i.e. the input images $A$ and $B$ are similar to each other and the motion between them is small.

In practice not all motions are small. One popular scheme to handle large motions is to embed the optical flow in a multi-scale ("pyramid") scheme. The idea is to create a coarse-to-fine structure that enables the estimation of the optical flow at a coarse

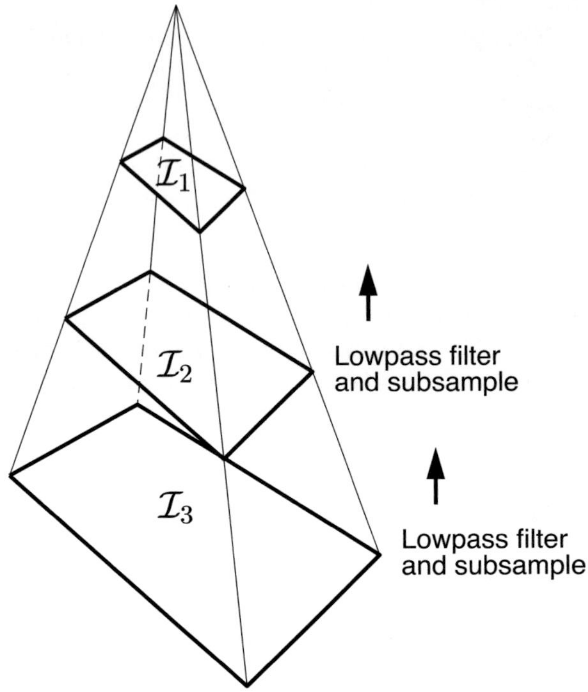


**Fig. 6.7** Shows the construction of a $K$-level pyramid representation of a $M \times N$ input image $I$. The input image $I$ is placed at the bottom of the pyramid and is labeled $\mathcal{I}_K$. Then $\mathcal{I}_K$ is low-pass filtered and sub-sampled by a factor of $\eta$ in each dimension. The resulting $(M/\eta) \times (N/\eta)$ image is labeled $\mathcal{I}_{K-1}$ and appears at the second level of the pyramid. The entire procedure is repeated $K - 1$ times. At the top of the pyramid lies the $(M/\eta^{K-1}) \times (N/\eta^{K-1})$ image which is labeled $\mathcal{I}_1$

scale and then gradually refines the solution at finer scales. In this scheme, we start by building a multi-scale pyramid representation [3] for each input image:

$$A \rightarrow \{\mathcal{A}_1, \mathcal{A}_2, \ldots, \mathcal{A}_K\} \quad \text{and} \quad B \rightarrow \{\mathcal{B}_1, \mathcal{B}_2, \ldots, \mathcal{B}_K\} \ , \tag{6.17}$$

where the (spatial) resolution of the pyramid images $\mathcal{A}_k, \mathcal{B}_k, k \in \{1, 2, \ldots, K\}$, increases with $k$. See Fig. 6.7.

Starting at the top of the pyramid, the optical flow equations are solved at each pyramid level $k$ to get successive approximations to the optical flow. In solving the optical flow equations, every intermediate solution is used as an initialization of the optical flow at the next level.

We now give the mathematical equations which underlie the coarse-to-fine algorithm. Let $\downarrow$ and $\uparrow$ denote, respectively, a downsampling and upsampling function. When we apply these functions to a $M \times N$ image $I$ we obtain images $\downarrow(I)$ and $\uparrow(I)$ of sizes $\frac{M}{\eta} \times \frac{N}{\eta}$ and $\eta M \times \eta N$. In many applications, we let $\eta = 2$.

For each pyramid level $k$, we perform the following three calculations (see Fig. 6.8):

1. Warp the pyramid image $\mathcal{B}_k$ using the optical flow $\mathbf{V}_{k-1} = \big(u_{k-1}, v_{k-1}\big)$ found at the previous level $(k-1)$:

$$\widetilde{\mathcal{B}}_k = \text{warp}(\mathcal{B}, \uparrow(\mathbf{V}_{k-1}) \ . \tag{6.18}$$

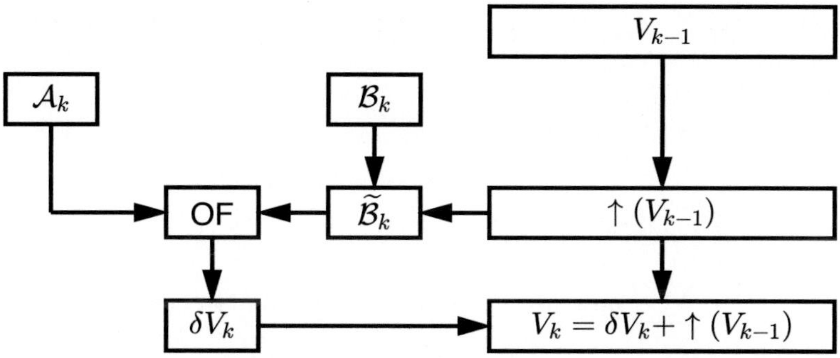


**Fig. 6.8** Shows the optical flow algorithm for the $k$th pyramid level. The input images at this level are $\mathcal{A}_k$ and $\mathcal{B}_k$. Image $\mathcal{B}_k$ is first warped using the optical flow calculated at the previous pyramid level: $\widetilde{\mathcal{B}}_k = \text{warp}(\mathcal{B}_k, \uparrow(V_{k-1}))$. Next the *residual* optical flow, $\delta \mathbf{V}_k$, is calculated between $\mathcal{A}_k$ and the warped image $\widetilde{\mathcal{B}}_k$: $\delta \mathbf{V}_k = OF(\mathcal{A}_k, \widetilde{\mathcal{B}}_k)$. Finally, the *total* optical flow, $\mathbf{V}$, is calculated: $\mathbf{V}_k = \uparrow(\mathbf{V}_{k-1}) + \delta \mathbf{V}_k$

2. Calculate *residual* optical flow for the level $k$, i.e. the optical flow calculated between $\mathcal{A}_k$ and the warped image $\widetilde{\mathcal{B}}_k$:

$$\delta \mathbf{V}_k = OF(\mathcal{A}_k, \widetilde{\mathcal{B}}_k) . \tag{6.19}$$

3. Estimate the *total* optical flow estimate for the level $k$, i.e. the optical flow between the images $\mathcal{A}_k$ and $\mathcal{B}_k$:

$$\mathbf{V}_k = \uparrow(\mathbf{V}_{k-1}) + \delta \mathbf{V}_k . \tag{6.20}$$

## 6.6.1 Dense Descriptor Flow (SIFT Flow*)

In dense descriptor flow [4] we replace each pixel gray-level $I(x, y)$ with a keypoint descriptor (handcrafted or deep learning) and reformulate optical flow estimation as an energy minimization problem. An important and early example of dense descriptor flow is SIFT flow [22] in which we replace $I(x, y)$ with a SIFT keypoint descriptor. To keep our discussion focused we shall initially limit ourselves to SIFT flow.

Given an gray-scale input image $I(x, y)$ we apply SIFT keypoint descriptor to every pixel in the image. The result is a "SIFT" image $\mathcal{S}$ which is of the same size as the input image but 128-deep. Then, given two SIFT images $\mathcal{A}$ and $\mathcal{B}$ we may match them using a technique similar to that used for optical flow. Thus we want SIFT descriptors to be matched along the flow vectors, and the flow field to be smooth, with discontinuities agreeing with object boundaries.

Based on these criteria we may define an energy function $\mathcal{E}$ which we minimize for SIFT flow [22], where

$$\begin{aligned}\mathcal{E}(u, v) = &\sum_{(x,y)} \min\Big(\|\mathcal{A}(x, y) - \mathcal{B}\big(x + u(x, y), y + v(x, y)\big)\|\Big) \\ &+\eta \sum_{(x,y)} \big(|u(x, y)| + |v(x, y)|\big) \\ &+\alpha \sum_{\substack{(x,y)\\(x',y')}} \Big(\min\big(|u(x, y) - u(x', y')|\big) + \min\big(|v(x, y) - v(x', y')|\big)\Big)\,.\end{aligned} \tag{6.21}$$

Note the last summation in Eq. (6.21) is taken over all pairs pixels $(x, y)$ and $(x', y')$ for which $|x' - x| \leq \epsilon$ and $|y' - y| \leq \epsilon$. To guard against outliers, [22] recommend using truncated $L_1$ norms in Eq. (6.21).

The three summations Eq. (6.21) are, respectively, the data, the small displacement and the smoothness terms. The data term minimizes the difference between SIFT descriptors along the flow vector. The small displacement term keeps the displacements as small as possible when no information is available. The smoothness term constrains the flow vectors of adjacent pixels to be similar. The coefficients $\alpha$ and $\eta$ are small displacement and smoothness coefficients, respectively.

**Example 6.19** *SIFT Flow for Kinship Verification** [33]. Anthropology studies show that genetic features are inherited by children from their parents resulting in visual resemblance between them. Nevertheless the similarity between images of children and their parents ("kinship" images) may be low. To enhance the genetic features between parent and child we apply SIFT operator to each pixel in a given kinship image. The result is a SIFT image of the same (spatial) size as the original image but 128-deep. Given two SIFT images $\mathcal{A}$ and $\mathcal{B}$ we align them by minimizing the energy $\mathcal{E}(\mathbf{w})$.

The following example illustrates human action recognition by fusing SIFT flow with CNN.

**Example 6.20** *Human Recognition by Fusing SIFT flow and CNN** [30]. Two-stream CNN is used for human action recognition. Two-stream CNN consists of a spatial stream and a temporal stream. The spatial stream through which the RGB image passes, extracts the shape features of human motion. The temporal stream which the optical flow images pass, extracts the sequence features of the listed motions.

However there are limitations of conventional optical flow-based pixel matching for recognizing human actions. Optical flow is derived assuming constant brightness and piecewise smoothness. In real-world videos these assumptions are often violated resulting in degraded optical flow performance. However feature-based SIFT flow is robust in situations violating brightness constancy and piecewise smoothness. By replacing optical flow with SIFT flow we therefore obtain high performance in real human action recognition when there are light changing and drastic movement situations.

### *6.6.2 CNN Optical Flow**

In CNN optical flow we apply a CNN to each input image $A$ and $B$.[3] At each pixel $(x, y)$ the CNN extracts feature vectors $\mathbf{f}_A(x, y)$ and $\mathbf{f}_B(x, y)$ *and* then estimates the corresponding optical flow vector $\mathbf{V}(x, y) = (u(x, y), v(x, y))$. In modern CNN optical flow the CNN is trained *end-to-end*. The architecture of the CNN [15, 34, 43, 44] is based on the classical optical flow algorithms and in particular on the coarse-to-fine scheme shown in Fig. 6.8.

For the sake of concreteness we shall consider SPyNet [34] (see Fig. 6.9) in which the classical coarse-to-fine approach for optical flow estimation is combined with deep learning CNN's. At each level of the pyramid, a CNN is trained independently on the optical flow at that level. Each level of the pyramid deals with motion within certain limits placed on the displacement. The final optical flow is produced by iteratively upsampling the coarse optical flow estimates and warping them with a higher resolution frame.

In SPyNet we first build a image pyramid for each of the RGB input images:

$$A \rightarrow \{\mathcal{A}_1, \mathcal{A}_2, \ldots, \mathcal{A}_K\} \quad \text{and} \quad B \rightarrow \{\mathcal{B}_1, \mathcal{B}_2, \ldots, \mathcal{B}_K\} ,$$

and then calculate the optical flow $\mathbf{V}$ as follows. Let $\{G_1, \ldots, G_K\}$ denote a set of trained CNN models, where $G_k$ computes the *residual* flow $\delta\mathbf{V}_k$ between $\mathcal{A}_k$ and $\widetilde{\mathcal{B}}_k$:

$$\delta\mathbf{V}_k = G_k(\mathcal{A}_k, \widetilde{\mathcal{B}}_k, \uparrow(\mathbf{V}_{k-1})) , \tag{6.22}$$

and $\widetilde{\mathcal{B}}_k$ denotes the image $\mathcal{B}_k$ after warping. The corresponding (full) flow, $\mathbf{V}_k$, for the $k$th level is found by adding $\delta\mathbf{V}_k$ to the upsampled flow, $\uparrow(\mathbf{V}_{k-1})$, obtained from the previous pyramid level:

$$\mathbf{V}_k = \uparrow(\mathbf{V}_{k-1}) + \delta\mathbf{V}_k . \tag{6.23}$$

[3] In contrast to classical optical flow, we shall assume $A$ and $B$ are color (RGB) images.

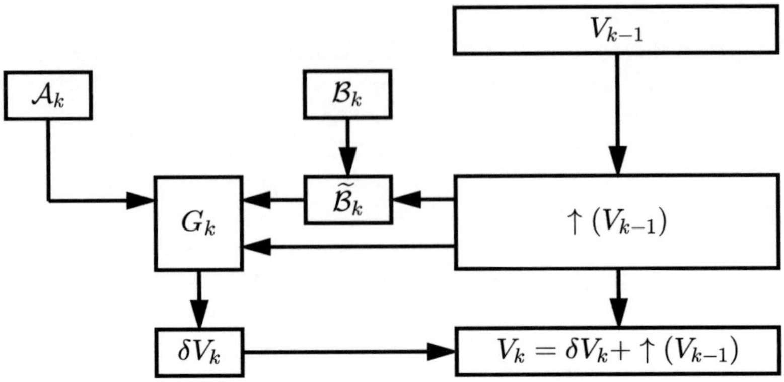


**Fig. 6.9** Shows the optical flow SPyNet. The SPyNet uses a CNN $G_k$ for each level $k$. Input to $G_k$ are: (1) Downsampled RGB image $\mathcal{A}_k$; (2) Warped downsampled RGB image $\widetilde{\mathcal{B}}_k$; (3) Upsampled optical flow velocity $\uparrow (\mathbf{V}_{k-1})$. Output from $G_k$ is the residual optical flow velocity $\delta\mathbf{V}_k = (\delta u_k, \delta v_k)$. The *full* optical flow velocity for the $k$th level is: $\mathbf{V}_k = \delta\mathbf{V}_k + \uparrow (\mathbf{V}_{k-1})$

In Eq. (6.22) the warped image $\widetilde{\mathcal{B}}_k$ is given by:

$$\widetilde{\mathcal{B}}_k = \text{warp}\big(\mathcal{B}, \uparrow (\mathbf{V}_{k-1})\big) . \tag{6.24}$$

Figure 6.9 shows the action of $G_k, k \in \{1, 2, \dots, K\}$. The input to $G_k$ are the RGB input image $\mathcal{A}_k$, the warped RGB image $\widetilde{\mathcal{B}}_k$ and the upsampled optical flow $\uparrow (\mathbf{V}_{k-1})$. The output is the 2D residual optical flow $\delta\mathbf{V}_k$.

We train SPynet end-to-end to minimize the average End Point Error (EPE).

## 6.7 Multiple Image Registration. Bundle Adjustment

Thus far we have considered the problem of registering a pair of images. In some applications we are interested in building a single panoramic or "mosaic" image $I^*$ from multiple images $I_m, m \in \{1, 2, \dots, M\}$. To do this we need to find functions $\mathcal{T}_m$ which transform each input image $I_m$ onto the mosaic image $I^*$.

Building a mosaic image from a sequence of partial views is a powerful means of obtaining a broader view of a scene than is available with a single view. Research on automated mosaic construction is ongoing with a wide range of different applications.

**Example 6.21** *Mosaic Fingerprint Image* [2]. Fingerprint-based verification systems have gained immense popularity due to the high level of uniqueness attributed to fingerprints and the availability of compact solid-state fingerprint sensors. However, the solid-state sensors sense only a limited portion of the fingerprint pattern and this may limit the accuracy of the user verification. To deal with this problem we construct a mosaic fingerprint image from multiple fingerprint impressions.

At first sight we may assume that the ability to register a pair of images is sufficient to solve the problem of forming a mosaic of the entire scene from multiple partial views. Ideally, if one image can be established as an "anchor image" $I_0$ on which to base the mosaic image $I^*$, then the transformation of each remaining images onto this anchor may be estimated using pairwise registration. The mosaic image $I^*$ is then formed by "stitching" the transformed images $T_m(I_m)$ (see Sect. 4.2). Unfortunately, in practice, this approach may not work for the following reasons:

**Overlap**. Some images may not overlap with $I_0$ at all. This makes a direct computation of the transformation impossible. In other cases, images may have insufficient overlap with $I_0$ to compute a stable transformation. The straightforward solution is to compose transformations using an "intermediate" image: Thus if $\mathcal{T}_{ij}$ is the pairwise transformation between images $i$ and image $j$, then

$$\mathcal{T}_{ik} = \mathcal{T}_{ij} \circ T_{jk} ,$$
$$\mathcal{T}_{ij} = \mathcal{T}_{ji}^{-1} ,$$

where $\circ$ denotes the composition operator. This process may, however, cause errors in the pairwise transformations $\mathcal{T}_{ij}$ to propagate.

**Consistency**. The transformations $\mathcal{T}_m$ may *not* be mutually consistent. This may happen even if all the image-to-anchor transformations have been accurately estimated.

A standard approach to solving these problems is *bundle adjustment* [28]. The idea is to globally adjust the transformations $\mathcal{T}_{ij}$ so that the overall re-projection error $\mathcal{E}$ is minimized, where $\mathcal{E}$ is defined as follows.

Consider two images $I_i$ and $I_j$. We assume they have $K$ corresponding (matching) keypoints where $(x_i^{(k)}, y_i^{(k)})$ and $(x_j^{(k)}, y_j^{(k)})$ denote, respectively, the locations of the $k$th keypoint in $I_i$ and $I_j$. We use the transformations $\mathcal{T}_{ij}$ and $\mathcal{T}_{ji}$ to, respectively, project $(x_i^{(k)}, y_i^{(k)})$ onto $I_j$ and $(x_j^{(k)}, y_j^{(k)})$ onto $I_i$:

$$(\hat{x}_j^{(k)}, \hat{y}_j^{(k)}) = \mathcal{T}_{ij}(x_i^{(k)}, y_i^{(k)}) ,$$
$$(\hat{x}_i^{(k)}, \hat{y}_i^{(k)}) = \mathcal{T}_{ji}(x_j^{(k)}, y_j^{(k)}) ,$$

Then the re-projection error for the images $I_i$ and $I_j$ is

$$\mathcal{E}_{ij} = \sum_k \left( |x_j^{(k)} - \hat{x}_j^{(k)}|^2 + |y_j^{(k)} - \hat{y}_j^{(k)}|^2 \right) .$$

and $\mathcal{E}$ is simply the sum of $\mathcal{E}_{ij}$ taken over all image pairs used in creating the mosaic image $I^*$.

## 6.8 Software

**MATLAB**. COMPUTER VISION TOOLBOX, IMAGE PROCESSING TOOLBOX and STATISTICS AND MACHINE LEARNING TOOLBOX include m-files for image registration (area matching and sparse parametric matching), optical flow (traditional and deep learning), handcrafted features (LBP, HOG, SIFT) and subspace models (PCA, LDA, CCA). A set of m-files for SIFT flow is: SIFT-flow (Author: Chienerh).

**PYTHON**. OPENCV includes python modules for image registration (and keypoint matching) and optical flow.

## 6.9 Further Reading

The subject of image registration has been intensely investigated for many years. Modern reviews/surveys include: [26] (image matching), [27, 40] (mutual information registration), [16, 48, 51] (multimodal registration), [12] (deep learning image registration). Other useful reviews include: [10] (image warping) and [29] (image mosaic).

Apart from SPyNet [34], powerful ÇNN's for optical flow are PWC-Net [44], LiteFlowNet2 [15] and LiteFlowNet3 [14]. A comprehensive review of optical flow methods is [1].

## Problems

**6.1** Define mutual information and explain the different approaches to measuring it.

**6.2** What are the main drawbacks to using mutual information for image registration? Explain how neighborhood and regional MI help alleviate these drawbacks.

**6.3** Resampling is an important step in spatial alignment. Explain what resampling is and why it is necessary.

**6.4** Develop the optical flow equations for gray-scale images. Describe the two approaches to introduce spatial coherency into the equations.

**6.5** What is a keypoint detector/descriptor. How do we use the these detector/descriptor to perform image registration.

# References


1. Alfarano A, Maiano L, Papa L, et al. (2024) Estimating optical flow: a comprehensive review of the state of the art. Comp Vis Im Under 249:104160
2. Alkhathami M, Han F, Van Schnydel R (2014) A mosaic approach to touchless fingerprint image with multiple views. In: Proceedings of the ACM/IEEE international conference on distributed smart cameras
3. Amiaz T, Lubetzky E, Kiryati N (2007) Coarse-to-over fine optical flow estimation. Patt Recogn 40:2496–2503
4. Baghaie A, D'Souza RM, Yu Z (2017) Dense descriptors for optical flow estimation: a comparative study. J Imaging 3:3010012
5. Botev ZI, Kroese DP (2011) The generalized cross-entropy method with applications to probability density estimation. Methodol Comp App Prob 13:1–27
6. Can A, Stewart CV, Roysam B et al (2002) A feature-based technique for joint, linear estimation of high-order image-to-mosaic transformations: mosaicing the curved human retina. IEEE Trans Patt Anal Mach Intell 24:412–419
7. Dusmanu M, Rocco I, Pajdla T, et al (2019) D2-Net: a trainable CNN for joint description and detection of local features. In: IEEE international conference on computer vision and pattern recognition (CVPR), pp 8092–8101
8. Elgammal A, Duraiswami R, Harwood D et al (2002) Background and foreground modeling using nonparametric kernel density estimation for visual surveillance. Proc IEEE 90:1151–1163
9. Fortun D, Bouthemy P, Kervrann C (2015) Optical flow modeling and computation: a survey (2015). Comp Vis Image Under 134:1–21
10. Glasbey CA, Mardia KV (1998) A review of image warping methods. J App Stat 25:155–171
11. Govindu VM (2006) Revisiting the brightness constraint: probabilistic formulation and algorithms. LNCS 3953:177–188
12. Haskins G, Kruger U, Yan P (2022) Deep learning in medical image registration: a survey. Mach Vis App 31:Article No. 8
13. Hseih C-K, Lai S-H, Chen Y-C (2010) An optical flow-based approach to robust face recognition under expression variations. IEEE Trans Im Proc 19:233–240
14. Hui TW, Loy CC (2020) LiteNetFlow3: resolving correspondence ambiguities for more accurate optical flow estimation. In: Proceedings of the European conference on computer vision (ECCV)
15. Hui T-W, Tang X, Loy CC (2021) A ligtweight optical flow cnn-revisiting data fidelity and regularization. IEEE Trans Patt Analysis Mach Intell 43:2555–2569
16. Jiang X, Ma J, Xiao G et al (2021) A review of multimodal image matching: methods and applications. Inf Fusion 73:22–71
17. Jones MC, Marron JS, Sheather SJ (1996) A brief survey of bandwidth selection for density estimation. J Am Stat Assoc 91:401–407
18. Knuth KH (2019) Optimal data-based binning for histograms and histogram-based probability density models. Dig Sig Proc 95:102581
19. Kybic J (2010) Bootstrap resampling form image registration uncertainty estimation without ground truth. IEEE Trans Im Proc 19:64–73
20. Lehmann TM, Gonner C, Spitzer K (1999) Survey: interpolation methods in medical image processing. IEEE Trans Med Imag 18:1049–1075
21. Li J, Li X, Wei Y, et al (2022) Multi-level feature aggregation-based joint keypoint detection and description. Comp Materials Continua
22. Liu C, Yuen J, Torralba A (2011) Sift flow: dense correspondence across scenes and its applications. IEEE Trans Patt Analy Mach Intell 33:978–994
23. Liu C, Xu J, Wang F (2021) A review of keypoints' detection and feature description in image registration. Sci Prog 2021:8509164
24. Long Y, Morris D, Liu X, et al (2021) Full-velocity radar returns by radar-camera fusion. In: IEEE international conference on computer vision (ICCV), pp 16178–16187

25. Ma J, Zhao J, Jiang J et al (2019) Locality preserving matching. Int J Comp Vis 127:512–531
26. Ma J, Jiang X, Fan A et al (2021) Image matching from handcrafted to deep features: a survey. Int J Comp Vis 129:23–79
27. Maes F, Vandermeulen D, Suetens P (2003) Medical image registration using mutual information. Proc IEEE 91:1699–1722
28. Mehrdad S, Satari M, Safdary M et al (2016) Toward real time uavs' image mosaicking. Int Arch Photogramm Remote Sense Spat Inf Sci 41:941–946
29. Pandey A, Pati UC (2019) Imag mosaicing: a deeper insight. Image Vis Comp 89:236–257
30. Park SK, Chung JH, Kang TK et al (2021) Binary dense sift flow based two stream cnn for human action recognition. Multimed Tools Appl 80:35697–35720
31. Park J-I, Lee Y, Suh E et al (2021) Improvement of optical flow estimation by using the hampel filter for low-end embedded systems. IEEE Robots Automat Lett 6:7233–7239
32. Perperidis D, Mohiaddin R, Rueckert D (2005) Spatio-temporal free-form registration of cardiac mr image sequences. Med Imag Anal 9:441–456
33. Puthenputhussery A, Liu Q, Liu C (2016) Sift flow based genetic fisher vector feature for kinship verification. In: Proceedings of international conference on image processing, pp 2921–2925
34. Ranjan A, Black MJ (2017) Optical flow estimation using a spatial pyramid network. In: Proceedings of IEEE conference on computer vision and pattern recognition
35. Ross A, Dass S, Jain A (2003) A deformable model for fingerprint matching. Patt Recogn 38:95–103
36. Russakoff DB, Tomasi C, Rohlfing T et al (2004) Image similarity using mutual information of regions. LNCS 3023:596–607
37. Savian S, Elahi M, Tillo T (2020) Optical flow estimation with deep learning, a survey on recent advances. In: Jiang R et al (eds) Deep Biometrics: unsupervised and Semi-Supervised. Published by Springer
38. Scott DW (1992) Multivariate density estimation. Published by Wiley
39. Senaratne R, Jap B, Lal S et al (2011) Comparing two video-based techniques for driver fatigue detection: classification versus optical flow approach. Mach Vis App 22:597–618
40. Sengupta D, Gupta P, Biswas A (2022) A survey on mutual information based medical image registration algorithms. Neurocomput 486:174–188
41. Silverman B (1986) Density estimation for statistical data analysis. Published by Chapman-Hall. Bioinform 18(Supplement 2):S231–S240
42. Sulewski P (2020) Equal-bin-width histogram versus equal-bin-count histogram. J Appl Stat 48:2092–2111
43. Sun D, Yang X, Liu M-Y, et al (2018) PWC-Net: CNNs for optical flow using pyramid, warping and cost volume. In: Proceedings of the IEEE/CVF conference on computer vision and pattern recognition, pp 8034–8043
44. Sun D, Yang X, Liu M-Y et al (2020) Models matter, so does training: an empirical study of cnns for optical flow estimation. IEEE Trans Patt Analysis Mach Intell 42:1408–1423
45. Thevenaz P, Blu T, Unser M (2000) Interpolation revisited. IEEE Trans Med Imag 19:739–758
46. Tian Y, Fan B, Wu F (2017) L2-net: deep learning of discriminative patch descriptor in euclidean space. In: IEEE conference on computer vision and pattern recognition, pp 6128–6136
47. Tu Z, Xie W, Zhang D et al (2019) A survey of variational and cnn-based optical flow techniques. Sig Proc Image Comm 72:9–24
48. Velesaca HO, Bastidas G, Rouhani M et al (2024) Multimodal image registration techniques: a comprehensive survey. Multimedia Tools App 83:63919–63947
49. Zagorchev L, Goshtasby A (2006) A comparative study of transformation functions for nonrigid image registration. IEEE Trans Imag Proc 15:529–538
50. Zhao X, Wu X, Miao J et al (2023) Alike: accurate and lightweight keypoint detection and descriptor extraction. IEEE Trans Multimedia 25:3101–3112
51. Zhang X, Leng C, Hong Y et al (2021) Multimodal remote sensing image registration methods and advancements: a survey. Remote Sens 13:5128

# Chapter 7
# Temporal Alignment

**Abstract** In this chapter we consider the conversion of the local time axis to a common time axis. This is the second step in the formation of a common representational format. To keep the discussion focused we shall concentrate on the methods based on dynamic time warping (DTW) including soft-DTW, drop-DTW and shape-DTW. We also consider the alignment of time series of different dimensionality (canonical time warping) and the alignment of multiple time series (dynamic barycenter average).

## 7.1 Introduction

The subject of this chapter is *temporal alignment*, or registration, which we define as the transformation $T(t)$ which maps local sensor observation times $t$ to a common time axis $t'$. Temporal alignment is one of the basic processes required for creating a common representational format. It often plays a critical role in applications involving in many multi-sensor data fusion applications. This is especially true for applications operating in real-time (see Sect. 2.4.2).

The following example illustrates the use of temporal registration in medical imaging.

**Example 7.1** *Cardiac MR Image Sequences* [11]. Cardiovascular diseases are a very important cause of death in the developed world. Their early diagnosis and treatment is crucial in order to reduce mortality and to improve patient's quality of life. The recent advances in the development of cardiac imaging modalities have led to an increased need for cardiac registration, whose aim is to bring $K$ sequences of cardiac images acquired from the same subject, or from the same subject at different times, into the same spatial-temporal coordinate system. To do this we must perform temporal registration which corrects for any *temporal misalignment* caused by different acquisition parameters, different lengths of cardiac cycles and different motion patterns of the heart.



H. B. Mitchell, *Data Fusion*, https://doi.org/10.1007/978-3-662-71023-4_7

Let $O = \langle E, \mathbf{u}, t, \mathbf{y}, \Delta\mathbf{y} \rangle$ denote a given sensor observation. Then, strictly speaking the transformation $T(t)$ is a function of the position $\mathbf{u}$ and the time $t$. However, it is common practice to assume (at least as a first approximation) that $T(t)$ is independent of $\mathbf{u}$.

The following example illustrates an application in which we perform temporal alignment on a sequence of input images. In this application we reduce each input image to a single point in space which reduces the errors involved in assuming $T(t)$ is independent of $\mathbf{u}$.

**Example 7.2** *Visualization of Dysphagia (Disorders Associated with Swallowing)* [19]. We consider an application in which we form a high resolution video sequence of a patient swallowing. In this application, a patient repeatedly swallows. For a given swallow $S$, a magnetic resonance imaging (MRI) camera records the swallow in a sequence of MRI images $I_i, i \in \{1, 2, \ldots, N\}$, taken at times $t_i$, where $t_i < t_{i+1}$. For each sequence (swallow) we segment the bolus and track its path in the space-time volume of the swallow. We do this by extracting the centroid, $(\bar{x}, \bar{y})$, of the segmented region from each MRI image and creating a path of the centroid motion for each swallow (see Fig. 7.1).

After performing temporal registration, we form a high resolution video of the patient swallowing by combining the multiple swallows. The temporal registration is performed by matching the spatial-temporal paths of the bolus in multiple swallows.

To keep our discussion focused we shall limit ourselves to the temporal alignment of a time series, $\mathbf{P}$, to a common time-axis $t'$, where $\mathbf{P} = \{P_1, P_2, \ldots, P_M\}$ consists of $M$ sensor observations, $P_i = \langle E_i, *, t_i, y_i, * \rangle, i \in \{1, 2, \ldots, M\}$, made with the same sensor and ordered in time, i. e. $t_i < t_{i+1}$. Furthermore, we shall assume (as is common in many multi-sensor data fusion systems) that we do not have an independent time-axis $t'$. Instead, we shall define the time-axis $t'$ by a second time series $\mathbf{Q}$, where $\mathbf{Q}$ consists of $N$ sensor observations, $Q_j = \langle E_j, *, t'_j, y'_j, * \rangle, j \in \{1, 2, \ldots, N\}$. *Note*. Sometimes we normalize the time series $\mathbf{P} = \{P_1, P_2, \ldots, P_M\}$ such that $\sum_{i=1}^{M} y_i = 1$. In this case the time series is, in fact, a *probability* distribution.

In many data fusion applications temporal alignment is the primary fusion algorithm. In Table 7.1 we list some of these applications together with their classification.

## 7.2 Dynamic Time Warping (DTW)

*Dynamic Time Warping* (DTW) [14, 15] is a general technique for performing temporal alignment between two time series, $\mathbf{P}$ and $\mathbf{Q}$. DTW was originally developed for use in speech recognition. However, recently, it has been used in many different areas, including data mining, pattern and shape recognition and biometric identification.

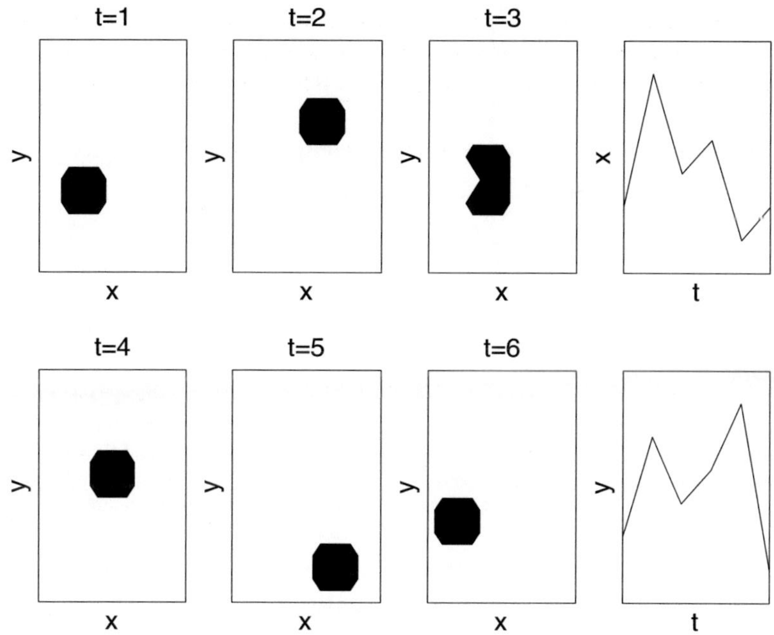


**Fig. 7.1** Shows six binary images, $I_i$, of the bolus (shown as a black irregularly shaped area) taken at times $t_i$, $i \in \{1, 2, \ldots, 6\}$. To the right of the images $I_i$ we show the average $x$ and $y$ position of the bolus as a function of time

**Table 7.1** Applications in which temporal alignment is primary fusion algorithm

| Class | | Application |
|---|---|---|
| Early-DaO | Example 7.1 | Cardiac MR Image Sequences |
| Early-FeO | Example 7.2 | Visualization of the Dysphagia (Disorders Associated with Swallowing) |
| | Example 7.5 | Speech recognition in an embedded speech recognition system |
| | Example 7.6 | Dynamic Temporal Alignment of Speech to Lips |
| | Example 7.7 | Word spotting in historical manuscripts |
| | Example 7.8 | Generating a digital elevation map using DTW |
| | Example 7.9 | Identifying a face profile using the DDTW algorithm |
| | Example 7.11 | Time Series Classification based on 1-NN and DBA |

DTW finds the *optimal* alignment between two time series **P** and **Q** in the sense that it minimizes the sum of the local distances $d(i, j)$ between the aligned observation pairs $(P_i, Q_j)$. In many applications we define $d(i, j)$ as the square of the Euclidean distance:

$$d(i, j) = (P_i - Q_j)^2 .$$

However, in fact, any appropriate local distance measure may be used instead.

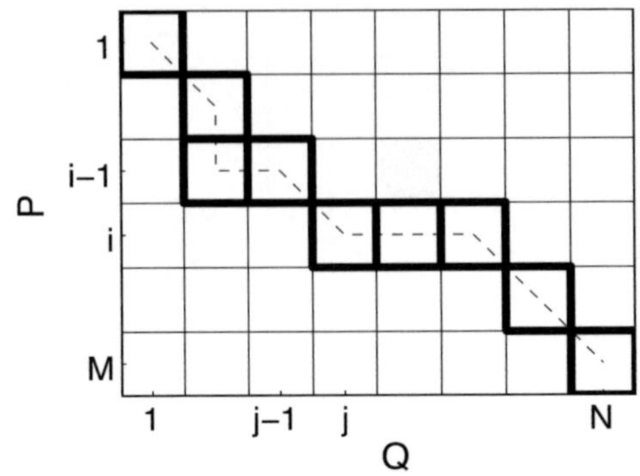


**Fig. 7.2** Shows the action of the DTW algorithm. We start at $w_1 = (1, 1)$ where $D(1, 1) = d(1, 1)$. We then proceed top-to-bottom and left-to-right updating $D(i, j)$ using the cumulative distances of the neighbouring cells. The optimal cost of matching **P** and **Q** is $DTW(\mathbf{P}, \mathbf{Q}) = D(M, N)$. The corresponding optimal path is $W_{\text{DTW}}$ and is represented by the dashed line and the thick-bordered cells, which can be recovered by tracing back the sequence of optimal neighbours

The entire procedure is called "time warping" because it warps the time axes, $t$ and $t'$, in such a way that corresponding sensor observations appear at the same location on a common time axis. In general, the time axes, $t$ and $t'$, are aligned in a non-linear way.

In the DTW algorithm we define a given alignment of the time series, **P** and **Q**, by means of a "warping path" **W** (Fig. 7.2). This is a set of matrix elements $w_k = (i_k, j_k), k \in \{1, 2, \ldots, K\}$, where $w_k$ defines a mapping between the observations $P_{i_k}$ and $Q_{j_k}$. Formally,

$$\mathbf{W} = \{w_1, w_2, \ldots, w_K\} ,$$

where $\max(M, N) \leq K \leq M + N - 1$.

In general, we are only interested in warping paths (i.. e. alignments) which satisfy the boundary constraints:

**End-point Conditions**. $w_1 = (1, 1)$ and $w_K = (M, N)$. This requires the warping path to start by matching the first observation of **P** with the first observation of **Q** and end by matching the last observation of **P** with the last observation of **Q**.

**Continuity**. Given $w_k = (a, b)$, then $w_{k-1} = (a', b')$, where $a - a' \leq 1$ and $b - b' \leq 1$. This restricts the allowable steps in the warping path to adjacent matrix elements (including diagonally adjacent elements).

**Monoticity**. Given $w_k = (a, b)$ then $w_{k-1} = (a', b')$, where $a - a' \geq 0$ and $b - b' \geq 0$. This forces the warping path sequence to increase monotonically in time.

There are exponentially many warping paths that satisfy the above conditions, however, we are interested only in the path which minimizes the sum of the local distances $d(i_k, j_k) \equiv d(w_k)$. This is the optimal ("DTW") warping path and is defined as

$$\mathbf{W}_{\text{DTW}} \equiv \mathbf{W}_{\text{DTW}}(\mathbf{P}, \mathbf{Q}) = \arg\min_{\mathbf{W}} \sum_k d(w_k) . \tag{7.1}$$

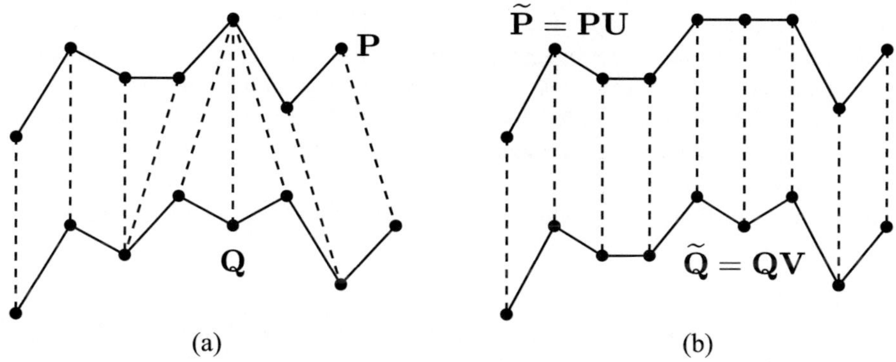


**Fig. 7.3** Shows two aligned series $\mathbf{P} = [P_1, P_2, \ldots, P_M]$ $(M = 7)$ and $\mathbf{Q} = [Q_1, Q_2, \ldots, Q_N]$ $(N = 8)$ and the corresponding "stretched" series $\widetilde{\mathbf{P}} = [\widetilde{P}_1, \widetilde{P}_2, \ldots, \widetilde{P}_L] = \mathbf{PU}$ and $\widetilde{\mathbf{Q}} = [\widetilde{Q}_1, \widetilde{Q}_2, \ldots, \widetilde{Q}_L] = \mathbf{QV}$ $(L = 9)$. **a** Shows $\mathbf{P}$ and $\mathbf{Q}$. **b** Shows $\widetilde{\mathbf{P}} = \mathbf{PU}$ and $\widetilde{\mathbf{Q}} = \mathbf{QV}$

It is sometimes convenient to write the optimal alignment path $\mathbf{W}_{\mathrm{DTW}}(\mathbf{P}, \mathbf{Q})$ as a *one-to-one* alignment between two "stretched" time series $\widetilde{\mathbf{P}}$ and $\widetilde{\mathbf{Q}}$ of length $L$. In this case,

$$\widetilde{\mathbf{P}} = [\widetilde{P}_1, \widetilde{P}_2, \ldots, \widetilde{P}_L] = \mathbf{PU} \quad \text{and} \quad \widetilde{\mathbf{Q}} = [\widetilde{Q}_1, \widetilde{Q}_2, \ldots, \widetilde{Q}_L] = \mathbf{QV} , \tag{7.2}$$

where $\mathbf{U} \in \mathbb{R}^{M \times L}$, $\mathbf{V} \in \mathbb{R}^{N \times L}$ are binary matrices which correspond to the optimal alignment $W_{DTW}$ and $L \geq \max(M, N)$. See Fig. 7.3.

### 7.2.1 Dynamic Programming

A brute force method for finding the optimal warping path requires the evaluation of an exponential number of warping paths. Fortunately dynamic programming can be used to find the optimal warping path, $\mathbf{W}_{\mathrm{DTW}}$, and its cost, $DTW(\mathbf{P}, \mathbf{Q})$, in $O(MN)$ time using the recursion:

$$D(i_k, j_k) = d(i_k, j_k) + \min(D(i_k, j_k - 1), D(i_k - 1, j_k), D(i_k - 1, j_k - 1)) , \tag{7.3}$$

where $D(i_k, j_k)$ is the cost of the optimal warping path from $(1, 1)$ to $(i_k, j_k)$ and $D(1, 1) = d(1, 1)$.

The DTW algorithm starts by building a $M \times N$ table $D(m, n)$, $m \in \{1, 2, \ldots, M\}$, $n \in \{1, 2, \ldots, N\}$, column by column. We start with $n = 1$ and compute $D(1, 1)$ using (7.3) and the boundary conditions

$$\begin{aligned} D(0, 0) &= 0 , \\ D(0, j_k) &= \infty = D(i_k, 0) . \end{aligned}$$

We then proceed from $m = 2$ to $m = M$ and compute $D(2, 1), D(3, 1), \ldots, D(M, 1)$. We then increase $n$ by one and proceed from $m = 1$ to $m = M$ computing $D(1, 2), D(2, 2), \ldots, D(M, 2)$. We continue the process until the last column $n = N$ and row $m = M$ are reached. By definition $D(M, N)$ is equal to $DTW(\mathbf{P}, \mathbf{Q})$, the cost of the optimal alignment of the sequences **P** and **Q**. The optimal warping path, $\mathbf{W}_{\mathrm{DTW}}$, is obtained by tracing the recursion backwards from $D(M, N)$. The following example illustrates the construction of the cumulative cost matrix $D(m, n)$ used in Fig. 7.2.

**Example 7.3** *Dynamic Programing*. Given two time series $\mathbf{P} = (1.7, 4.1, 3.1, 2.2, 3.4, 1.6)^T$ and $\mathbf{Q} = (2.4, 3.5, 2.7, 3.2, 4.5, 4.7, 4.8, 5.9)^T$, we create a $6 \times 8$ matrix of squared Euclidean distances $d(i, j) = (P_i - Q_j)^T$:

$$
\begin{array}{cccccccc}
0.49 & 3.24 & 1.00 & 2.25 & 7.84 & 9.00 & 9.61 & 17.64 \\
2.89 & 0.36 & 1.96 & 0.81 & 0.16 & 0.36 & 0.49 & 3.24 \\
0.49 & 0.16 & 0.16 & 0.01 & 1.96 & 2.56 & 2.89 & 7.84 \\
0.04 & 1.69 & 0.25 & 1.00 & 5.29 & 6.25 & 6.76 & 13.69 \\
1.00 & 0.01 & 0.49 & 0.04 & 1.21 & 1.69 & 1.96 & 6.25 \\
0.64 & 3.61 & 1.21 & 2.56 & 8.41 & 9.61 & 10.24 & 18.49
\end{array}
$$

The corresponding cumulative matrix $D$ is

$$
\begin{array}{cccccccc}
0.49 & 3.73 & 4.73 & 6.98 & 14.82 & 23.82 & 33.43 & 51.07 \\
3.39 & 0.85 & 2.81 & 3.62 & 3.73 & 4.14 & 4.63 & 7.87 \\
3.87 & 1.01 & 1.01 & 1.02 & 2.98 & 5.54 & 7.03 & 12.47 \\
3.91 & 2.70 & 1.26 & 2.01 & 6.31 & 9.23 & 12.30 & 20.72 \\
4.91 & 2.71 & 1.75 & 1.30 & 2.51 & 4.20 & 6.16 & 12.41 \\
5.50 & 6.32 & 2.96 & 3.86 & 9.71 & 12.12 & 14.44 & 24.65
\end{array}
$$

The cost of the optimal alignment of **P** and **Q** is $DTW(\mathbf{P}, \mathbf{Q}) = 24.65$. By tracing back (see Fig. 7.2) we find the corresponding optimal warp path:

$$
\mathbf{W}_{\mathrm{DTW}} = \big((1, 1), (2, 2), (3, 3), (4, 3), (5, 4), (5, 5), (5, 6), (5, 7), (6, 8)\big)^T .
$$

### 7.2.2 Noise Smoothing

The DTW algorithm is sensitive to noise and/or outliers in the inputs **P** and **Q**. We therefore often pre-process the inputs by smoothing them with a sliding filter. If **P** and **Q** are *uniformly sampled* then an effective smoothing filter is the Savitzky-Golay (SG) filter [17, 20]. The original SG filter is (maximum-likelihood) optimum for Gaussian noise.

The following example illustrates the idea underlying the SG filter.

**Example 7.4** *Savitzky-Golay (SG) Filter** [17, 20]. SG filters are widely used to smooth noisy data points in a time series. They work on the principle of local polynomial regression as follows. Consider an unknown clean signal $x$ that is smooth enough to be locally approximated by a polynomial and $w$ be the random noise that corrupts it. Then the observed noisy version of $x$ at the $k$th data point is $y_k = x_k + w_k$, where $x_k$ and $w_k$ are, respectively, the values of $x$ and $w$ at the $k$th data point. Given the noisy data $y$, we choose a symmetric window about a data point, perform polynomial approximation over that window, and assign the value of the polynomial function at the central index of the window to the selected data point. The process is repeated for all data points $y_i$.

Consider the data points $k \in \{-K, -K+1, \dots, K\}$ over which $y$ is approximated by an $N$th order polynomial function whose value at $k$ is $p_k = \sum_{n=0}^{N} a_n k^n$, where $a_n$, are the coefficients of the polynomial. The optimum coefficients $a_n^*$ are found by minimizing the square error:

$$\mathbf{a}^* = \{a_0^*, a_1^*, \dots, a_N^*\} = \arg\min \sum_{k=-K}^{K} (y_k - p_k)^2 \, .$$

*Analytical* expressions have been tabulated for the smoothed values $x_i$ for different polynomials and sliding-window sizes. E.g. for a cubic polynomial and 5 point sliding-window, the formula for $x_i$ is: $x_i = \frac{1}{35}\big(y_{i-1} + 17y_i + 12y_{i+1} - 3y_{i+2}\big)$.

If the times series are thought to contain outliers, then the time series may be smoothed using a robust SG filter (see Example 12.10) or a Hempel filter (see Example 12.6). Alternatively the drop-DTW algorithm (see Sect. 7.3) may be used. This algorithm is a DTW algorithm *specifically* designed to allow for the presence of outliers.

### *7.2.3 Singularities*

One drawback to the DTW is that it can produce pathological results. The reason is the algorithm may try to explain a difference between **P** and **Q** by unnaturally warping the two time axes $t$ and $t'$. This can lead to counter-intuitive alignments. For example, a single point on one time series may map onto a large subsection of another time series (Fig. 7.4a).

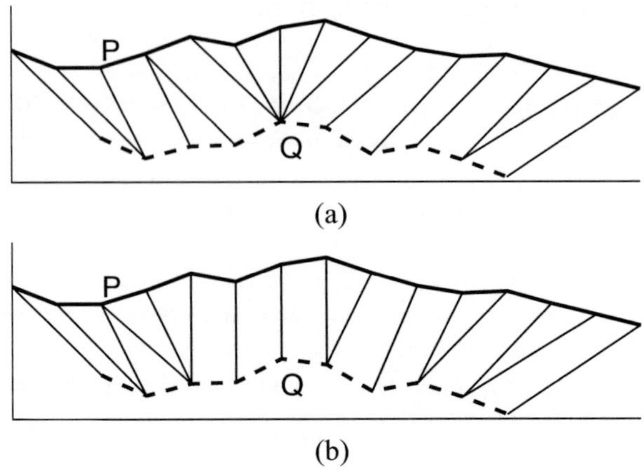


**Fig. 7.4** Shows matching of two time series **P** and **Q** using DTW and DDTW algorithms. **a** DTW algorithm with Euclidean local cost function. **b** DDTW algorithm

One way of dealing with these *singularities* is to place an additional constraint on **W**. Two constraints which are sometimes used for this purpose are:

**Windowing**. Locally constrain **W** so it does not deviate too far from the diagonal path (as defined by the end points $w_1$ and $w_K$). The distance that the path is allowed to deviate is the window length $R$.

**Slope Constraint**. Locally constrain **W** so it is neither too steep nor too shallow. We may do this by replacing Eq. (7.3) with $D(i_k, j_k) = d(i_k, j_k) + \min(\alpha D(i_k, j_k - 1), \alpha D(i_k - 1, j_k), D(i_k - 1, j_k - 1))$, where $\alpha$ is positive real number.

An alternative approach is to use a different local cost function (cf. Sect. 7.3).

## 7.2.4 Examples

In this section we give several examples, over a wide range of applications, in which two input series **P** and **Q** are fused together using the DTW.

### Speech Recognition

The original use of DTW was in speech recognition and although nowadays better performing (deep-learning) speech recognition techniques are available, the DTW is still widely used in systems which require minimal hardware e.g. small-scale embedded-speech recognition systems in cellular telephones.

**Example 7.5** *Speech-Recognition in an Embedded-Speech Recognition System.* Although nowadays high performance deep-learning speech-recognition techniques are available, the DTW is widely used in systems which require minimal hardware e.g. in cellular telephones.

A DTW speech-recognition system works as follows. Let $\omega^{(l)}$ and $\mathbf{Q}^{(l)}$, $l \in \{1, 2, \ldots, L\}$, denote, respectively, a set of $L$ reference words and their corresponding time series. Given an unknown input time series $\mathbf{P}$ we use the DTW algorithm to optimally match $\mathbf{P}$ to each time series $\mathbf{Q}^{(l)}$, $l \in \{1, 2, \ldots, L\}$, in turn. If $DTW(\mathbf{P}, \mathbf{Q}^{(l)})$ denotes the cost of matching $\mathbf{P}$ with $\mathbf{Q}^{(l)}$, then we assign $\mathbf{P}$ to the word $\Omega = \omega_{\text{OPT}}$, where

$$\omega_{\text{OPT}} = \arg\min_{l} DTW(\mathbf{P}, \mathbf{Q}^{(l)}) .$$

### Alignment of Speech and Lips

In filming a movie, poor sound quality is common for speech recorded on location. When this occurs, the speech is re-recorded in a studio during post-production using a process called "Automated Dialogue Replacement" (ADR). In ADR the actor watches his original performance in a loop and re-performs each line to match the wording and lip movements.

**Example 7.6** *Dynamic Temporal Alignment of Speech to Lips* [4]. Recently an algorithm for automatic ADR has been developed [4]. The algorithm automatically aligns audio and video of a speaking person using DTW. The algorithm consists of three parts:

**Feature Extraction.** Extract audio-visual features from both reference and unaligned signals.

**Temporal Alignment.** Use DTW to optimally align the two signals.

**Warped Signal.** Synthesize a warped version of the unaligned signal so it temporally matches the reference signal.

### Word Spotting in Historical Manuscripts

Easy access to handwritten historical manuscripts requires an index similar to that found in the back of this book. The current approach—manual transcription followed by index generation from the transcript—is extremely expensive and time-consuming.

For collections of handwritten manuscripts written by a single author the images of multiple instances of the same word are likely to look similar. For such collections, the *word spotting* algorithm provides an alternative approach to index generation.

**Example 7.7** *Word-Spotting in Historical Manuscripts* [16]. The word spotting algorithm works as follows. Each page in the collection of manuscripts is segmented into words. Then different instances of a given word are clustered together using an image matching algorithm. By annotating the clusters we create an index which links words to the locations where they occur.

A critical issue in word spotting is matching the word images. We may use the DTW algorithm to match the word images as follows [16]: We divide each word image into columns of equal width $\Delta$. In each column we extract $K$ features, $\mathbf{p}_i = \left(p_i^{(1)}, p_i^{(2)}, \ldots, p_i^{(K)}\right)^T$, where $i$ denotes the corresponding index of the column. In this manner we convert each word image into a *multivariate* series of feature values in which the horizontal image axis functions as a time axis. Once the conversion is complete we match any two word images $\mathbf{P} = [\mathbf{p}_1, \mathbf{p}_2, \ldots, \mathbf{p}_M]$ and $\mathbf{Q} = [\mathbf{q}_1, \mathbf{q}_2, \ldots, \mathbf{q}_N]$ by matching their multivariate feature values using the DTW algorithm. Here local distance between the $i$th multivariate feature in $\mathbf{P}$ and the $j$th multivariate feature in $\mathbf{Q}$ is the squared Euclidean distance

$$d(i, j) = \sum_k \left(p_i^{(k)} - q_j^{(k)}\right)^2 .$$

## Stereovision Depth Estimation

A simple model for a stereo-vision system, consists of two pinhole cameras $C_1$ and $C_2$ which observe a 3D scene from two slightly different viewing points. The scene is displayed on two 2D images $A$ and $B$. It is standard practice, to separately warp these images to a common representational format consisting of two epipolar-rectified images $\widetilde{A}$ and $\widetilde{B}$ (see Example 4.3).

**Example 7.8** *Generating a Digital Elevation Map (DEM) using DTW* [8]. Let $A$ and $B$ denote two 2D images in a stereovision system. We separately warp $A$ and $B$ to form a common representational format consisting of two epipolar-rectified images $\widetilde{A}$ and $\widetilde{B}$. In this format, corresponding points in $\widetilde{A}$ and $\widetilde{B}$ lie in the same row (see Fig. 7.5). Let $\widetilde{\mathbf{a}}_i = \left(\widetilde{A}(i, 1), \widetilde{A}(i, 2), \ldots, \widetilde{A}(i, M)\right)^T$ and $\widetilde{\mathbf{b}}_i = \left(\widetilde{B}(i, 1), \widetilde{B}(i, 2), \ldots, \widetilde{B}(i, M)\right)^T$ denote, respectively, the pixel gray-values in the $i$th row in $\widetilde{A}$ and $\widetilde{B}$. Then we use the DTW algorithm to find the

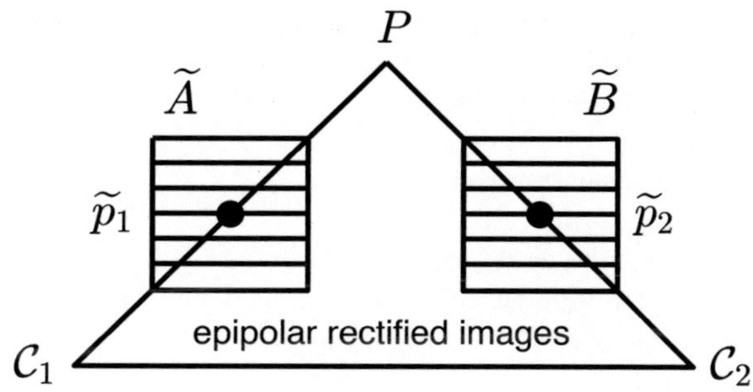


**Fig. 7.5** Shows a 3D scene captured by two (pinhole) cameras $C_1$ and $C_2$ and displayed on two "epipolar-rectified" images $\widetilde{A}$ and $\widetilde{B}$. The points $\widetilde{p}_1$ and $\widetilde{p}_2$ both correspond to the 3D point $P$ and therefore lie on rows with the *same row number*

optimal warping path $W_{\text{DTW}}$ between $\widetilde{\mathbf{a}}_i$ and $\widetilde{\mathbf{b}}_i$, where the distance matrix $D(m,n)$ is defined as

$$D(m,n) = |\widetilde{a}_i(m) - \widetilde{b}_i(n)| \ .$$

The path $D_{\text{DTW}}$ defines the *disparity* between *corresponding* pixels in $\widetilde{\mathbf{a}}_i$ and in $\widetilde{\mathbf{b}}_i$ from which the DEM may be calculated.

## 7.3 DTW Variants

Many different variants of the DTW algorithm have been developed. We discuss a few of the most important variants.

### Derivative Dynamic Time Warping (DDTW)

In derivative dynamic time warping (DDTW) algorithm [6] we use a local distance $d(i,j)$, which is defined as the square of the difference between the slopes of the curves $\mathbf{P}$ and $\mathbf{Q}$ at the times $t_i$ and $t'_j$. See Fig. 7.4b. Mathematically, the local distance is

$$d(i,j) = \left( \frac{dP}{dt}\bigg|_i - \frac{dQ}{dt'}\bigg|_j \right)^2 ,$$

where $dP/dt|_i$ and $dQ/dt'|_j$ are, respectively, the local slopes of $\mathbf{P}$ and $\mathbf{Q}$ at the times $t_i$ and $t'_j$.

**Example 7.9** *Identifying a Face Profile Using the DDTW Algorithm* [1]. Most current face profile recognition algorithms rely on the detection of fiducial points. Unfortunately features such as a concave nose, protrubing lips or a flat chin make the detection of such points difficult and unreliable. Also the number and position of the fiducial points vary when the expression changes even for the same person. As an alternative we may perform face profile recognition by matching an unknown face profile $\mathbf{P}$ to a data base of known face profiles $\mathbf{Q}^{(l)}$, $l \in \{1, 2, \ldots, L\}$, using the DDTW algorithm.

### Soft-DTW*

Although DTW is a powerful mechanism for time series fusion, DTW is not differentiable which makes it difficult to incorporate the DTW into a deep learning network. Soft-DTW [2] is a recent DTW algorithm which has been made differentiable by replacing the $min$ function in Eq. (7.1) with a generalized $min$ operator:

$$\min^{\gamma}(a_1, a_2, \ldots, a_N) = \begin{cases} \min(a_1, a_2, \ldots, a_N) & \text{if} \quad \gamma = 0\,, \\ -\gamma \log \sum_{n=1}^{N} \exp -a_n/\gamma & \gamma \geq 0\,. \end{cases} \tag{7.4}$$

where $\gamma \geq 0$ is an adjustable smoothing parameter.

### Drop-DTW*

We consider the problem of DTW for signals containing outliers. Without outliers then DTW efficiently computes the optimal alignment between two sequences. However, in the presence of outliers the DTW fails to align the sequences in a meaningful way. To address this problem we may use drop-DTW [3] which enables us to discard ("drop") elements from $\mathbf{P}$ and/or $\mathbf{Q}$ which are regarded as outliers. The drop-DTW algorithm works by defining a cost function $C_{\text{drop}}$. Then an element from one time series may be discarded if the cost of associating it with any element of the second time series is greater than $C_{\text{drop}}$.

### PSI-DTW

In the DTW algorithm we assume the end-points are constrained such that the first pairs of points is $(P(1), Q(1))$ and the last pair of points is $(P(M), Q(N))$.

In the prefix and suffix invariant DTW (PSI-DTW) [18] we relax this constraint (see Fig. 7.6a) as follows: Given an integer $r$, we allow the alignment

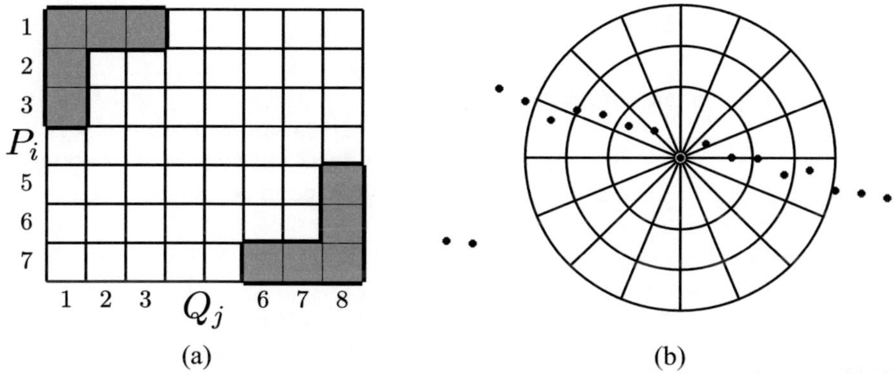


**Fig. 7.6** Two DTW variants. **a** PSI-DTW. Shows relaxation of end points in PSI-DTW for $r = 2$. Alignment path starts at any shaded box in top left corner and ends at any shaded box in bottom right corner. **b** SC-DTW. Shows time series $\mathbf{P} = \{P_1, P_2, \ldots\}$. For each point $P_i$ construct a local set of $B$ histogram bins which are centered on $P_i$. The shape context for $P_i$ is the number of points $P_{i'}, i' \neq i$ which fall in the $b$th bin, $b \in \{1, 2, \ldots, B\}$

path between $\mathbf{P} = (P(1), P(2), \ldots, P(M))^T$ and $\mathbf{Q} = (Q(1), Q(2), \ldots, Q(N))^T$ to start at any pair of points in $\{(P(1), Q(c_1 + 1)\} \cup \{(P(c_1 + 1), Q(1))\}$ and to end at any pair of points in $\{(P(M - c_2), Q(N))\} \cup \{(P(M), Q(N - c_2))\}$, where $c_1, c_2 \in \{0, 1, \ldots, r\}$. The value of $r$ is typically a small integer and is set by the user.

The relaxed constraint helps avoid undesirable matches at the end points of a time series by removing the obligation for the alignment path to start and end in the first and last pairs of observations.

### Shape DTW

Shape-DTW [24] is a class of hybrid dynamic time warping algorithms which attempt to match points which have *locally similar structures*.

**Example 7.10** *Shape-Context DTW* [25]. In SC-DTW we use the difference in shape contexts (see Example 4.12) to align two series $\mathbf{P}$ and $\mathbf{Q}$. We take each time series in turn. For each point $P_i$ construct a set of $B$ bins centered on $P_i$. Let $h_i(b)$, $b \in \{1, 2, \ldots, B\}$, be the number of points $P_{i'}, i' \neq i$ which fall in the $b$th bin (Fig. 7.6b). Repeat for $\mathbf{Q}$: For each point $Q_j$ let $H_j(b)$, $b \in \{1, 2, \ldots, B\}$, be the number of points $Q_{j'}, j' \neq j$ which fall in the $b$th bin. The cost function between $P_i$ and $Q_j$ is

$$C(i, j) = \frac{1}{2} \sum_{b=1}^{B} \frac{|h_i(b) - H_j(b)|^2}{h_i(b) + H_j(b)} .$$

## 7.4 Canonical Time Warping*

Given two multivariate time series $\mathbf{P} = [\mathbf{p}_1, \mathbf{p}_2, \ldots, \mathbf{p}_M] \in \mathbb{R}^{d_P \times M}$ and $\mathbf{Q} = [\mathbf{q}_1, \mathbf{q}_2, \ldots, \mathbf{q}_N] \in \mathbb{R}^{d_Q \times N}$, the DTW algorithm requires the dimensions $d_P$ and $d_Q$ to be equal. In canonical time warping (CTW) [26] we relax this constraint by adding a feature weighting mechanism to DTW that is able to align two signals $\mathbf{P}$ and $\mathbf{Q}$ of different dimensionality.

The method for doing this is to combine DTW with canonical correlation analysis (CCA) (see Sect. 4.7.3). Let $\mathbf{X} = [\mathbf{x}_1, \mathbf{x}_2, \ldots, \mathbf{x}_M] \in \mathbb{R}^{K_X \times M}$ and $\mathbf{Y} = [\mathbf{y}_1, \mathbf{y}_2, \ldots, \mathbf{y}_N] \in \mathbb{R}^{K_Y \times N}$, denote two *zero-mean* multivariate time series. Then, assuming $K_X = K = K_Y$, we represent the DTW alignment path as the "stretched" time series $\widetilde{\mathbf{X}}$ and $\widetilde{\mathbf{Y}}$ (see Eq. 7.2).

We now allow for $K_x$ and $K_y$ to be different by applying a linear transformation $\mathbf{V}_X \in \mathbb{R}^{K_X \times K}$ to $\widetilde{\mathbf{X}}$ and a linear transformation $\mathbf{V}_Y \in \mathbb{R}^{K_Y \times K}$ to $\widetilde{\mathbf{Y}}$:

$$\widetilde{\widetilde{\mathbf{X}}} = \mathbf{V}_X^T \mathbf{X} \mathbf{W}_X \quad \text{and} \quad \widetilde{\widetilde{\mathbf{Y}}} = \mathbf{V}_Y^T \mathbf{X} \mathbf{W}_Y \ , \tag{7.5}$$

where $\widetilde{\widetilde{\mathbf{X}}}$ and $\widetilde{\widetilde{\mathbf{Y}}}$ are two multi-variate matching time series with the *same* dimensionality $K$ and are solutions of the *Canonical Time Warping (CTW)* equation:

$$(\widetilde{\widetilde{\mathbf{X}}}, \widetilde{\widetilde{\mathbf{Y}}}) = \min_{\mathbf{V}_X, \mathbf{V}_Y, \mathbf{W}_X, \mathbf{W}_Y} \left\| \mathbf{V}_X^T \mathbf{X} \mathbf{W}_X - \mathbf{V}_Y \mathbf{Y} \mathbf{W}_Y \right\|_F^2 . \tag{7.6}$$

## 7.5 Multiple Time Series. Dynamic Barycenter Average

Until now we have limited ourselves to the temporal alignment of two time series $\mathbf{P}$ and $\mathbf{Q}$. We now consider how we may perform the simultaneous temporal alignment of $N$ time series $\mathbf{P}_1, \mathbf{P}_2, \ldots, \mathbf{P}_N$. Specifically, our aim is to create an average time series $\overline{\mathbf{Q}}$ such that the sum of the squared DTW distances from $\mathbf{Q}$ to $\mathbf{P}_1, \mathbf{P}_2, \ldots, \mathbf{P}_N$ is minimal.

One algorithm for doing this is the *Dynamic Barycenter Averaging* (DBA) algorithm [12]. This works as follows.

Let $\mathcal{P} = \{\mathbf{P}_1, \mathbf{P}_2, \ldots, \mathbf{P}_N\}$ denote a set of $N$ one-dimensional time series, where $P_n(i)$ denotes the $i$th observation in $\mathbf{P}_n$. Let $\overline{\mathbf{Q}}$ be an estimate of the average time series, where $\overline{Q}(j)$ denotes the $j$th observation in $\overline{\mathbf{Q}}$. We compute the DTW between each time series $\mathbf{P}_n, n \in \{1, 2, \ldots, N\}$, and $\overline{\mathbf{Q}}$. Each DTW gives the associations between $\mathbf{P}_n(i)$ and $\overline{\mathbf{Q}}(j)$. Then we update $\overline{Q}(j)$ as the average of those observations, $P_n(i)$, which are associated with $\overline{Q}(j)$. The process is repeated until convergence. Pseudo-code for the DBA algorithm is given in Algorithm 3.

**Algorithm 3:** BARYCENTER DYNAMIC AVERAGE ALGORITHM

**Input**: Set of $N$ time series: $\mathcal{P} = \{\mathbf{P}_1, \mathbf{P}_2, \ldots, \mathbf{P}_N\}$; Initial average sequence (of length $L$): $\overline{\mathbf{Q}}$ which is selected randomly from $\mathcal{P}$; number of iterations: $niter$

**Output**: Average of time sequence $\overline{\mathbf{Q}}$

```
for iter ← 1 to niter do
    alignedSet(k) = 0; k = 1, 2, ..., L
    for i ← 1 to N do
        use DTW to align Q̄ and P_i
        for k ← 1 to L do
            Identify elements C_k in P_i which align with element Q̄(k)
            alignedSet(k) = alignedSet(k) ∪ C_k
        end
    end
end
for k ← 1 to L do
    Q̄(k) = mean(alignedSet(k))
end
return Q̄
```

**Example 7.11** *Time Series Classification based on* 1-*NN and DBA* [13, 21]. The classification of multiple time series is an important issue in time series data mining. Let $\mathcal{P} = \{\mathbf{P}_1, \mathbf{P}_2, \ldots, \mathbf{P}_N\}$ denote a set of $N$ one-dimensional time series. Each time series $\mathbf{P}_n, n \in \{1, 2, \ldots, N\}$, has a label $l_n$ which specifies to which class $\mathbf{P}_n$ belongs. We suppose there are $K$ unique classes $c_k, k \in \{1, 2, \ldots, K\}$. For each $c_k$ we calculate the corresponding DBA $\mathbf{Q}^{(k)}$, then given an unknown time series $\mathbf{R}$, we match $\mathbf{R}$ to each DBA $\mathbf{Q}^{(k)}, k \in \{1, 2, \ldots, K\}$. Let $DTW(\mathbf{R}, \mathbf{Q}^{(k)})$ denote the corresponding match distance, then we classify $\mathbf{R}$ as belonging to class $c_{k^*}$, where

$$k^* = \arg\min_k \left( DTW(\mathbf{R}, \mathbf{Q}^{(k)}) \right) .$$

## 7.6 Further Reading

Temporal alignment has been intensely investigated over the years. In recent years it has been increasingly used in data fusion applications in automotive applications [10], medical imaging [7, 9] and speech recognition [5, 14]. The DTW has also been used in applications which do not explicitly involve temporal alignment but require curve matching in one form or another. Examples include: word spotting, identifying facial profiles and signature verification. Recently [22] has described a flight maneuver recognition methods using canonical time warping. Methods for reducing the computational load for performing DTW includes [23].

## 7.7 Software

**MATLAB**. MATLAB SIGNAL PROCESSING toolbox includes m-file for DTW. Additional m-files are: Dynamic Barycenter Average (Author: Petitjean) and Canonical Time Warping) (Author: Zhou).

**PYTHON**. DTW-PYTHON includes python modules for DTW.

## Problems

**7.1** Describe the process of dynamic time warping. Explain the role the boundary conditions and the constraints of continuity and monoticity play in the algorithm.

**7.2** Describe the derivative DTW algorithm. Suggest when DDTW should be used instead of DTW.

## References

1. Bhabu B, Zhou X (2004) Face recognition from face profile using dynamic time warping. In: Proceedings of the 17th international conference on pattern recognition (ICPR '04), vol 4, pp 499–502
2. Cuturi M, Blondel M (2017) Soft-DTW: a differentiable loss function for time-series. In: International conference on machine learning (ICML)
3. Dvornik N, Hadji I, Derpanis KG, Garg A, et al (2021) Drop-DTW: aligning common signal between sequences while dropping outliers. In: 35th conference on neural information processing systems (NeurIPS)
4. Halperin T, Ephrat A, Peleg S (2019) Dynamic temporal alignment of speech to lips. In: Proceedings of the IEEE international conference acoustics, speech signal processing, pp 3980–3984
5. Jiang S, Chen Z (2023) Application of dynamic time warping optimization algorithm in speech recognition of machine translation. Heliyon 9:e21625
6. Keogh EJ, Pazzani MJ (2001) Derivative dynamic time warping. In: 1st SIAM international conference on data mining
7. Klein GJ (2002) Four-dimensional processing of deformable cardiac pet data. Med Image Anal 6:29–46
8. Krauss T, Reinartz P, Lehner M, et al (2005) DEM generation from very high resolution stereo satellite data in urban areas using dynamic programming. Proc ISPRS
9. Ledesma-Carbayo MJ, Kybic J, Desco M et al (2005) Spatio-temporal nonrigid registration for ultrasound cardiac motion estimation. IEEE Trans Med Imag 24:1113–1126
10. Moser U, Schramm D (2019) Multivariate dynamic time warping in automotive applications: a review. Intell Data Anal 23:535–553
11. Perperidis D, Mohiaddin R, Rueckert D (2005) Spatio-temporal free-form registration of cardiac mr image sequences. Med Imag Anal 9:441–456
12. Petitjean F, Ketterlin A, Gancarski P (2011) A global averaging method for dynamic time warping with applications to clustering. Patt Recogn 44:678–693
13. Peritjean F, Foresrier G, Webb GI, et al (2014) Dynamic time warping averaging of time series allows faster and more accurate classification. Int Conf Data Mining

14. Rabiner L, Juang B (1993) Fundamentals of speech processing. Published by Prentice-Hall
15. Ratanahatano CA, Keogh E (2005) Three myths about dynamic time warping. In: SIAM 2005 data mining conference, CA, USA
16. Rath TM, Manmatha R (2003) Word spotting for historical documents. Int J Doc Anal Recogn 9:139–152
17. Savitzky Ã, Golay MJ (1964) Smoothing and differentiation of data by simplified least squares procedure. Anal Chem 36:1627–1639
18. Silva DF, Batista GE, Keogh E (2016) Prefix and suffix invariant dynamic time warping. In: IEEE international conference on data mining, pp 1209–1214
19. Singh M, Thompson R, Basu A, Rieger J, Mandal M (2006) Image based temporal registration of MRI data for medical visualization. In: IEEE international conference on image processing, Atlanta, Georgia
20. Steiner J, Termonia Y, Deltour J (1972) Comments on smoothing and differentiation of data by simplified least square procedure. Anal Chem 44:1906–1909
21. Tran TM, Le X-MT, Nguyen HT, Huynh V-N (2019) A novel non-parametric method for time series classification based on $k$-nearest neighbors and dynamic time warping barycenter averaging. Engng Appl Art Intell 78:173–185
22. Wei Z, Ding D, Zhou H et al (2020) A fight maneuver recognition method based on multi-strategy affine canonical time warping. App Soft Comp J 95:106527
23. Wu J, Wang R, Zhang Y, et al. Adaptive sequential three-way decision for dynamic time warping. Inf Sci 690:121541
24. Zhao J, Itti L (2018) Shapedtw: shape dynamic time warping. Patt Recogn 74:171–184
25. Zhang Z, Tang P, Rubing D (2015) Dynamic time warping under pointwise shape context. Inform Sci 315:88–101
26. Zhou F, Torre F (2016) Generalized canonical time warping. IEEE Trans Patt Anal Mach Intell 38:279–294

# Chapter 8
# Semantic Alignment

**Abstract** The subject of this chapter is semantic alignment and deals with the conversion of multiple input data or measurements which do not refer to the same object, or phenomena, to a common object or phenomena. This is a core element in any data fusion algorithm since different inputs can only be fused together if the inputs refer to the same object or phenomena. In the chapter we first introduce the concept of an assignment matrix and its use in matching shapes and contours. Then we consider clustering algorithms considering in depth the $K$-means algorithm and its variants. This includes automatically estimating the number of clusters and robustifying the algorithm when outliers are present. In the last part of the chapter we consider cluster ensembles and how they may be fused together using the co-association matrix.

## 8.1 Introduction

The subject of this chapter is semantic alignment. This is the conversion of multiple input data or measurements which do not refer to the same object, or phenomena, to a common object or phenomena. The reason for performing semantic alignment is that different inputs can only be fused together if the inputs refer to the same object or phenomena. In general, if the observations have been made by sensors of the same type, then the observations should refer to the same object or phenomena. In this case, no semantic alignment is required, although radiometric normalization may be required.

**Example 8.1** *Multiple Edge Maps*. Consider an input image $I$ on which we separately apply a $K$ edge detectors. The detectors may work on different physical principles but all of them measure the presence, or absence, of an edge in $I$. The corresponding edge images, $F_k, k \in \{1, 2, \ldots, K\}$, are said to be "semantically equivalent" and no semantic alignment is required. In general, $F_k, k \in \{1, 2, \ldots, K\}$, do not have the same radiometric scale. Radiometric normalization may therefore be required before they can be fused together.



H. B. Mitchell, *Data Fusion*, https://doi.org/10.1007/978-3-662-71023-4_8

However, when the sensors are of different types, the observations may refer to different phenomena. If this is true, then it is not possible to perform data fusion unless the input data are semantically aligned to a common object or phenomena. From this point-of-view the thought experiment in Example 6.6 may be regarded as semantic alignment of two multi-modal images. The following example illustrates an example of semantic alignment of several inputs, each of which refers to a different phenomena.

**Example 8.2** *Multi-Feature Infra-Red Target Detection in an Input Image* [29]. We consider the detection of a small target in an infra-red input image $I$. At each pixel $(m, n)$ in $I$ we test for the presence of a target by extracting the following local features:

- $\mathcal{M}$: Maximum Gray Level
- $\mathcal{C}$: Contrast Mean Difference.
- $\mathcal{G}$: Average Gradient.
- $\mathcal{V}$: Gray-level Variation.
- $\mathcal{E}$: Entropy.

The features $\mathcal{M}, \mathcal{C}, \mathcal{G}, \mathcal{V}, \mathcal{E}$, clearly do not measure the same phenomena. However, according to the theory of target detection in an infra-red image, they are all causally linked to the presence of a target. In this case, we may semantically align the features by converting each of the features into evidence, $E(x, y)$, that a target is present at $(x, y)$.

In some cases, the inputs may refer to the same phenomena but each sensor uses a different set of names or symbols. In this case, we must semantically align the inputs so that for each phenomena all the sensors will use the same name or symbol. This is commonly known as "solving the (label) correspondence problem".

**Example 8.3** *Label Correspondence*. We consider an input image $\mathbf{X} = (x_1, x_2, \ldots, x_N)^T$, which we segment using two different algorithms. The first algorithm, $\mathfrak{A}$, partitions the pixels $\{x_1, x_2, \ldots, x_N\}$ into $K$ clusters which are labeled $\{\alpha_1, \alpha_2, \ldots, \alpha_K\}$. The corresponding segmented image is $\mathbf{A} = (a_1, a_2, \ldots, a_N)^T$, where $a_i = k$ if $\mathfrak{A}$ assigns $x_i$ to the $k$th cluster $\alpha_k$. Similarly, the second algorithm, $\mathfrak{B}$, partitions $\{x_1, x_2, \ldots, x_N\}$ into $L$ clusters which are labeled $\{\beta_1, \beta_2, \ldots, \beta_L\}$. The corresponding segmented image is $\mathbf{B} = (b_1, b_2, \ldots, b_N)^T$, where $b_i = l$ if $\mathfrak{B}$ assigns $x_i$ to $l$th cluster $\beta_l$. Then, one way of semantically aligning $\mathbf{A}$ and $\mathbf{B}$ is to find which labels $\alpha_k$ in $\mathbf{A}$ are associated with labels $\beta_l$ in $\mathbf{B}$ and vice versa. A convenient way of defining the associations is through an assignment matrix $\lambda$, where

$$\lambda(k, l) = \begin{cases} 1 \text{ if labels } \alpha_k \text{ and } \beta_l \text{ are associated with each other} , \\ 0 \text{ otherwise.} \end{cases}$$

**Table 8.1** Applications in which semantic alignment is primary fusion algorithm

| Class | | Application |
|---|---|---|
| Inter-FeO | Example 6.10 | Global image registration using RANSAC |
| | Example 6.11 | L2-Net Keypoint Descriptors |
| | Example 6.12 | Joint Keypoint Detection and Description |
| | Example 6.13 | ALIKE: Accurate and Lightweight Keypoint Detection and Descriptor |
| | Example 6.14 | Local Patch Matching |
| | Example 8.4 | Handwritten character recognition |
| | Example 8.5 | Closed contour matching using the shape context |
| | Example 8.7 | Lesion Detection in Lung Computed Tomography Images |
| | Example 8.9 | Image Segmentation via the Method of Cluster Ensembles |
| | Example 8.11 | Multiple-Atlas Segmentation (MAS) |
| | Example 12.14 | Target tracking initialization using the Hough transform |
| | Example 13.9 | Tracking pedestrians across non-overlapping cameras |

In many data fusion applications, semantic alignment is the primary fusion algorithm. In Table 8.1 we list some of these applications together with their classification. We now consider the concept of an assignment matrix in detail and the essential role it plays in semantic alignment.

## 8.2 Assignment Matrix

If $C(k, l)$ denotes the cost of matching a label $\alpha_k$ with a label $\beta_l$, then the *optimal* assignment matrix $\lambda^*$ is defined as the assignment matrix with minimal overall cost:

$$\lambda^* = \arg\min \sum_{k,l} C(k, l)\lambda(k, l) \ . \tag{8.1}$$

In applications in which each label represents a different physical object, it is common practice to use a distance measure for $C(k, l)$, and to solve Eq. (8.1) subject to the following one-to-one constraints:

$$\sum_k \lambda(k, l) \leq 1 \ , \quad \forall l \quad \text{and} \quad \sum_l \lambda(k, l) \leq 1 \ , \quad \forall k \ . \tag{8.2}$$

Fast algorithms for finding the optimal assignment matrix $\lambda^*$ are available. Among them is the *Hungarian algorithm* [16]. Traditionally these algorithms are used when $K = L$ and we require each label $\alpha_k$, $k \in \{1, 2, \ldots, K\}$, is matched to a corresponding label $\beta_l$, $l \in \{1, 2, \ldots, L\}$, and vice versa. The algorithms may, however, be used when $K \neq L$, or when we wish to make the assignment robust against outliers, by finding the best $M$ associations, where $M \leq \min(K, L)$. In this case we use an enlarged cost matrix $C_e$:

$$C_e(k,l) = \begin{cases} C(k,l) \text{ if } k \in \{1,2,\ldots,K\} \textbf{ and } l \in \{1,2,\ldots,L\} , \\ P \quad \text{otherwise} , \end{cases}$$

where $P$ is the cost, or penalty, of not associating a label $\alpha_k$ with any label $\beta_l$ or of not associating a label $\beta_l$ with any label $\alpha_k$.

In some applications it may difficult (or not even possible) to define a penalty $P$. In this case, we may use the following algorithm to find a sub-optimal assignment matrix $\lambda^*$. Note: For simplicity Algorithm 4 does *not* include the one-to-one constraints (Eq. 8.2).

**Algorithm 4:** Sub-Optimal Assignment Algorithm

**Input** : Cost matrix $C(k,l), k \in \{1,2,\ldots,K\}, l \in \{1,2,\ldots,L\}$
**Output** : Assignment matrix $\lambda^*(k,l), k \in \{1,2,\ldots,K\}, l \in \{1,2,\ldots,L\}$

Initialize: $\lambda^*(k,l) \leftarrow 0, \quad \forall k,l$
$m \leftarrow 0$
**while** $m < M$ **do**
  $m \leftarrow m+1$
  $(k',l') \leftarrow \arg\min_{(k,l)} C(k,l)$
  $\lambda^*(k',l') = 1$
  $C(k',l') = 0$
**end**

In many applications, solving the assignment problem is the fusion algorithm itself. This is illustrated in the following example.

**Example 8.4** *Handwritten Character Recognition* [2]. Shape matching is a powerful method for automatic handwritten character recognition. In this technique we match the contour of an unknown character with the contour of a known character. Let **A** and **B** denote the two contours. We sample **A** at $K$ points $a_k, k \in \{1,2,\ldots,K\}$, where we regard each point $a_k$ as a "physical" object with its own unique label $\alpha_k$. Similarly, **B** is sampled at $L$ points $b_l, l \in \{1,2,\ldots,L\}$, each of which has its own label $\beta_l$. For each pair of labels $(\alpha_k, \beta_l)$, i.e. points $(a_k, b_l)$, we let

$$C(k,l) = d(S_k, S_l) ,$$

where $S_k$ and $S_l$ are, respectively, the shape contexts (see Example 4.12) of the points $a_k$ and $b_l$ and $d(S_k, S_l)$ is the $\chi^2$ distance metric. If $\lambda^*$ denotes the optimal one-to-one assignment matrix, then we may use

$$C^* = \sum_{(k,l)} C(k,l)\lambda^*(k,l) ,$$

as a similarity measure: the smaller $C^*$ the more likely **A** and **B** are the same character.

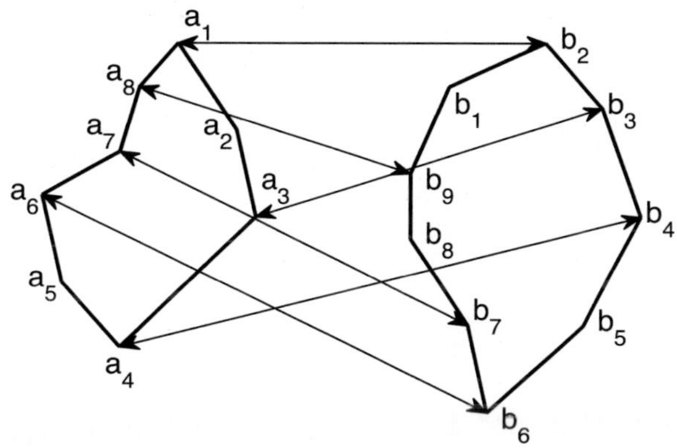


**Fig. 8.1** Shows matching of closed contours $\mathbf{A} = (a_1, a_2, \ldots, a_8)^T$ and $\mathbf{B} = (b_1, b_2, \ldots, b_9)^T$. The optimum matching is $\pi(1, 2, 3, 4, 5, 6, 7, 8) = (2, 0, 3, 4, 0, 6, 7, 9)$, i.e. $a_1 \leftrightarrow b_2$, $a_3 \leftrightarrow b_3, \ldots, a_8 \leftrightarrow b_9$, while points $a_2, a_5$ are not associated with any points on **B** and points $b_1, b_5, b_8$ are not associated with any points on **A**

In some applications we may impose additional constraints on Eq. (8.1). For example, in Example 8.4, we may require the order of the points to be preserved [25].

**Example 8.5** *Closed Contour Matching using Shape Context* [25]. Suppose the points $a_k$ and $b_l$ in Example 8.4 are labeled in increasing order (see Fig. 8.1). Then, we match the two contours **A** and **B** assuming:

**One-to-one Constraint.** No point $a_k$ may be paired with more than one point $b_l$ and no point $b_l$ may be paired with more than one point $a_k$. (Eq. 8.2).

**Increasing-order constraint.** If $b_l = \pi(a_k)$ and $b_{l'} = \pi(a_{k'})$, then $k' > k$ if $l' > l$, where $b_l = \pi(a_k)$ denotes that the point $a_k$ is matched with $b_l$ and $\pi(a_k) = 0$ denotes that the point $a_k$ is not matched to any of the points $b_l$.

Sometimes we may solve Eq. (8.1) subject to only one of the constraints listed in Eq. (8.2). This may occur when the sensors have different resolutions: For example, if sensor $A$ has a lower resolution than sensor $B$, then a point $a_m$ may be associated with several points $\{b_{n'}, b_{n''}, \ldots\}$.

In Examples 8.4 and 8.5, the labels $\alpha_k$ and $\beta_l$ represented physical objects. We now turn our attention to the case when the labels represent symbolic class names, and not specific physical objects.

## 8.3 Clustering Algorithms

Clustering algorithms constitute the most popular method for partitioning a set of objects, $\{O_1, O_2, \ldots, O_N\}$, into $K$ classes. Mathematically, we write the partition of objects into $K$ classes by using the assignment matrix $\lambda$:

$$\lambda(i,k) = \begin{cases} 1 & \text{if } O_i \text{ is assigned to cluster } k \ , \\ 0 & \text{otherwise} \ . \end{cases} \tag{8.3}$$

In traditional "hard" cluster algorithms, each object can only belong to one cluster. In this case, the assignment matrix $\lambda$ satisfies the following constraint:

$$\sum_k \lambda(i,k) = 1 \quad \forall i \ . \tag{8.4}$$

There are also "soft" cluster algorithms. In these algorithms each object may belong to several clusters. Broadly speaking there are two classes of "soft" cluster algorithms:

**Probabilistic**. These algorithms use a probabilistic assignment matrix $\lambda_{\text{prob}}(i,k)$, where $\lambda_{\text{prob}}(i,k)$ is the probability of assigning object $O_i$ to cluster $k$. The matrix $\lambda_{\text{prob}}(i,k)$ satisfies the constraints: $\lambda_{\text{prob}}(i,k) \in [0,1]$ and $\sum_{k=1}^{K} \lambda_{\text{prob}}(i,k) = 1$.

**Fuzzy**. These algorithms use a fuzzy assignment matrix $\lambda_{\text{fuzzy}}(i,k)$, where $\lambda_{\text{fuzzy}}(i,k)$ is the degree to which $O_i$ belongs to cluster $k$. Unlike the probabilistic assignment matrix, $\lambda_{\text{fuzzy}}(i,k)$ only satisfies one constraint:$\lambda_{\text{fuzzy}}(i,k) \in [0,1]$.

In this chapter we shall only consider hard cluster algorithms. However, in Chap. 10 we introduce a probabilistic model (Gaussian mixture model) which may be interpreted as a probabilistic cluster algorithm.

Of the different hard clustering algorithms which are available, the $K$-means algorithm [13] is the workhorse. This algorithm and its variants are described next.

## 8.4 *K*-Means Cluster Algorithm

Suppose $\{O_1, O_2, \ldots, O_N\}$ denotes a set of $N$ objects, where each object is characterized by a vector $\mathbf{x}_i = \left(x_i^{(1)}, x_i^{(2)}, \ldots, x_i^{(L)}\right)^T$ in $L$-dimensional feature space. The $K$-means algorithm partitions the objects into $K$ *disjoint* clusters $C_1, C_2, \ldots, C_K$, such that each object $O_i$ belongs to its nearest cluster and objects belonging to different clusters are far from each other. We represent the center of each cluster $C_k$ by a vector $\mathbf{y}_k = \left(y_k^{(1)}, y_k^{(2)}, \ldots, y_k^{(L)}\right)^T$, in $L$-dimensional feature space. Mathematically the action of the $K$-means algorithm is to minimize the cost function $J$:

$$J = \sum_i \min\left(\|\mathbf{x}_i - \mathbf{y}_k\|\right)^2 , \tag{8.5}$$

where $\|\mathbf{x}_i - \mathbf{y}_k\| = \sqrt{\sum_l (x_i^{(l)} - y_k^{(l)})^2}$ is the Euclidean distance between $O_i$ and the cluster $C_k$. Mathematically we may rewrite Eq. (8.5) as

$$
\begin{aligned}
&\min \Big( \sum_i \sum_k \lambda(i,k) \|\mathbf{x}_i - \mathbf{y}_k\|^2 \Big) , \\
&\text{subject to} \quad \lambda(i,k) \in \{0,1\} \quad \text{and} \quad \sum_k \lambda(i,k) = 1 ,
\end{aligned} \tag{8.6}
$$

where the cluster center $\mathbf{y}_k$ is computed by taking the average over all objects $O_i$ in cluster $C_k$.

We solve Eq. (8.6) iteratively as follows. We first randomly select $K$ objects as the initial cluster centers $\mathbf{y}_1, \mathbf{y}_2, \ldots, \mathbf{y}_K$. Then for each iteration: calculate the distance from each object to each cluster center and place each object in its nearest cluster. The center of each cluster is updated by averaging all the objects in the cluster. The process is repeated until the cluster centers stop moving (convergence).

The $K$-means cluster algorithm is shown in pseudo-code in Algorithm 5.

**Algorithm 5:** $K$-means Algorithm

**Input**: Set of Objects $O_i, i \in \{1, 2, \ldots, N\}$, where $O_i$ is characterized by a vector $\mathbf{x}_i = (x_i(1), x_i(2), \ldots, x_i(L))^T$
Number of clusters $K$
**Output**: Final cluster assignment matrix $\lambda(i,k)$ and final cluster centers $\mathbf{y}_k$

**Initialize:** Randomly select $K$ objects as cluster centers $\mathbf{y}_k, k \in \{1, 2, \ldots, K\}$
**while** *no convergence* **do**
  Initialize assignment matrix:
  $\lambda(i,k) \leftarrow 0, \quad i \in \{1, 2, \ldots, N\}, \quad k \in \{1, 2, \ldots, K\}$
  For each object $O_i$ find nearest cluster:
  **for** $i \leftarrow 1$ ***to*** $N$ **do**
    Compute $k^* = \arg\min_k \|\mathbf{x}_i - \mathbf{y}_k\|^2$
    update assignment matrix (in effect "placing" $O_i$ in cluster $C_{k^*}$):
    $\lambda(i,k^*) \leftarrow 1$
  **end**
  Update each cluster center:
  **for** $k \leftarrow 1$ ***to*** $K$ **do**
    $\mathbf{y}_k = \sum_i \lambda(i,k)\mathbf{x}_i / \sum_i \lambda(i,k)$
  **end**
**end**
**return** Final assignment matrix $\lambda(i,k)$ and cluster centers $\mathbf{y}_k, k \in \{1, 2, \ldots, K\}$

The following example illustrates an important use of $K$-means clustering to *over*-segment an input image into $K$ *superpixels* [1, 26].

**Example 8.6** *SLIC: A Superpixel Algorithm* [1]. SLIC (Simple Linear Iterative Clustering) is a widely used algorithm for over-segmenting an input image $I$ into *superpixels*. Let $(x_i, y_i)$ and $(R_i, G_i, B_i), i \in \{1, 2, \ldots, N\}$, denote, respectively, the position and color of the $i$th pixel in the input image $I$. Note:

Different color spaces may be used instead of RGB. For example, in the original SLIC algorithm, the LAB color space is used instead of RGB.

We cluster the tuples $z_i = (x_i, y_i, R_i, G_i, B_i)$ into $K$ clusters using a weighted Euclidean distance $d_{ik}$:

$$d_{ij} = \sqrt{\alpha\big((x_i - \overline{x}_k)^2 + (y_i - \overline{y}_k)^2\big) + \big((R_i - \overline{R}_k)^2 + (G_i - \overline{G}_k)^2 + (B_i - \overline{B}_k)^2\big)}$$

where the tuple $\overline{z}_k = (\overline{x}_k, \overline{y}_k, \overline{R}_k, \overline{G}_k, \overline{B}_k)$ represents the centroid of the $k$th cluster. The purpose of the parameter $\alpha$ is to ensure that the same scale is used to measure spatial proximity (i.e. distance between $(x_i, y_i)$ and $(\overline{x}_k, \overline{y}_k)$) and color proximity (i.e. distance between $(R_i, G_i, B_i)$ and $(\overline{R}_k, \overline{G}_k, \overline{B}_k)$).

The clusters are known as *superpixels*. Physically, they represent small spatially compact groups of pixels which have similar color values. The reader should note, however, that the superpixels do *not* represent physical objects in the image. Instead they simply represent an over-segmentation of the input image into a large number of spatially compact regions with similar color values.

Superpixels are increasingly used in computer vision applications where they help to reduce the computational workload and remove redundant information.

**Deep Learning Superpixels**

Traditional superpixel algorithms use only spatial distance and color features for segmentation. This may, however, limit their performance in low contrast or monchromatic (gray-scale) images. To alleviate we may augment the SLIC algorithm using deep features extracted from different layers in a convolutional neural network (CNN) (cf. Sect. 5.6). In [28] the authors found that the inclusion of these deep features consistently improved the quality of the superpixel segmentation.

We now consider an important computer vision application of $K$-means clustering to image segmentation.

### 8.4.1 Computer Vision Applications of K-Means Algorithm

In this section we consider the use of $K$-means clustering for image segmentation. The aim of image segmentation is to divide an input image $I$ into *several disjoint and* **meaningful** *regions*.

We may use the $K$-means algorithm *directly* for this purpose by letting each pixel in $I$ be an object $O_i$ in the $K$-means algorithm. In the case of an input grey image

$I$, $L = 1$ and $\mathbf{x}_i$ is a scalar value: $x_i = I(r, c)$, where pixel $i$ lies in row $r$, column $c$. For some *specific* types of images this method works well as shown in the following example.

**Example 8.7** *Lesion Detection in Lung Computed Tomography Images* [19]
The $K$-means cluster algorithm was used in developing a method for the early identification and quantification of infected lung areas. Given a CT (computed tomography) image of a patients lung area, we construct a histogram of the gray-levels in the image. Typically the histogram has $P = 4$ or $P = 5$ prominent peaks $G_p$, $p \in \{1, 2, \ldots, P\}$, corresponding to $P$ specific regions: background, lungs, lesion region, breast and spinal cord, ribs and shoulder blade, where background is dark, lungs are dark gray and lesions are light gray. We set the number of clusters $K$ equal to the number of histogram peaks $P$ and use the gray-levels $G_p$ to initialize the cluster centers.

The algorithm is simple, fast and gives reliable good quality segmentation.

However, for general input images, the segmentation results are often less satisfactory. One reason for this is that the $K$-means algorithm does *not* include any spatial information describing the relationship between a pixel and its neighbors. The absence of this information may lead to a segmentation of the input image that bears no relationship to the objects in the image.

### Image Segmentation Using Histogram-Based $K$-Means

The traditional approach to incorporating spatial information into the $K$-means algorithm is to let $O_i$ refer to a small *patch* of pixels and *not* to a single pixel [18]. Let $P_i$ denote a $p \times p$ image patch centered on pixel $i$, then we could simply represent $\mathbf{x}_i$ as the concatenation of the $p^2$ grey values $\{I(r', c'), I(r'', c''), \ldots, \}$ which lie in the patch $P_i$.

However, in practice, [18] found it best to let $\mathbf{x}_i$ be the *histogram* $\mathbf{H}_i$ of the gray-levels $\{I(r', c'), I(r'', c''), \ldots, \}$. In this case, suppose the histogram has $L$ bins, then $\mathbf{x}_i = \mathbf{H}_i$, where $\mathbf{H}_i = (H_i^{(1)}, H_i^{(2)}, \ldots, H_i^{(L)})^T$ and

$$H_i^{(l)} = \begin{cases} \text{\# gray-levels } \{I_i(r', c'), I_i(r'', c''), \ldots, \} \\ \text{in patch } P_i \text{ which fall in } l\text{th histogram bin} . \end{cases} \tag{8.7}$$

Very good segmentation results are obtained by running the $K$-means algorithm using the histogram $\mathbf{H}$. To emphasize the role played by the histogram $\mathbf{H}$ we shall refer to the segmentation algorithm as *histogram-based $K$-means*.

The histogram-based $K$-means algorithm is found to be robust against noise and has been successfully used to segment color images (see Example 8.10).

### Image Segmentation Using Superpixels

An alternative approach to incorporating spatial information into the $K$-means algorithm is to first oversegment the input image into $K'$ superpixels, where typically $K' \geq N/30$ and $N$ denotes the number of pixels in the input image. Then cluster the superpixels into $K$ clusters using the $K$-means algorithm (or similar cluster algorithm), where $K \ll K'$. See [24] for a modern review of this approach.

## 8.4.2 Robust Variants of the K-Means Algorithm

Although widely used, the $K$-means algorithm is sensitive to the presence of outliers. Many variants of the $K$-means algorithm have therefore been developed which tend to be more robust against outliers. Among them are the $K$-median, the $K$-medoid and the $K$-means-sharp algorithms, each of which use a different approach to improving the robustness of the algorithm.

### $K$-Median

The $K$-median algorithm is a robustified version of the $K$-means algorithm in which we replace the Euclidean distance (Eq. 8.5) by the $L_1$ (Manhatten) distance:

$$|\mathbf{x}_i - \mathbf{y}_k| = \sum_l |x_i^{(l)} - y_k^{(l)}| \,. \tag{8.8}$$

In this case, the $K$-median algorithm is defined as the solution of

$$\begin{aligned} &\min \Big( \sum_i \sum_k \lambda(i,k) |\mathbf{x}_i - \mathbf{y}_k| \Big) \,, \\ &\text{subject to} \;\; \lambda(i,k) \in \{0,1\} \;\; \text{and} \;\; \sum_k \lambda(i,k) = 1 \,. \end{aligned} \tag{8.9}$$

We solve Eq. (8.9) iteratively as follows. We first randomly select $K$ objects as the initial cluster centers $\mathbf{y}_1, \mathbf{y}_2, \ldots, \mathbf{y}_K$. Then for each iteration calculate the distance from each object to each cluster center and place each object in its nearest cluster. The center of each cluster $\mathbf{y}_k$, $k \in \{1, 2, \ldots, K\}$, is computed by taking the median value for each dimension over all objects in the cluster. Pseudo-code for the $K$-median algorithm is given in Algorithm 6.

**Algorithm 6:** $K$-median Algorithm

**Input**: Set of Objects $O_i, i \in \{1, 2, \ldots, N\}$, where $O_i$ is characterized by a vector $\mathbf{x}_i = (x_i(1), x_i(2), \ldots, x_i(L))^T$
Number of clusters $K$
**Output**: Final Cluster Assignment Matrix $\lambda(i, k)$ and Final Cluster Centers $\mathbf{y}_k$

**Initialize:** Randomly select $K$ objects as cluster centers $\mathbf{y}_k^{(0)}, k \in \{1, 2, \ldots, K\}$
**while** *no convergence* **do**
  Initialize assignment matrix:
  $\lambda(i, k) \leftarrow 0, \quad i \in \{1, 2, \ldots, N\}, \quad k \in \{1, 2, \ldots, K\}$
  For each object $O_i$ find nearest cluster:
  **for** $i \leftarrow 1$ *to* $N$ **do**
    Compute $k^* = \min_k |\mathbf{x}_i - \mathbf{y}_k|$
    Place $O_i$ into cluster $C_{k^*}$ by updating assignment matrix: $\lambda(i, k^*) \leftarrow 1$
  **end**
  Take each cluster in turn:
  **for** $k \leftarrow 1$ *to* $K$ **do**
    For cluster index $k$ find all objects in cluster and place them temporary list temp:
    $ntemp \leftarrow 0$
    $temp \leftarrow 0$
    **for** $i \leftarrow 1$ *to* $N$ **do**
      **if** $\lambda(i, k) = 1$ **then**
        $ntemp \leftarrow ntemp + 1$
        $temp(ntemp, 1 : L) \leftarrow (x_i(1), x_i(2), \ldots, x_i(L))$
      **end**
    **end**
    For cluster index $k$ take each component $l$ in turn. For each component update corresponding cluster center $y_k(l)$:
    **for** $l \leftarrow 1$ *to* $L$ **do**
      $y_k(l) = \text{med}(temp(1, l), temp(2, l), \ldots, temp(ntemp, l))$
    **end**
  **end**
**end**
**return** Final assignment matrix $\lambda(i, k)$ and cluster centers $\mathbf{y}_k, k \in \{1, 2, \ldots, K\}$

### *K*-Mediod

The $K$-mediod algorithm is similar to $K$-median. However instead of *calculating* the cluster centers $\mathbf{y}_k, k \in \{1, 2, \ldots, K\}$, we simply choose a different object $O_i$ to represent each cluster. The objects are chosen to minimize the $L_1$ distance.

### *K*-Means-Sharp

The $K$-means-sharp algorithm is a robustified version of the $K$-means algorithm in which we use an outlier identification test to exclude outliers. In this case, instead of using the assignment matrix $\lambda(i, k)$ (Eq. 8.3) we use a modified assignment matrix $\lambda_{\text{mod}}(i, k)$:

$$\lambda_{\text{mod}}(i,k) = \begin{cases} 1 & \text{if } O_i \text{ is assigned to cluster } k \textbf{ and } d_{ik} \leq T \ , \\ 0 & \text{otherwise} \ . \end{cases} \tag{8.10}$$

The $K$-means-sharp algorithm works in an iterative manner as follows. We randomly select $K$ objects as the initial cluster centers $\mathbf{y}_1, \mathbf{y}_2, \ldots, \mathbf{y}_K$. For each iteration: calculate the Euclidean distance $d_{ik}$ between each object $O_i$ and each cluster center $\mathbf{y}_k$. If the minimum distance for a given object $O_i$ is less than the threshold $T$, then $O_i$ is classified as an *inlier* and placed in its nearest cluster $C_{k^*}$, where $k^* = \arg\min_k(d_{i1}, d_{i2}, \ldots, d_{iK})$. The cluster centers are then updated by averaging over inlier objects. The process is repeated until the cluster centers stop moving (convergence). The threshold $T$ is calculated by performing a preliminary run using the standard $K$-means algorithm. If in the preliminary run, $d_i^*$ is the distance between object $O_i$ and its corresponding cluster, then [21] recommend using $T = 3 \times \sigma$, where $\sigma$ is a robust estimate of the standard deviation of the $\{d_1^*, d_2^*, \ldots, d_N^*\}$ (see Eq. 12.3).

Pseudo-code for the $K$-means-sharp cluster algorithm is shown in Algorithm 7.

**Algorithm 7:** $K$-means-sharp Algorithm

**Input**: Set of Objects $O_i, i \in \{1, 2, \ldots, N\}$, where $O_i$ is characterized by a vector $\mathbf{x}_i = (x_i(1), x_i(2), \ldots, x_i(L))^T$
Number of clusters $K$ and object-cluster threshold $T$
**Output**: Final cluster assignment matrix $\lambda_{\text{mod}}(i,k)$ and final cluster centers $\mathbf{y}_k$

**Initialize:** Randomly select $K$ objects as cluster centers $\mathbf{y}_k, k \in \{1, 2, \ldots, K\}$
**while** *no convergence* **do**
    Initialize assignment matrix:
    $\lambda_{\text{mod}}(i,k) \leftarrow 0, \quad i \in \{1, 2, \ldots, N\}, \quad k \in \{1, 2, \ldots, K\}$
    For each object $O_i$ find index of the nearest cluster and its distance to the cluster:
    **for** $i \leftarrow 1$ ***to*** $N$ **do**
        $d^* = \min_k \|\mathbf{x}_i - \mathbf{y}_k\|^2$
        $k^* = \arg\min_k \|\mathbf{x}_i - \mathbf{y}_k\|^2$
        If $d^*$ is less than threshold $T$ then object $O_i$ is an inlier. Place it in cluster $C_{k^*}$ and update assignment matrix:
        **if** $d^* \leq T$ **then**
            $\lambda_{\text{mod}}(i,k^*) \leftarrow 1$
        **end**
    **end**
    Update each cluster center:
    **for** $k \leftarrow 1$ ***to*** $K$ **do**
        $\mathbf{y}_k = \sum_i \lambda_{\text{mod}}(i,k)\mathbf{x}_i / \sum_i \lambda_{\text{mod}}(i,k)$
    **end**
**end**
**return** Final assignment matrix $\lambda_{\text{mod}}(i,k)$ and cluster centers $\mathbf{y}_k, k \in \{1, 2, \ldots, K\}$

### 8.4.3 Determining the Number of Clusters K

An important drawback to using the $K$-means algorithm and its variants is the need to specify the number of clusters $K$. Many different techniques have been developed for automatically estimating $K$. The following example illustrates one such technique [22] which is based on Chebyshev inequalities (see also Algorithm 8).

**Example 8.8** *Determining the Number of Clusters Using Chebyshev Inequalities* [22]. Assume a set of objects $O_1, O_2, \ldots, O_N$ are clustered into $K$ clusters using the $K$-means algorithm. Let $d_i$ denote the corresponding distance of object $O_i$ to its nearest cluster center. Then [22] shows that when $K$ is chosen correctly, the distances $d_i, i \in \{1, 2, \ldots, N\}$, satisfy two Chebyshev inequalities:

- The number of objects whose $d_i$ is less than $\sigma$ should be less than $8N/9$
- The number of objects whose $d_i$ is less than $2\sigma$ should be less than $5N/9$.

where $\sigma$ is the standard deviation of the $d_i$.

**Algorithm 8:** Number of Clusters Using Chebyshev Inequalities

```
Input  : Set of objects {O1, O2, ..., ON}
Output : Estimated Number of clusters K
Initialize: K ← 1
Test ← FALSE
while (Test ← FALSE and K < N) do
    (λ, d1, d2, ..., dN) ← K-means(O1, O2, ..., ON, K)
    σ ← standard deviation of {d1, d2, ..., dN}
    M1 ← 0 and M2 ← 0
    for i ← 1 to N do
        if (di < σ) then
            M1 ← M1 + 1
        end
        if (di < 2σ) then
            M2 ← M2 + 1
        end
    end
    if (M1 < 8N/9 and M2 < 5N/9) then
        Test ← TRUE
    else
        K ← K + 1
    end
end
return K
```

The reader should note that generally speaking the techniques developed for automatically estimating $K$ have limited applicability. It is therefore important that a preliminary test be made to determine the suitability of any technique.

## 8.5 NJW Spectral Clustering Algorithm*

In recent years spectral clustering algorithms have become increasingly popular. These techniques are known to perform well in cases where classical clustering methods such as the $K$-means fail. The idea behind spectral clustering is to cluster the observations $O_i, i \in \{1, 2, \ldots, N\}$, based on their similarity or affinity. The algorithm uses the eigenvalues and eigenvectors of a similarity matrix to project the observations into a $K$-dimensional feature space in which the observations are easily partitioned into $K$ groups.

Let $K$ be the desired number of clusters. Then, given a set of observations, $\{O_1, O_2, \ldots, O_N\}$, we calculate a $N \times N$ affinity matrix

$$\mathcal{A}(i, j) = \exp - \frac{dist^2(O_i, O_j)}{\sigma^2}$$

where $dist(O_i, O_j)$ is the distance between $O_i$ and $O_j$ and $\sigma$ is a global scale parameter. A convenient formula for $\sigma$ is

$$\sigma = \frac{1}{N} \sum_i dist(O_i, \widetilde{O}_i) ,$$

where $\widetilde{O}_i$ denotes the nearest neighbour of $O_i$ [3]. In place of a global scale parameter, a local scale $\sigma_i$ may also be used [32]:

$$\sigma_i = dist(O_i, \widetilde{O}_i) .$$

In this case,

$$\mathcal{A}(i, j) = \exp - \left( \frac{dist^2(O_i, O_j)}{\sigma_i \sigma_j} \right) .$$

The following algorithm (Algorithm 9) gives the pseudo-code for the NJW spectral clustering algorithm.

A major drawback to spectral clustering is the computational cost and storage requirements although methods have been developed for reducing the cost and storage [8]. In addition, spectral clustering algorithms require as input the number of clusters $K$. One method for estimating $K$ is [32].

We now consider the segmentation of a histogram. This is an important application which is used in many data fusion algorithms.

**Algorithm 9:** NJW Spectral Clustering Algorithm

**Input** : Set of objects $O_i, i \in \{1, 2, \ldots, N\}$, Affinity matrix $\mathcal{A}(i, j), i, j \in \{1, 2, \ldots, N\}$
**Output** : Cluster assignment matrix $\lambda(i, k)$, where $\lambda(i, k) = 1$ if $i$th object belongs to $k$th cluster; otherwise $\lambda(i, k) = 0$

1 Set diagonal elements of $\mathcal{A}$ to zero: $\mathcal{A}(i, i) = 0$
2 Calculate $D(i) = \sum_{h=1}^{N} \mathcal{A}(i, h), i \in \{1, 2, \ldots, N\}$
3 Construct normalized affinity matrix: $\mathcal{N}(i, j) = \mathcal{A}(i, j)/\sqrt{D(i) \times D(j)}$
4 Compute eigenvector solutions of: $\mathcal{N}\mathbf{u} = \lambda\mathbf{u}$
5 Form $N \times K$ matrix $U = (\mathbf{u}_1, \mathbf{u}_2, \ldots, \mathbf{u}_K)$
6 Calculate $V(i, j) = U(i, j)/\sqrt{\sum_{h=1}^{K} U^2(i, h)} \quad \forall i, j$
7 Treat each row of $V$ as $K$-dimensional point. Cluster rows of $V$ with $K$-means algorithm.
8 If (and only if) $i$th row of $V$ was assigned to cluster $k$, then $\lambda(i, k) = 1$

## 8.6 Histogram Segmentation

In this section we discuss clustering 1D data which is in the form of a 1D-histogram. One of the features that describes a 1D-histogram is the list of its modes. These are intervals around which the data is concentrated. If the number of modes is known, then we can find the locations of the modes using the $K$-means algorithm. Unfortunately, in very many cases, the number of modes is not known. When this happens we may use the following non-parametric approach.

### Non-parametric Histogram Segmentation

We describe an effective non-parametric histogram segmentation algorithm [4, 17]. Suppose the histogram has a finite number of $M$ points distributed among a set of $L$ equal-width bins. Let $H_{ab}$ denote the number of points which lie in the interval $[a, b]$. Then the aim of the algorithm is to find a set of discrete intervals $[a, b]$ over which it is "highly likely" that the histogram $H_{ab}$ is unimodal, or nearly so (which we refer to as "statistically" unimodal).

In each interval $[a, b]$ we apply a statistical test which measures the likelihood that $H_{ab}$ is unimodal. The statistical test used in [4] is: For each discrete interval $[a, b]$, let $r_{ab} = H_{ab}/M$ represent the proportion of points in the interval. If the points were randomly distributed among the $L$ bins, then the expected number of points in $[a, b]$ is $p_{ab} = (b - a + 1)/L$.

An interval $[a, b]$ is statistically unimodal if there exists $c \in [a, b]$ such that

- In the interval $[a, c]$: $r_{ac} \geq p_{ac}$ and $\mathcal{E}_{ac} \geq \frac{1}{M} \log \frac{L(L+1)}{2}$.
- In the interval $[c, b]$: $r_{cb} \leq p_{cb}$ and $\mathcal{E}_{cb} \geq \frac{1}{M} \log \frac{L(L+1)}{2}$.

where $\mathcal{E}_{xy} = r_{xy} \log \frac{r_{xy}}{p_{xy}} + (1 - r_{xy}) \log \frac{1-r_{xy}}{1-p_{xy}}$.

Using these Eqs. Delon et al. [4] developed an automatic algorithm for 1D histogram segmentation.

## 8.7 Cluster Ensembles

Cluster ensembles is a subject of increasing importance and involves the semantic alignment, or label correspondence, of symbolic labels.

**Example 8.9** *Image Segmentation via the Method of Cluster Ensembles* [9, 14]. Image segmentation is a critical task in image fusion. In order to deal with the great variability of features encountered in different images, specific segmentation methods have been designed for different types of images. In [9] we apply several different segmentation algorithms (or the same segmentation algorithm with different parameters) to a single input image. We then combine the multiple segmented images into a single *consensus* segmented image. In general the consensus segmented image is found to have better overall quality than any of the input segmented images.

Formally, we may define the method of cluster ensembles as follows. Let $O = \{O_1, O_2, \ldots, O_N\}$ denote a set of $N$ objects. We partition the objects using an ensemble of $M$ cluster algorithms $\{\mathfrak{A}, \mathfrak{B}, \ldots, \mathfrak{D}\}$. The algorithm $\mathfrak{A}$ partitions the $N$ objects into clusters: $\{\alpha_1, \alpha_2, \ldots\}$. The corresponding identity vector is $\mathbf{A} = (a_1, a_2, \ldots, a_N)^T$, where $a_n = \alpha_k$ if $O_n$ is assigned to the $k$th cluster $\alpha_k$, $k \in \{1, 2, \ldots\}$. Similarly, algorithms $\mathfrak{B}$ and $\mathfrak{D}$ partition the objects into clusters $\{\beta_1, \beta_2, \ldots\}$ and $\{\delta_1, \delta_2, \ldots\}$. The corresponding identity vectors are $\mathbf{B} = (b_1, b_2, \ldots)^T$ and $\mathbf{D} = (d_1, d_2, \ldots)^T$, where $b_n = \beta_l$ if $\mathfrak{B}$ assigns $O_n$ to cluster $\beta_l$ and $d_n = \delta_h$ if $\mathfrak{D}$ assigns $O_n$ to $\delta_h$. *Note*: Each cluster algorithm may generate a different number of clusters. Finally, we combine the identity vectors $\{\mathbf{A}, \mathbf{B}, \ldots, \mathbf{D}\}$ into a single "consensus" vector $\widetilde{\mathbf{Y}} = (\widetilde{y}_1, \widetilde{y}_2, \ldots, \widetilde{y}_N)^T$.

In order to find $\widetilde{\mathbf{Y}}$ we need to transform the individual vectors $\{\mathbf{A}, \mathbf{B}, \ldots, \mathbf{D}\}$, into a common representational format. One way to do this is to designate one of the cluster algorithms as a "reference" algorithm $\mathfrak{R}$. If $\{\phi_1, \phi_2, \ldots\}$ denotes the clusters associated with $\mathfrak{R}$, then we calculate the optimum assignment matrices $\{\lambda_A(k, r), \lambda_B(k, r), \ldots, \lambda_D(k, r)\}$ for the remaining algorithms $\{\mathfrak{A}, \mathfrak{B}, \ldots, \mathfrak{D}\}$, where

$$\lambda_A(k, r) = \begin{cases} 1 \text{ if } \alpha_k \text{ corresponds to } \phi_r \ , \\ 0 \text{ otherwise} \ , \end{cases}$$

$$\lambda_B(k, r) = \begin{cases} 1 \text{ if } \beta_k \text{ corresponds to } \phi_r \ , \\ 0 \text{ otherwise} \ , \end{cases}$$

$$\vdots$$

$$\lambda_D(k, r) = \begin{cases} 1 \text{ if } \delta_k \text{ corresponds to } \phi_r \ , \\ 0 \text{ otherwise} \ . \end{cases}$$

A convenient cost function which may be used to calculate $\{\lambda_A(k,r), \lambda_B(k,r), \ldots, \lambda_D(k,r)\}$, is to maximize the overlap of $\{\mathbf{A}, \mathbf{B}, \ldots, \mathbf{D}\}$ with $\mathbf{R} = (r_1, r_2, \ldots, r_N)^T$. Alternatively, we may side-step the need to calculate the pairwise assignment matrices $\{\lambda_A(k,r), \lambda_B(k,r), \ldots, \lambda_D(k,r)\}$ (and the need to define an appropriate cost function) as follows:

**Concatenation.** For each object $O_n, n \in \{1, 2, \ldots, N\}$, we form a feature vector $\mathbf{f}_n$ by concatenating the identities $a_n, b_n, \ldots, d_n$:

$$\mathbf{f}_n = (a_n, b_n, \ldots, d_n)^T \ . \tag{8.11}$$

We may then cluster the vectors $\mathbf{f}_n, n \in \{1, 2 \ldots, N\}$, to give a consensus identity vector $\widetilde{\mathbf{Y}}$. See Example 8.10.

**Co-Association Matrix.** For each cluster algorithm $\{\mathfrak{A}, \mathfrak{B}, \ldots, \mathfrak{D}\}$, we rewrite the identity vectors $\{\mathbf{A}, \mathbf{B}, \ldots, \mathbf{D}\}$ as a set of $N \times N$ co-association matrices $\{\mathcal{A}, \mathcal{B}, \ldots, \mathcal{D}\}$. The co-association matrices are special similarity matrices which all use the same binary language. We then fuse together the matrices $\{\mathcal{A}, \mathcal{B}, \ldots, \mathcal{D}\}$ and generate the fused co-association matrix $\widetilde{\mathcal{Y}}$. Finally, we perform spectral analysis on $\widetilde{\mathcal{Y}}$ to generate the corresponding consensus identity vector $\widetilde{\mathbf{Y}}$. See Example 8.12.

The following example describes using a cluster ensemble for high performance color image segmentation.

**Example 8.10** *Color Image Segmentation* [15, 18]. Given a color RGB input image we transform it into $M = 6$ different color spaces (RGB, HSV, YIQ, XYZ, LAB and LUV) [5]. In each color space we segment the pixels into $K_m, m \in \{1, 2, \ldots, M\}$, clusters using the histogram-based $K$-means algorithm (Sect. 8.4). Then we represent each pixel $(i, j)$ by a $M$-dimensional vector $\mathbf{f}(i, j)$:

$$\mathbf{f}(i,j) = \left(y^{(1)}(i,j), y^{(2)}(i,j), \ldots, y^{(M)}(i,j)\right) ,$$

where $y^{(m)}(i,j) = k_m$ if in the $m$th color space, the cluster algorithm assigns the pixel $(i, j)$ to the $k_m$th cluster. The $M$-dimensional feature map $\mathbf{f}(i, j)$ is then segmented using a $K$-means cluster algorithm (see Example 8.4). Using this technique, we obtain high quality color image segmentation [18].

The following example describes multiple atlas segmentation using a cluster ensemble. In this example the cluster labels are not arbitrary but represent given physical phenomena. In this case we may use a simple majority vote as the fusion operation.

**Example 8.11** *Multiple-Atlas Segmentation (MAS)* [23, 31]. With the development of modern imaging techniques in medicine, imaging-based studies are becoming more important. In particular, we consider the development of automatic image segmentation tools. One such technique is *multiple-atlas segmentation* (MAS).

The idea of MAS is to segment an unknown target image $I_0$ by transferring the labels from a database $\mathcal{D}$ of previously annotated images through image registration (see Sect. 6.2). The underlying idea is that if two anatomical structures have similar location and show similar variation in intensity, they should have the same labels (or tissue types).

Suppose the database $\mathcal{D}$ contains $N$ images $I_1, I_2, \ldots, I_N$, which have been annotated by hand. Let $S_1, S_2, \ldots, S_N$ denote the corresponding annotated images. Each annotated image $S_i$ consists of a set of disjoint regions $\{R_1^{(i)}, R_2^{(i)}, \ldots\}$ which are given a label $k$ according to their anatomical structure or tissue type. Thus, if regions $R_r^{(i)}$, $R_s^{(i)}$ in $S_i$ and $R_t^{(j)}$ in $S_j$, represent the same tissue type, then all three regions will be assigned the same label $k$. Altogether there are $K$ possible labels $k$, $k \in \{1, 2, \ldots, K\}$.

We spatially align (and warp) $I_i$, $i \in \{1, 2, \ldots, N\}$, by separately registering the target image $I_0$ with each $I_i$ (see Fig. 8.2). Mathematically, the spatially aligned (and warped) images are found by applying the inverse transformation $\mathcal{T}_i^{-1}$ to the pixel coordinates $(x', y')$ in $I_i$:

$$\widetilde{I}_i(x, y) = I_i(x', y') , \tag{8.12}$$

where $\mathcal{T}_i$ is the transformation which optimally maps coordinates $(x, y)$ in $I_0$ to coordinates $(x', y')$ in $I_i$ (see Sect. 6.2):

$$(x, y) = \mathcal{T}_i(x', y') . \tag{8.13}$$

We now apply the inverse transformation $\mathcal{T}_i^{-1}$ to the coordinates $(x', y')$ in the annotated image $S_i$. The result is a spatially aligned and warped image $\widetilde{S}_i$, where

$$\widetilde{S}_i(x, y) = S_i(x', y') . \tag{8.14}$$

The result of applying Eqs. (8.12–8.14) is: For each pixel $(x, y)$ in $I_0$ we have $N$ cluster labels $\widetilde{S}_1(x, y), \widetilde{S}_2(x, y), \ldots, \widetilde{S}_N(x, y)$. We then fuse these labels together by choosing the label which appears most frequently. In this way we obtain an annotated image $\widetilde{S}_0$ for the unknown image $I_0$.

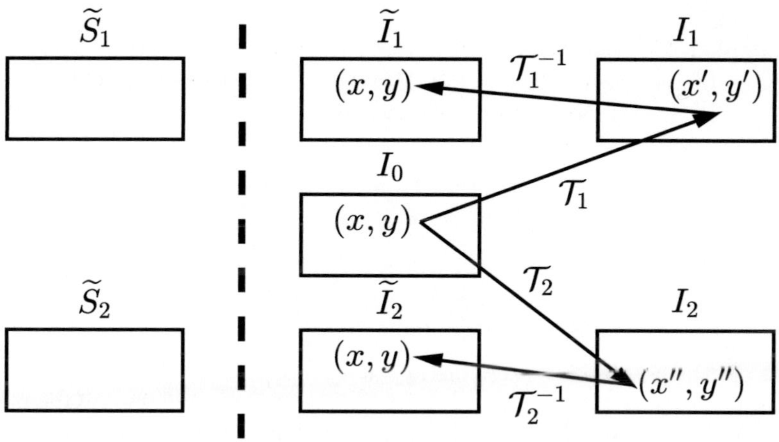


**Fig. 8.2** Shows MAS algorithm. Database contains images $I_1$, $I_2$ and the corresponding annotated images $S_1$, $S_2$ (not shown). Target image $I_0$ is separately registered with $I_1$ and $I_2$. The corresponding transformations $\mathcal{T}_1$, $\mathcal{T}_2$ map coordinates $(x, y)$ in $I_0$ to $(x', y')$, $(x'', y'')$. Apply inverse transformations $\mathcal{T}_1^{-1}$, $\mathcal{T}_2^{-1}$ to $I_1$, $I_2$ gives images $\widetilde{I}_1$, $\widetilde{I}_2$ which are spatially aligned with $I_0$. Similarly, apply $\mathcal{T}_1^{-1}$, $\mathcal{T}_2^{-1}$ to $S_1$, $S_2$. This gives spatially aligned annotated images $\widetilde{S}_1$, $\widetilde{S}_2$. Fuse $\widetilde{S}_1$ and $\widetilde{S}_2$ to generate annotated image $\widetilde{S}_0$ for the target image $I_0$

## Co-association Matrix

Co-association matrices are special similarity measures which enable us to fuse together multiple identity vectors $\{\mathbf{A}, \mathbf{B}, \ldots, \mathbf{D}\}$ *without* calculating an optimum assignment matrix. In this method, we convert each identity vector into a corresponding *co-association* matrix [10, 11] as follows:

$$\mathcal{A}(i, j) = \begin{cases} 1 \text{ if } a_i = a_j \ , \\ 0 \text{ otherwise} \ . \end{cases}$$

$$\mathcal{B}(i, j) = \begin{cases} 1 \text{ if } b_i = b_j \ , \\ 0 \text{ otherwise} \ . \end{cases}$$

$$\vdots$$

$$\mathcal{D}(i, j) = \begin{cases} 1 \text{ if } d_i = d_j \ , \\ 0 \text{ otherwise} \ . \end{cases}$$

The co-association matrices all use the same binary language and may therefore be fused together without further processing as shown in the following example.

**Example 8.12** *Mean co-association matrix* [30]. Let $\mathbf{X} = (x_1, x_2, \ldots, x_7)^T$ denote an input vector. We partition $\mathbf{X}$ using three different cluster algorithms $\{\mathfrak{A}, \mathfrak{B}, \mathfrak{C}\}$. The resulting identity vectors are:

$$\begin{aligned}
\mathbf{A} &= (\alpha_1\ \alpha_1\ \alpha_2\ \alpha_2\ \alpha_2\ \alpha_3\ \alpha_3)^T \\
\mathbf{B} &= (\beta_3\ \beta_3\ \beta_2\ \beta_2\ \beta_3\ \beta_3\ \beta_1)^T\ , \\
\mathbf{C} &= (\gamma_2\ \gamma_3\ \gamma_2\ \gamma_2\ \gamma_1\ \gamma_1\ \gamma_1)^T\ ,
\end{aligned}$$

and the corresponding co-association matrices are:

$$\mathcal{A} = \begin{pmatrix} 1\,1\,0\,0\,0\,0\,0 \\ 1\,1\,0\,0\,0\,0\,0 \\ 0\,0\,1\,1\,1\,0\,0 \\ 0\,0\,1\,1\,1\,0\,0 \\ 0\,0\,1\,1\,1\,0\,0 \\ 0\,0\,0\,0\,0\,1\,1 \\ 0\,0\,0\,0\,0\,1\,1 \end{pmatrix}, \quad \mathcal{B} = \begin{pmatrix} 1\,1\,0\,0\,1\,1\,0 \\ 1\,1\,0\,0\,1\,1\,0 \\ 0\,0\,1\,1\,0\,0\,0 \\ 0\,0\,1\,1\,0\,0\,0 \\ 1\,1\,0\,0\,1\,1\,0 \\ 1\,1\,0\,0\,1\,1\,0 \\ 0\,0\,0\,0\,0\,0\,1 \end{pmatrix}, \quad C = \begin{pmatrix} 1\,0\,1\,1\,0\,0\,0 \\ 0\,1\,0\,0\,0\,0\,0 \\ 1\,0\,1\,1\,0\,0\,0 \\ 1\,0\,1\,1\,0\,0\,0 \\ 0\,0\,0\,0\,1\,1\,1 \\ 0\,0\,0\,0\,1\,1\,1 \\ 0\,0\,0\,0\,1\,1\,1 \end{pmatrix}.$$

The co-association matrices all use the same binary language and may therefore be fused by summing them together element-by-element:

$$\overline{\mathcal{Y}} = \begin{pmatrix} 1 & \frac{2}{3} & \frac{1}{3} & \frac{1}{3} & \frac{1}{3} & \frac{1}{3} & 0 \\ \frac{2}{3} & 1 & 0 & 0 & \frac{1}{3} & \frac{1}{3} & 0 \\ \frac{1}{3} & 0 & 1 & 1 & \frac{1}{3} & 0 & 0 \\ \frac{1}{3} & 0 & 1 & 1 & \frac{1}{3} & 0 & 0 \\ \frac{1}{3} & \frac{1}{3} & \frac{1}{3} & \frac{1}{3} & 1 & \frac{2}{3} & \frac{1}{3} \\ \frac{1}{3} & \frac{1}{3} & 0 & 0 & \frac{2}{3} & 1 & \frac{2}{3} \\ 0 & 0 & 0 & 0 & \frac{1}{3} & \frac{2}{3} & 1 \end{pmatrix},$$

We may regard $\overline{\mathcal{Y}}$ as a "fuzzy" co-association, or affinity, matrix whose elements vary continuously between 0 and 1. Physically, the closer $\overline{\mathcal{Y}}(i, j)$ is to 1, the more the original decision maps put the elements $x_i$ and $x_j$ into the same cluster and the stronger the bond between the elements $x_i$ and $x_j$.

**Example 8.13** *Bagged co-association matrix* [6]. Tumor classification in DNA microarray experiments requires both the partition of the tumor samples into clusters and the confidence of the cluster assignments. For this purpose we may use a bagging procedure [6]. Given $N$ DNA observations $\mathbf{X} = \{x_1, x_2, \ldots, x_N\}$ we create $B$ bootstrap sets of observations: $\mathbf{X}^{(b)}$, $b \in$

$\{1, 2, \ldots, B\}$, by sampling $\mathbf{X}$ with replacement (see Sect. 15.4.2). Let $\mathbf{A}^{(b)} = \left(a_1^{(b)}, a_2^{(b)}, \ldots, a_N^{(b)}\right)^T$ denote the identity vector obtained as a result of clustering $\mathbf{X}^{(b)}$, where

$$a_i^{(b)} = \begin{cases} k \text{ if } x_i \in \mathbf{X}^{(b)} \text{ and } x_i \text{ has been placed in the } k\text{th partition} , \\ 0 \text{ if } x_i \notin \mathbf{X}^{(b)} . \end{cases}$$

If $\mathcal{A}^{(b)}$ is the corresponding co-association matrix, then

$$\mathcal{A}^{(b)}(i, j) = \begin{cases} 1 \text{ if } a_i^{(b)} = a_j^{(b)} \text{ and } a_i^{(b)} > 0 , \\ 0 \text{ otherwise} , \end{cases}$$

and the "bagged" co-association matrix is $\overline{\mathcal{Y}}$, where

$$\overline{\mathcal{Y}}(i, j) = \frac{1}{M(i, j)}\left(\mathcal{A}^{(1)}(i, j) + \mathcal{A}^{(2)}(i, j) + \ldots, \mathcal{A}^{(B)}(i, j)\right) ,$$

and $M(i, j)$ is # of bootstrap samples for which both $x_i \in \mathbf{X}^{(b)}$ and $x_j \in \mathbf{X}^{(b)}$.

By viewing the fused co-association matrix $\widetilde{\mathcal{Y}}$ as a special similarity matrix $\mathcal{S}$, we may use the NJW algorithm to generate the corresponding identity vector, $\widetilde{\mathbf{y}}$ [20].

## 8.8 Software

**MATLAB**. MATLAB IMAGE PROCESSING and STATISTICS AND MACHINE LEARNING toolboxes include m-files on clustering (K-means, K-mediods, spectral). An additional toolbox for cluster ensemble is: CLUSTER ENSEMBLE (Author: Sapio). A m-file for histogram segmentation [4] is: HistogramSegmentation (Author: Delon).

**PYTHON**. SCIKIT-LEARN includes python modules for clustering (K-means, spectral). OPENENSEMBLE is a python package for cluster ensembles.

## 8.9 Further Reading

A comprehensive survey of clustering algorithms in general is given in [7]. Surveys which deal only with $k$-means clustering and spectral clustering are given in [12, 27].

## References


1. Achanta R, Shaji A, Smith K et al (2012) SLIC superpixels compared to state-of-the-art superpixel methods. IEEE Trans Patt Analy Mach Intell 34:2274–2282
2. Belongie S, Malik J, Puzicha J (2002) Shape matching and object recognition using shape contexts. IEEE Trans Patt Anal Mach Intell 24:509–522
3. Brand M, Huang K (2002) A unifying theorem for spectral embedding and clustering. Ninth Int Conf Art Intell Stat
4. Delon J, Desolneux A, Lisani J-L et al (2007) A nonparametric approach for histogram segmentation. IEEE Trans Img Proc 16:253–261
5. Dubois E (2010) The structure and properties of color spaces and the representation of color images. Published by Springer
6. Dudoit S, Fridlyand J (2003) Bagging to improve the accuracy of a clustering procedure. Bioinformatics 19:1090–1099
7. Ezugwu AE, Ikotun AM, Oyelade OO, et al (2022) A comprehensive survey of clustering algorithms: state-of-the-art machine learning applications, taxonomy, challenges and future research prospects. Engng App Artif Intell 11:104743
8. Fowlkes C, Belongie S, Chung F et al (2004) Spectral grouping using the Nystrom method. IEEE Trans Patt Anal Mach Intell 26:214–225
9. Franek L, Abdala DD, Vega-Pons S, Jiang X (2011) Image segmentation fusion using general ensemble clustering methods. LNCS 6495:373–384
10. Fred ALN (2001) Finding consistent clusters in data partitions. LNCS 2096:309–318
11. Fred ALN, Jain AK (2005) Combining multiple clusterings using evidence accumulation. IEEE Trans Patt Anal Mach Intell 27:835–850
12. Ikotun AM, Ezugwu AE, Abualigah L et al (2023) K-means clustering algorithms: a comprehensive review, variants analysis and advances in the era of big data. Inform Sci 622:178–210
13. Jain AK (2010) Data clustering: 50 years beyond $K$-means. Patt Recogn Lett 31:651–666
14. Jia J, Liu B, Jiao L (2004) Soft spectral clustering ensemble applied to image segmentation. Front Comp Sci China 5:66–78
15. Khelifi L, Mignotte M (2017) EFA-BMFM: a multi-criteria framework for the fusion of colour image segmentation. Inf Fusion 38:194–121
16. Kuhn HW (2005) The Hungarian method for the assignment problem. Nav Res Logist 52:7–21
17. Lisani J-L, Petro AB (2021) Automatic 1D histogram segmentation and application to the computation of color palettes. Img Proc On Line (IPOL) 11:76–104
18. Mignotte M (2008) Segmentation by fusion of histogram-based $K$-means clusters in different color spaces. IEEE Trans Image Proc 17:780–787
19. Nivetha S, Inbarani HH (2023) Automated lesion image segmentation based on novel histogram-based K-means clustering using Covid-19 chest CT images. LNNS 613:763–776
20. Ng A, Jordan M, Weiss Y (2001) On spectral clustering: analysis and an algorithm. Adv Neural Inform Proc Sys 14:849–856
21. Olukanmi P, Twala B (2017) K-means-sharp: modified centroid update for outlier-robust k-means clustering. Int Conf Patt Recogn Assoc of S. Africa and Robotics Mechtronics, pp 14–19
22. Olukanmi P, Nelwamondo F, Marwala T et al (2022) Automatic detection of outliers and the number of clusters in $k$-means clustering via Chebyshev-type inequalities. Neural Comput Applicat 34:5939–5958
23. Sanroma G, Wu G, Gao Y et al (2014) Learning to rank atlases for multiple-atlas segmentation. IEEE Trans Med Imag 33:1939–1953
24. Sasmal B, Dhal KG (2023) A survey on the utilization of superpixel image for clustering based image segmentation. Multimedia Tools App 82:35493–35555
25. Scott C, Nowak R (2006) Robust contour matching via the order preserving assignment problem. IEEE Trans Image Process 15:1831–1838
26. Stutz D, Hermans A, Liebe B (2018) Superpixels: an evaluation of the state-of-the-art. Comp Vis Image Under 166:1–27

27. von Luxburg U (2007) A tutorial on spectral clustering. Stat Comp 17:395–416
28. Verelst T, Blaschko M, Berman M (2019) Generating superpixels using deep image representations. arXiv:1903.04586
29. Wang Z, Gao C, Tian J, et al (2005) Multi-feature distance map based feature detection of small infra-red targets with small contrast in image sequences. Proc SPIE 5985
30. Wang X, Yang C, You J (2008) Spectral aggregation for clustering ensemble. In: Proceedings of international conference on pattern recognition, pp 1–4
31. Zaffino P, Ciardo D, Raudaschi P, et al (2018) Multi atlas based segmentation: should we prefer the best atlas group over the group of best atlases? Phys Med Biol 63(12)
32. Zelnik-Manor L, Perona P (2005) Self-tuning spectral clustering. Adv Neural Inform Proc Syst 17:1601–1608

# Chapter 9
# Radiometric Normalization

**Abstract** In this chapter we consider radiometric normalization. This is defined as the conversion of all sensor values to a common scale and is the fourth function required for the formation of a common representational format. In this chapter we consider different methods of normalization including whitening, histogram matching, binarization, fuzzy logic, ranking and isotonic regression. We also include methods for similarity matching including Kullback-Leibler, Earth Movers Distance and Weighted Distribution Matching.

## 9.1 Introduction

The subject of this chapter is radiometric calibration, or normalization, which we define as the conversion of all sensor values to a common scale. This is the fourth and last function listed in Sect. 4.1 which is required for the formation of a common representational format. Although conceptually the radiometric normalization and semantic alignment are very different, we often use the same probabilistic transformation for both semantic alignment and radiometric normalization. The reader should be careful not to confuse the two terms.

In many data fusion applications radiometric normalization is the primary fusion algorithm. In Table 9.1 we list some of these applications together with their classification.

## 9.2 Scales of Measurement

Physically, the scale of measurement of a variable describes how much information the values associated with the variable contains. Different mathematical operations on variables are possible, depending on the scale at which a variable is measured.

Four scales of measurement are usually recognized:

**Nominal scale**. The variable values are names or labels. The only operations that can be meaningfully applied to the variable values are "equality" and "inequality".



H. B. Mitchell, *Data Fusion*, https://doi.org/10.1007/978-3-662-71023-4_9

**Table 9.1** Applications in which radiometric normalization is primary fusion algorithm

| Class | | Application |
|---|---|---|
| Early-DaO | Example 2.2 | Approximate Lesion Localization in Demoscopy Images |
| | Example 7.1 | Cardiac MR Image Sequences |
| | Example 9.14 | Change Detection Thresholding using Mutual Information |
| Early-FeO | Example 9.1 | Similarity Measures for Virtual Screening |
| Inter-FeO | Example 9.11 | Histogram Equalization for Robust Speech Recognition |
| | Example 9.12 | Midway Histogram Equalization |
| | Example 9.19 | Fusing Similarity Coefficients in a Virtual Screening Programme |
| Inter-DeO | Example 9.4 | $\chi^2$-distance and Probability Binning |

**Ordinal scale**. The variable values have all the features of a nominal scale and are numerical. The values represent the rank order (first, second, third, etc.) of the variables. Comparisons of "greater" and "less" can be made, in addition to "equality" and "inequality".

**Interval Scale**. The variable values have all the features of an ordinal scale and, additionally, are separated by the same interval. In this case, differences between arbitrary pairs of values can be meaningfully compared. Operations such as "addition" and "subtraction" are therefore meaningful. Additionally, negative values on the scale can be used.

**Ratio Scale**. The variable values have all the features of an interval scale and can also have meaningful ratios between arbitrary pairs of numbers. Operations such as "multiplication" and "division" are therefore meaningful. The zero value on a ratio scale is non-arbitrary. Most physical quantities, such as mass, length, or energy are measured on ratio scales.

In Table 9.2 we list the mathematical properties and types of statistics that can be applied to the different scales of measurement variables.

**Table 9.2** Scales of measurement

| Name | Properties | Mathematical operations | Descriptive statistics | Inferential statistics |
|---|---|---|---|---|
| Nominal | Identity | Count | Mode | Non-parametric, $\chi^2$ |
| Ordinal | Identity, magnitude | Rank order | Mode, median, range statistics | Non-parametric, $\chi^2$, Mann-Whiteney, Kruskal-Wallis, ANOVA |
| Interval | Identity, magnitude, equal interval | Add, subtract | Mode, median, mean, range statistics, variance | Non-parametric, parametric, $t$-test, ANOVA |
| Ratio | Identity, magnitude, equal interval, true zero | Add, subtract, multiply, divide | Mode, median, mean, range statistics, variance | Non-parametric, parametric, $t$-test, ANOVA |

## 9.3 Degree-of-Similarity Scales

Degree-of-similarity is a popular interval scale which is widely used for radiometric normalization. In fact most of the data fusion applications listed in Table 9.1 use a degree-of-similarity scale. The following example illustrates the use of different degree-of-similarity scales in virtual screening.

**Example 9.1** *Similarity Measures for Virtual Screening.* [6] "Virtual screening" uses a computational scoring scheme to rank molecules in decreasing order of probability of activity. It has been used for a long time in drug discovery where its popularity is due to its ease of use, speed and performance.

One approach to virtual screening is "similarity searching". This involves taking a molecule (the "target compound") with the required activity and then searching a database to find molecules that are structurally most similar to it.

A molecular similarity measure has two principal components: (1) Structural representation used to characterize the molecules; (2) Similarity coefficient used to compute degree of resemblance between pairs of such representations. Most systems for similarity-based virtual screening use molecular "fingerprints" which are binary vectors encoding the presence, or absence, of sub-structural fragments within the molecule. Table 9.3 lists several similarity coefficients for the comparison of pairs of molecular fingerprints.

In recent years, CNN's have also been used for virtual screening [21, 34] as described in the following example.

**Example 9.2** *Convolutional neural networks (CNN) for virtual screening* [21]. In Table 9.4 we list a simple 1D CNN for virtual screening. Input to the CNN is a molecular fingerprint of length 1024 bits. This represents a given

**Table 9.3** Similarity coefficients for virtual screening programme

| Name | Formula | Range |
|---|---|---|
| Jaccard/Tanimoto | $c/(a+b+c)$ | [0, 1] |
| Dice | $2c/(a+b)$ | [0, 1] |
| Euclidean distance | $\sqrt{a+b-2c}$ | $[0, \sqrt{N}]$ |
| City Block/Manhattan/Hamming distance | $a+b-2c$ | $[0, N]$ |
| Cosine similarity | $c/\sqrt{ab}$ | [0, 1] |

$N$ is total # of attributes which are present in molecules $A$ and $B$; $a$ is # attributes which are present in $A$, $b$ is # of attributes which are present in $B$, $c$ is # of attributes which are present in both $A$ and $B$

**Table 9.4** 1D-CNN for virtual screening

| Layer | Input | Output | Type | Kernel | Remarks |
|---|---|---|---|---|---|
| 1 | $1024 \times 1 \times 1$ | $1024 \times 1 \times 128$ | CNV | $3 \times 1(128)$ | Includes zero-pad + ReLU |
| 2 | $1024 \times 1 \times 128$ | $512 \times 1 \times 128$ | MP | $2 \times 1$ | |
| 3 | $512 \times 1 \times 128$ | $512 \times 1 \times 64$ | CNV | $3 \times 1(64)$ | Includes zero-pad + ReLU |
| 4 | $512 \times 1 \times 64$ | $256 \times 1 \times 64$ | MP | $2 \times 1$ | |
| 5 | $256 \times 1 \times 64$ | $256 \times 1 \times 32$ | CNV | $3 \times 1(32)$ | Includes zero-pad + ReLU |
| 6 | $256 \times 1 \times 32$ | $128 \times 1 \times 32$ | MP | $2 \times 1$ | |
| 7 | $128 \times 1 \times 32$ | $128 \times 1 \times 16$ | CNV | $3 \times 1(16)$ | Includes zero-pad + ReLU |
| 8 | $128 \times 1 \times 16$ | $64 \times 1 \times 16$ | MP | $2 \times 1$ | |
| 9 | $512 \times 1$ | $512 \times 1$ | DP | – | Dropout 50% of weights |
| 10 | $64 \times 1 \times 16$ | $1024 \times 1$ | FL | – | Flattening |
| 11 | $1024 \times 1$ | $512 \times 1$ | FC | – | Includes ReLU |
| 12 | $512 \times 1$ | $256 \times 1$ | FC | – | Includes ReLU |
| 13 | $256 \times 1$ | $256 \times 1$ | DP | – | Dropout 50% of weights |
| 14 | $256 \times 1$ | $1 \times 1$ | FC | – | With Sigmoid activation |

drug compound. The CNN has 4 convolutional layers and 3 fully-connected layers. Output from the last layer is a single neuron with sigmoid activation which classifies the input drug as bioactive or not.

## 9.4 Features

In many applications the object of interest is described by a vector of $M$ feature values. In this case, we may measure the degree of similarity between two objects by measuring the corresponding *distance* between their feature vectors. Note that in this case, the degree-of-similarity between two objects is high if the corresponding distance between the feature vectors is small and vice versa.

Let $\mathbf{A} = (a_1, a_2, \ldots, a_M)^T$ and $\mathbf{B} = (b_1, b_2, \ldots, b_M)^T$ denote the feature vectors of two objects. Then, in Table 9.5 we list several distance measures which we may use to compare the feature vectors.

**Table 9.5** Feature vector distance functions

| Name | Formula |
|---|---|
| Euclidean distance | $\sqrt{\sum_{m=1}^{M} \lvert a_m - b_m\rvert^2}$ |
| City-block distance | $\sum_{m=1}^{M} \lvert a_m - b_m\rvert$ |
| Chebyshev distance | $\max_m(\lvert a_m - b_m\rvert)$ |
| Minkowski distance | $\left[\sum_{m=1}^{M} \lvert a_m - b_m\rvert^p\right]^{1/p}$ |

### Deep Learning Features

Deep learning features are increasingly being used in data fusion applications. In these applications the features are extracted from one or more layers of a CNN (see Sect. 5.6). By definition, in each CNN layer, the features are minimally normalized using a ReLU, or similar, activation function. Furthermore, CNN features often undergo additional normalization when the CNN has a batch normalization layer (cf. Sect. 5.4). It is for this reason, that we may fuse CNN features, or compute the degree of similarity between CNN features, without (preliminary) radiometric normalization.

### Histograms

In some applications the feature vector is in fact a histogram of feature values. In this case, it is important to choose an appropriate set of histogram bins. The following method is a simple and effective method for optimally choosing the number of histogram bins in a *regular* histogram (see also Table 6.5).

**Example 9.3** *Optimum Number of Bins in a Regular Histogram: Birge and Rozenhol* [3]. Let $x_n, n \in \{1, 2, \ldots, N\}$, denotes a set of $N$ measurements taken from an unknown density $f$ on $[0, 1]$. Then Birge and Rozenholc [3] find the optimum number of histogram bins, $K^*$, by optimizing a penalized likelihood: Given $K$ histogram bins, let $H_k, k \in \{1, 2, \ldots, K\}$, denote the number of measurements which fall in the $k$th bin. Then, the optimum number of histogram bins is

$$K^* = \max_K \big(L_K - P_K\big) ,$$

where

$$L_K = \sum_k H_k \log_2 \left( \frac{K \times H_k}{N} \right) \quad \text{and} \quad P_K = D - 1 + \big(\log_2 K\big)^{2.5} .$$

If the bins are allowed to vary in size, then a popular rule-of-thumb is to choose bins which have at least 5 or 10 observations.

Given two histograms $F$ and $G$ we may measure their similarity using the $\chi^2$-distance.

**Example 9.4** *$\chi^2$-distance and Probability Binning* [24]. Suppose $f$ represents $M$ observations $x_1, x_2, \ldots, x_M$ and $g$ represents $N$ observations $y_1, y_2, \ldots, y_N$. We divide $f$ and $g$ into $K$ non-overlapping bins $B_k = [b'_k, b''_k)$,

$k \in \{1, 2, \ldots, K\}$, where $b'_{k+1} = b''_k$. If $m_k$ and $n_k$ represent, respectively, the number of observations which fall in the $k$th bin, then the corresponding histograms are

$$F = (m_1, m_2, \ldots, m_K)^T \quad \text{and} \quad G = (n_1, n_2, \ldots, n_K)^T \; .$$

We then compare $F$ and $G$ using the $\chi^2$ distance:

$$\chi^2 = \sum_k \frac{|m_k - n_k|^2}{m_k + n_k} \; .$$

In probability binning [24] we use a modified $\chi^2$ distance as follows. We use *variable* width bins $B_k$ such that each bin contains the same number of observations $m_k$. Then we convert the histograms $F$ and $G$ into probability distributions by dividing, respectively, through by $M$ and $N$. Let

$$\mathcal{F} = (\widetilde{m}_1, \widetilde{m}_2, \ldots, \widetilde{m}_K)^T \quad \text{and} \quad \mathcal{G} = (\widetilde{n}_1, \widetilde{n}_2, \ldots, \widetilde{n}_K)^T \; ,$$

where $\widetilde{m}_k = m_k/M$ and $\widetilde{n}_k = n_k/N$. Then the modified $\chi^2$ measure used for comparing $\mathcal{F}$ and $\mathcal{G}$ is

$$\widetilde{\chi}^2 = \sum_k \frac{|\widetilde{m}_k - \widetilde{n}_k|^2}{\widetilde{m}_k + \widetilde{n}_k} \; .$$

where $0 \leq \widetilde{\chi}^2 \leq 2$. By using $\widetilde{\chi}^2$ we can define a metric for $\widetilde{\chi}^2$:

$$T(\widetilde{\chi}^2) = \max\left(0, \frac{\widetilde{\chi}^2 - \mu}{\sigma}\right) \; ,$$

where $T(\widetilde{\chi}^2)$ represents the difference between $\mathcal{F}$ and $\mathcal{G}$ as the number of standard deviations above $\mu$, where $\mu = K/\min(M, N)$ is the minimum statistical significant value of $\widetilde{\chi}^2$ (i.e. the minimum value for which a confident decision of histogram difference can be made) and $\sigma = \sqrt{K}/\min(M, N)$ is an appropriate standard deviation for $\widetilde{\chi}^2$. Baggerly [1] suggests a more accurate metric is: $T(\widetilde{\chi}^2) = \frac{1}{\sqrt{2(K-1)}}\left(2\widetilde{\chi}^2\left(\frac{MN}{M+N}\right) - (K-1)\right)$.

Additional histogram similarity measures are listed in Table 9.6 [7]. The histogram similarity measures may be divided into two classes: bin-to-bin distances and cross-bin distances. The bin-to-bin distances compare corresponding bins in the two histograms. They thus ignore all correlations between neighbouring bins. As a consequence, these measures may be sensitive to changes in the positions and sizes of the histogram bins. In contrast, the cross-bin distances overcome these shortcomings,

**Table 9.6** Signature and histogram similarity measures

| Name | Formula |
|---|---|
| $L_p$ distance | $\left(\sum_{k=1}^{K} \|m_k - n_k\|^p\right)^{1/p}$ |
| $\chi^2$ distance | $\chi^2 = \sum_{k=1}^{K} \|m_k - n_k\|^2/(m_k + n_k)$ |
| Kullback-Leibler | $\sum_{k=1}^{K} \widetilde{m}_k \log(\widetilde{m}_k/\widetilde{n}_k)$; $\widetilde{m}_k = m_k/M, \widetilde{n}_k = n_k/N, M = \sum m_k, N = \sum n_k$ |
| Jeffrey divergence | $\sum_{k=1}^{K}\big(m_k \log(m_k/n_k) + n_k \log(n_k/m_k)\big)$ |
| EMD | See Example 9.5 (Earth Mover's Distance) |
| VSBD | See Example 9.6 (Variable-Size Bin Distance) |
| WBD | See Example 9.7 (Weighted Distribution Matching) |

by comparing both corresponding bins and non-corresponding bins in the two histograms. Among the different cross-bin measures, the Earth mover's distance (EMD) is recognized as a powerful and robust distance measure.

The following example explains the working of EMD.

**Example 9.5** *Earth Mover's Distance (EMD)* [25]. EMD defines the distance between two histograms $F = \{(m_1, u_1), (m_2, u_2), \ldots, (m_K, u_K)\}$ and $G = \{(n_1, v_1), (n_2, v_2), \ldots, (n_L, v_L)\}$ as the minimum "work" required to transform $F$ into $G$, where $m_k$ is the number of elements in $F$ which fall in the $k$th bin whose position is $u_k$ and $n_l$ is the number of elements in $G$ which fall in the $l$th bin whose position is $v_l$.

EMD treats the elements in $F$ as "supplies" located at $u_k$ and treats the elements in $G$ as "demands" located at $v_l$. The work expended in moving one element from $u_k$ to $v_l$ is defined as the ground distance $d_{kl}$ between $u_k$ and $v_l$. The EMD is then given by

$$EMD = \min_{F=\{f_{kl}\}} \left( \frac{\sum_k \sum_l f_{kl} d_{kl}}{\sum_k \sum_l f_{kl}} \right) ,$$

subject to

$$f_{kl} \geq 0 , \quad \sum_l f_{kl} \leq m_k , \quad \sum_k f_{kl} \leq n_l , \quad \sum_k \sum_l f_{kl} = \min\left(\sum_k m_k, \sum_l n_l\right) ,$$

where the flow $f_{kl}$ is number of elements moved from the $k$th supply at $u_k$ to the $l$th demand at $v_l$.

For 1D vectors, the formula for the earth mover's distance reduces to

$$EMD = \sum(|\mathrm{cdf}(F) - \mathrm{cdf}(G)|) ,$$

where cdf is the cumulative distribution function.

The main disadvantage of the EMD is its high computational complexity: $O(N^3 \log_2(N))$. For this reason, there is continuing interest in the development of bin-to-bin similarity measures with better similarity performance. Two examples are the Variable-Size Bin Distance and the Weighted Distribution Matching.

**Example 9.6** *Variable-Size Bin Distance (VSBD)* [19]. The VSBD is calculated iteratively as follows. First we calculate a bin-to-bin distance $d_1$ for the two histograms $F$ and $G$ using a refined histogram bin. Then the intersection $I = |F \wedge G|$ of the two histograms is subtracted from each histogram:

$$F' = F - I \quad \text{and} \quad G' = G - I \ .$$

The residual histograms $F'$ and $G'$ are recalculated using a coarser bin size and the corresponding bin-to-bin distance $d_2$ is calculated for the new coarser histograms. The entire process is repeated for several iterations. At each iteration $k$ we obtain a bin-to-bin distance $d_k$. The VSBD is then given by:

$$d_{\text{VSBD}} = \sum_k d_k \ .$$

The Weighted Distribution Matching inherits the low computational cost of a bin-to-bin similarity measure but also takes into account the relationship between consecutive bins from the same histogram.

**Example 9.7** *Weighted Distribution Matching* (WBD) [10]. Let $\mathbf{F} = \{F_1, F_2, \ldots, F_N\}$ and $\mathbf{G} = \{G_1, G_2, \ldots, G_N\}$ denote two histograms, where $F_i$ and $G_i$ are, respectively, the number of observations $\{x_1, x_2, \ldots\}$ and $\{y_1, y_2, \ldots\}$ which fall in the $i$th bin $B_i = [b'_i, b''_i]$. Then we calculate the WDM between $F$ and $G$ as follows. Let

$$\delta F_i = \begin{cases} 1 & \text{if } F_i > F_{i-1} \ , \\ -1 & \text{if } F_i < F_{i-1} \ , \\ 0 & \text{if } F_i = F_{i-1} \ , \end{cases} \quad \text{and} \quad \delta G_i = \begin{cases} 1 & \text{if } G_i > G_{i-1} \ , \\ -1 & \text{if } G_i < G_{i-1} \ , \\ 0 & \text{if } G_i = G_{i-1} \ . \end{cases}$$

Then WDM similarity match is defined as

$$WDM = \frac{1}{N} \sum_i \alpha_i |F_i - G_i|$$

where $w \in [0, 1]$ and $\alpha_i = w$ if $\delta F_i = \delta G_i$; otherwise $\alpha_i = (w - 1)$.

Although the histogram similarity measures can also be used to measure the similarity between two pdf's, statistical measures based on mutual information or the Kullback-Leibler (KL) distance, $d_{KL}$, are more common. Given two pdf's $f$ and $g$, the Kullback-Leibler (KL) distance is defined as:

$$d_{KL}(f|g) = \int f(x)\log\bigl(f(x)/g(y)\bigr)dx \ ,$$

(see Sect. 10.5).

### Hybrid Features*

In some applications, we use a "hybrid" representation or features, which requires a special purpose degree-of-similarity scale. The following example illustrates a hybrid representation and its similarity scale.

**Example 9.8** *Hue-Saturation Similarity Scale* [20]. For illumination-tolerant image processing we often use only hue and saturation images. Given a RGB color image, the hue and saturation images are given by:

$$H = \tan^{-1}\left(\frac{\sqrt{3}(G-B)}{2R-G-B}\right) \quad \text{and} \quad S = 1 - \frac{3}{R+G+B}\min(R,G,B)$$

The Hue-Saturation similarity (distance) measure between two pixels with hue and saturation values $(H_1, S_1)$ and $(H_2, S_2)$ is defined as:

$$d\bigl((H_1,S_1),(H_2,S_2)\bigr) = \frac{\sqrt{(H_1-H_2)^2+(S_1-S_2)^2}}{\sqrt{H_1^2+S_1^2}+\sqrt{H_2^2+S_2^2}} \ .$$

## 9.5 Radiometric Normalization

The process of converting the values of different pieces of information to a common scale is known as radiometric normalization. Suppose we use a set of logical sensors, or sources-of-information, $S_i, i \in \{1, 2, \ldots, N\}$, to make $N$ measurements $x_i$ on an object $O$. Then, we transform the $x_i$ into a common representational format by changing the statistical parameters of the $x_i$.[1] Mathematically we write the transformation of an input measurement $x$ to a normalized value $y$ as

[1] The most important statistical parameters are the scale and location parameters.

$$y = f(x|\alpha, \beta, \gamma, \ldots, \delta) \;,$$

where $f$ is a parametric *transformation function*, $\alpha$ and $\beta$ are the scale and location parameters and $\gamma, \delta$, etc. are the higher order statistical parameters.

In most cases the parameters $\alpha, \beta, \ldots, \delta$ are learnt from measurements made on a set of *training samples* $\mathcal{D} = \{x_1, x_2, \ldots, x_N\}$. In this case the procedure is referred to as *fixed sensor value normalization*. In a few cases, the transformation is estimated from the current measurement values. This is known as *adaptive sensor value normalization* and its main advantage is its ability to adapt to variations in the input data.

**Example 9.9** *Whitening: Adaptive Normalization*. In whitening we transform input measurements $x_i, i \in \{1, 2, \ldots, N\}$, into normalized values $y_i$:

$$y_i = \frac{x_i - \bar{x}}{\sigma}$$

where $\bar{x} = \frac{1}{N}\sum_{i=1}^{N} x_i$ and $\sigma^2 = \frac{1}{N-1}\sum_{i=1}^{N}(x_i - \bar{x})^2$. The $y_i$ are then distributed according to a Gaussian distribution $\mathcal{N}(\bar{x}, \sigma^2)$.

For a good normalization scheme, the parameters $\alpha, \beta, \ldots, \delta$ should be *robust*, where *robustness* refers to insensitivity in the presence of outliers (cf. Chap. 12). The next example describes a robust whitening procedure.

**Example 9.10** *Robust Whitening*. The whitening scheme in Example 9.9 is not robust against outliers since it uses the mean value $\bar{x}$ and the standard deviation $\sigma$. We may robustify it by replacing $\bar{x}$ with the median $m$ and replacing $\sigma$ with the median absolute deviation (MAD) $S$ (see Sect. 12.2). In this case, the robust whitened values are

$$y_i = \frac{x_i - m}{S} \;,$$

where $m = \text{median}(x_1, x_2, \ldots, x_N)$ and $S = 1.4826 \times \text{median}(|x_1 - m|, |x_2 - m|, \ldots, |x_N - m|)$.

Some common parametric transfer functions are listed in Table 9.7.[2]

[2] Only the trimmed min-max, robust z-transform and robust tanh normalization functions are robust against outliers (see Chap. 12). The remaining functions are not robust against outliers and should not be used if outliers are likely to be present.

**Table 9.7** Parametric normalization functions

| Function | Formula |
|---|---|
| Min-Max | $y = (x - a)/(b - a)$, where $a = \min_i(x_i)$, $b = \max_i(x_i)$. In trimmed min-max we replace $a$ and $b$ by the $l$th smallest and largest $x_i$ values. Normalized result is $y = \min(\max(0, (a - x)/(b - a)), 1)$. *Note*: In min-max, $y_i$ retains the same distribution as input $x_i$ |
| Z-Transform | $y = (x - \mu)/\sigma$, where $\mu = \sum_i x_i/N, \sigma^2 = \sum_i (x_i - \mu)^2/(N - 1)$. In robust Z-transform we replace $\mu$ by median$\{x_i\}$ and $\sigma$ by median absolute deviation $S$ or by interquartile distance $Q = (x_{(3N/4)} - x_{(N/4)})/2$, where $x_{(l)}$ is $l$th largest value in $\{x_i\}$ |
| Robust Tanh | $y = \frac{1}{2}\tanh(\alpha(x - \mu_H)/\sigma_H + 1)$, where $\alpha$ determines the spread of the normalized scores and $\mu_H$ and $\sigma_H$ are robust Hampel mean and standard deviation estimates of the $x_i$, $i \in \{1, 2, \ldots, N\}$ [17] (see Sect. 12.2) |

The parameters $\alpha, \beta, \ldots, \delta$ should also be *efficient*, where *efficiency* refers to the proximity of the obtained estimate to the optimal estimate when the distribution of the data is known.

The following example illustrates the normalization of the input data $x_i$ in which we match the pdf of the $x_i$ to a reference pdf.

**Example 9.11** *Histogram Equalization for Robust Speech Recognition* [28]. The performance of automatic speech recognition (ASR) systems degrades significantly if there is an acoustic mismatch between the training and test data. One method for compensating for the acoustic mismatch is histogram equalization (HEQ). *Note:* Another widely used method of reducing the mismatch is dynamic time warping (DTW) (see Chap. 7).

The goal of HEQ is to transform the speech features in such a way that the acoustic environment does not affect its probability distribution. This is achieved by transforming the distribution of each test feature into the corresponding training set, or reference, distribution. Suppose $x$ denotes a given test feature whose pdf and cumulative distribution function (cdf) are, respectively, $p_1(x)$ and $c_1(x)$, where $c_1(x) = \int_{-\infty}^{x} p_1(x)dx$. We seek a function $y = F(x)$ which maps $p_1(x)$ into the corresponding training pdf, $p_2(y)$. This is obtained by equating $c_1(x)$ and $c_2(y)$, where $c_2(y) = \int_{-\infty}^{y} p_2(y)dy$:

$$c_1(x) = c_2(y) = c_2(F(x)) ,$$

or

$$y = F(x) = c_2^{-1}(c_1(x)) ,$$

where $c_2^{-1}$ denotes the inverse of $c_2$ and is defined as follows: If $y = c_2(x)$, then $c_2^{-1}(y) = x$.

A critical issue in HEQ is the reliable estimation of the cdf's $c_1(x)$ and $c_2(y)$. In speech recognition applications, the cdf $c_2(y)$ can be estimated quite reliably by computing cumulative histograms with a large amount of training data. However, when short utterances are used as test data, the length of each utterance may be insufficient for a reliable estimation. In this case, we may use an order-statistic estimate for the test cdf, $c_1(x)$: Given a test sequence consisting of $K$ examples $\{x_1, x_2, \ldots, x_K\}$ of a given feature, the order-statistic estimate of the cdf $c_1(x)$ is:

$$\hat{c}_1(x_k) \approx \frac{r_k - \frac{1}{2}}{K} ,$$

where $r_k$ denotes the rank of $x_k$ and $r_k = j$ if $x_k$ is the $j$th largest input data.

The following example illustrates the normalization of two input vectors $\mathbf{x}$ and $\mathbf{y}$ by simultaneously matching their pdf's.

**Example 9.12** *Midway Histogram Equalization* (MHE) [11]. MHE is defined as any method which warps two input histograms $p_1(x)$ and $p_2(y)$ to a common "intermediate" histogram $p_3(z)$, such that $p_3(z)$ retains as much as possible of the shapes of $p_1(x)$ and $p_2(y)$. Mathematically, MHE may be implemented as follows: Given the two cumulative distribution functions (cdf) $c_1(x)$ and $c_2(y)$ we define the inverse cdf of the intermediate histogram $p_3(z)$ by:

$$c_3^{-1}(z) = \frac{c_1^{-1}(z) + c_2^{-1}(z)}{2}$$

We now warp $p_1(x)$ by matching it to $p_3(z)$. Let $c_3(z) = \int_0^z p_3(z)dz$ denote the cdf of $p_3(z)$, then the warped distribution is $p_1'(x')$, where $p_1'(x') = p_1(x)$ and

$$x' = c_3^{-1}(c_1(x)) .$$

Similarly we warp $p_2(y)$ by matching it to $p_3(z)$. The corresponding warped distribution is $p_2'(y')$, where $p_2'(y') = p_2(y)$ and

$$y' = c_3^{-1}(c_2(y)) .$$

## 9.6 Binarization

*Binarization* is probably the simplest radiometric normalization technique. It is robust and is widely used when multiple sensors $S_i, i \in \{1, 2, \ldots, \}$, each makes several measurements $\mathbf{x}_i = \left(x_i^{(1)}, x_i^{(2)}, \ldots, x_i^{(K)}\right)$, on the same object $O$: We threshold each measurement $x_i^{(k)}$, using a local threshold $T_i$, or a global threshold $T_G$. The corresponding normalized values are $y_i^{(k)}$, where

$$y_i^{(k)} = \begin{cases} 1 \text{ if } x_i^{(k)} \geq T_i \ , \\ 0 \text{ otherwise} \ . \end{cases} \tag{9.1}$$

The thresholds $T_i$, $i \in \{1, 2, \ldots, N\}$, may be established *a priori* on physical grounds or may be learnt on a training set $\mathcal{D}$ ("supervised" learning) or learnt on the current set of measurements ("unsupervised" learning).

## Otsu Threshold

The Otsu threshold [26] is the workhorse of unsupervised thresholding algorithms. It works as follows.

Assume the sensor $S$ is a camera generating a $M \times N$ gray-image image $I$ with $L$ gray levels $[0, 1, \ldots, (L-1)]$. The Otsu method finds a threshold $T \in \{0, 1, \ldots, (L-1)\}$ to segment $I$ into two classes $C_0$ and $C_1$. Let $(\mu_0, \sigma_0^2)$ and $(\mu_1, \sigma_1^2)$ denote, respectively, the mean and variance of $C_0$ and $C_1$. Let $p_0$ and $p_1$ be the probability of classes $C_0$ and $C_1$, respectively. Then the Otsu method minimizes the intra-class variance by searching through the histogram of $I$ to find the optimum threshold $T^*$:

$$T^* = \arg\min_T \left( p_0(T)\sigma_0^2(T) + p_1(t)\sigma_1^2(T) \right) , \tag{9.2}$$

When $T^*$ is found, the image is segmented into two classes, $C_0$ and $C_1$; and their sizes are $p_0 MN$ and $p_1 MN$, respectively. Mathematically, Eq. (9.2) is equivalent to maximizing the between-class variance

$$T^* = \arg\max_T \left( p_0(T)(\mu_0(T) - \mu)^2 + p_1(T)(\mu_1(T) - \mu)^2 \right) , \tag{9.3}$$

$$= \arg\max_T \left( p_0(T) p_1(T)(\mu_0(T) - \mu_1(T))^2 \right) , \tag{9.4}$$

where $\mu$ is the mean value of the image $I$.

In general the Otsu algorithm works well if the image histogram is bi-modal with two distinct peaks. When this condition is not met the segmentation is often inaccurate. In many real-world situations the condition for a bi-modal histogram is not met when the two classes of the image have very different sizes. For example, in Example 4.11, the class containing background pixels ($C_0$) is very large while the class containing the foreground pixels ($C_1$) is very small, i.e. $p_0 \gg p_1$. In this case, the histogram may have only one prominent peak.

### Weighted Otsu Threshold

One way of tackling the problem of histograms which are unimodal, or close to unimodal, is to weigh the term in Eq. (9.2) which has the largest probability. Thus, if $p_0 \gg p_1$, [18] suggest weighing the first term $p_0\sigma_0^2(T)$ with the probability $p_1$. In this case the modified threshold is:

$$T^* = \arg\min_T \left( p_0(T)p_1(T)\sigma_0^2(T) + p_1(T)\sigma_1^2(T) \right) .$$

### Regularized Otsu Threshold

A more principled approach [5] to solving the problem of unimodal histograms is to regularize the solution by adding two new terms to the right-hand side of Eq. (9.4), In this case the new threshold is

$$T^* = \arg\max_T \left( p_0(T)p_1(T)(\mu_0(T) - \mu_1(T))^2 + (\mu_0(T) - \mu)^2 + (\mu_1(T) - \mu)^2 \right) . \tag{9.5}$$

The motivation for introducing the two new terms in Eq. (9.5) is to correct the tendency of the Otsu method to penalize the class of smaller size. It does this by finding a threshold $T^*$ such that the two classes $C_0$ and $C_1$ are well separated from each other *and* are also well separated from the global mean $\mu$.

For additional thresholding techniques, see [26]. See also Example 12.4 for an Otsu-based algorithm which is robust against outliers.

We now consider a novel image thresholding technique which is useful when we wish to *jointly* threshold a set of spatially registered images $I_i, i \in \{1, 2, \ldots, N\}$, which were acquired with different imaging modalities.

### Image Binarization using Mutual Information

Given a pair of spatially registered gray-scale images $I_1$ and $I_2$ of the same object $O$. Then we may find the thresholds $T_1$ and $T_2$ by maximizing the mutual information (MI) between the two binary images.

**Example 9.13** *Multiple Image Binarization using MI* [23]. Let $A = \{a_1, a_2, \ldots, a_N\}$ and $B = \{b_1, b_2, \ldots, b_N\}$ be two spatially aligned images of the same object or scene. After thresholding with thresholds $t_A$ and $t_B$, we obtain binary images $\widetilde{A} = \{\widetilde{a}_1, \widetilde{a}_2, \ldots, \widetilde{a}_N\}$ and $\widetilde{B} = \{\widetilde{b}_1, \widetilde{b}_2, \ldots, \widetilde{b}_N\}$, where

$$\widetilde{a}_i = \begin{cases} 1 & \text{if } a_i \geq t_A , \\ 0 & \text{otherwise} , \end{cases} \quad \text{and} \quad \widetilde{b}_i = \begin{cases} 1 & \text{if } b_i \geq t_B , \\ 0 & \text{otherwise} . \end{cases}$$

The thresholds $t_A$ and $t_B$ are chosen to maximize the MI between the distributions of $\widetilde{a}_i$ and $\widetilde{b}_i$:

$$(t_a, t_b) = \arg\max \left( \sum_u \sum_v p_{ab}(u, v) \log \frac{p_{ab}(u, v)}{p_a(u) p_b(v)} \right) ,$$

where $p_{ab}(u, v)$ is the probability that $\widetilde{a}_i = u$ and $\widetilde{b}_i = v$, $p_a(u)$ is the probability that $\widetilde{a}_i = u$ and $p_b(v)$ is the probability that $\widetilde{b}_i = v$.

The following example describes an important application of this technique.

**Example 9.14** *Change Detection Thresholding using MI* [23]. The most common methodology to carry out an unsupervised change detection in remotely sensed imagery is to compare two co-registered multi-temporal remote sensing images $I_1(m, n)$ and $I_2(m, n)$ taken at two different dates over the same geographical area [4]. The result of the comparison is a binary image $C(m, n)$, where $C(m, n) = 1$ if the pixel $(m, n)$ is classified as "changed", otherwise $C(m, n) = 0$. One way of generating $C$ is to threshold the difference image $D = |I_1 - I_2|$. A recent alternative is to separately threshold $I_1$ and $I_2$ using thresholds $T_1$ and $T_2$. If $B_k, k \in \{1, 2\}$, are the corresponding thresholded images:

$$B_k(m, n) = \begin{cases} 1 & \text{if } I_k(m, n) > T_k \ , \\ 0 & \text{otherwise} \ , \end{cases}$$

then $T_1$ and $T_2$ are chosen such that $MI(B_1, B_2)$ is maximized.

The following example illustrates the use of a "soft" binarization procedure in data fusion.

**Example 9.15** *"Soft" Binarization.* In many data fusion applications it is not possible to establish the correct threshold $t_i$ exactly. To reduce the impact of choosing an incorrect threshold, we may replace Eq. (9.1) with a trimmed min-max transfer function (see Table 9.7):

$$y_i = \begin{cases} 1 & \text{if } x_i > t_i + \delta \ , \\ 0 & \text{if } x_i < t_i - \delta \ , \\ \frac{1}{2}(x_i - t_i + \delta)/\delta & \text{otherwise} \ . \end{cases}$$

or with a sigmoid-like transfer function (see Fig. 9.1).

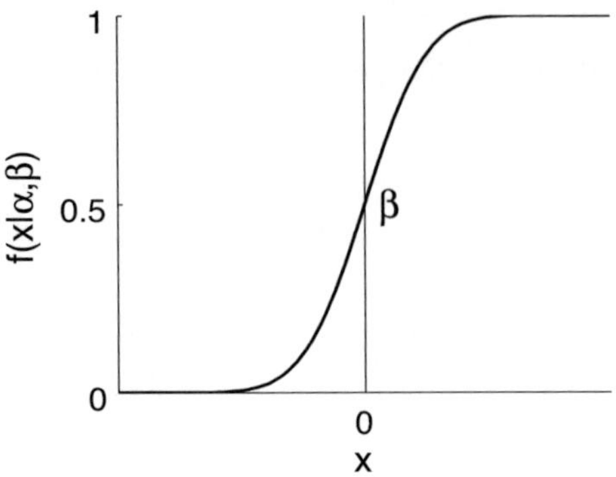


**Fig. 9.1** Shows a logistic sigmoid transfer function

**Table 9.8** Sigmoid-like transfer functions $f(x|\alpha, \beta)$ and slopes $df(x|\alpha, \beta)/dx$

| Name | $f(x\|\alpha, \beta)$ | $df(x\|\alpha, \beta)/dx$ |
|---|---|---|
| Logistic | $1/(1 + \exp(\alpha(\beta - x)))$ | $\alpha f(x\|\alpha, \beta) \times (1 - f(x\|\alpha, \beta))$ |
| Probit | $(\alpha/\sqrt{2\pi}) \int_{-\infty}^{x} \exp -\frac{1}{2}(\alpha(t - \beta))^2 dt$ | $(\alpha/\sqrt{2\pi}) \exp -(\alpha(x - \beta))^2/2)$ |
| Weibull | $1 - \exp(-(x/\beta)^{\alpha})$ | $\alpha x^{\alpha-1}(1 - f(x\|\alpha, \beta))/\beta^{\alpha}$ |

**Example 9.16** *Psychometric Transfer Functions* [12]. Psychophysics explores the connection between physical stimuli and subjective human responses. The psychometric transfer function, $f(x|\alpha, \beta)$, relates the intensity of the stimulus, $x$, to an observers's response, $y$. It is a parametric function with two parameters: a scale parameter $\alpha$ and a location parameter $\beta$. The psychometric function is commonly chosen to have a sigmoid-like shape. Three sigmoid-like psychometric transfer functions are listed, together with their slopes in Table 9.8.

Binarization is often used as a preliminary step where it functions as a *detection* step. For example, in calculating the similarity coefficients in Example 9.1 we assumed that all the features had undergone a preliminary binarization, or detection, process.

Although very effective, problems may arise if, for a given object $O$, we do not have a complete set of $N$ measurement values. The reason is that when we perform binarization we do not differentiate between (1) $y_i = 0$ because $x_i \leq t_i$ and (2) $y_i = 0$ because, for one reason or another, no measurement value $x_i$ is available. For example, the sensor $S$, which is responsible for making the measurement $x_i$, may have malfunctioned when the object $O$ was being processed. To handle such situations we use a *censored* binary scale which is defined as follows:

**Table 9.9** Questionnaire used for detecting acute gastrointestinal syndrome

| Question | Medical diagnosis | Possible answer |
|---|---|---|
| Diarrhea present? | Yes, No | $[1, 0]$ |
| Diarrhea duration? | Acute, Chronic, Unknown | $[1, 0, *]$ |
| Diarrhea etiology? | Infectious, Non-infectious, Unknown | $[1, 0, *]$ |
| Stool culture taken? | Yes, No | $[1, 0]$ |
| Vomiting present? | Yes, No | $[1, 0]$ |
| Vomiting duration? | Acute, Chronic, Unknown | $[1, 0, *]$ |
| Vomiting etiology? | Infectious, Non-infectious, Unknown | $[1, 0, *]$ |
| Acute GI disorder? | Yes, No | $[1, 0]$ |

$$y_i = \begin{cases} 1 \text{ if } x_i \geq t_i \ , \\ 0 \text{ if } x_i < t_i \ , \\ * \text{ if } x_i \text{ is missing} \ , \end{cases}$$

where $*$ is a special symbol which is handled differently from 0 and 1 by the data fusion algorithm. Censored binary scales are widely used when the information concerning the $x_i$ are *subjective* in nature. The following example illustrates the use of a censored binary scale when the $x_i$ are subjective replies to questions contained in a questionnaire.

**Example 9.17** *Automatic Disease Classification: Syndromic Surveillance* [15, 16]. The concept of *syndromic surveillance* was developed to ensure early detection of an epidemic in a given population. Syndrome surveillance works by monitoring the temporal and spatial trends of a syndrome, or a group of related symptoms or diseases, in the local population. For example we may use a censored questionnaire, similar to that shown in Table 9.9, to monitor cases of acute gastrointestinal syndrome which were admitted to the local hospital emergency departments.

## 9.7 Ranking

*Ranking* is a robust normalization technique which has enjoyed a growing popularity in recent years. Ranking is an example of *adaptive score normalization*, i.e. the normalization is based on the current input measurements and it does not require the use of a training set $\mathcal{D}$. Its main advantages are: it is simple and fast to implement, linear, robust against outliers and can be applied to all kinds of input information without the need to make additional assumptions regarding their distributions.

Mathematically, rank normalization is defined as follows. Suppose the input data consists of $N$ objects $O_i, i \in \{1, 2, \ldots, N\}$. On each object $O_i$ we make a measurement $x_i$. Then the corresponding rank is $r_i$, where

$$r_i = N - j + 1 \text{ if } x_i \text{ is } j\text{th largest measurement} . \tag{9.6}$$

If there are no ties among the $x_i$, then the ranks $r_i$ take on the values $\{1, 2, \ldots, N\}$, where $r_i = 1$ denotes that the object $O_i$ has the smallest $x$ value and $r_i = N$ denotes that the object $O_i$ has the largest $x$ value. If there are ties, then the ranks $r_i$ are chosen so that all $x_i$ with the same value have the same rank. See Example 9.18.

**Example 9.18** *Ranking: An example.* Given a set of $N = 10$ objects $O_i, i \in \{1, 2, \ldots, N\}$. We make a measurement $x_i$ on each object $O_i$. Then given the vector $\mathbf{x} = (x_1, x_2, \ldots, x_N)$, we calculate the corresponding rank vector $\mathbf{r}$ using Eq. (9.6):

| | | | | | | | | | | | |
|---|---|---|---|---|---|---|---|---|---|---|---|
| **O** | | $O_1$ | $O_2$ | $O_3$ | $O_4$ | $O_5$ | $O_6$ | $O_7$ | $O_8$ | $O_9$ | $O_{10}$ |
| **x** | or | 1 | 3 | 2 | 2 | 5 | 6 | 6 | 6 | 8 | 7 |
| ↓ | | ↓ | ↓ | ↓ | ↓ | ↓ | ↓ | ↓ | ↓ | ↓ | ↓ |
| **r** | | 1.0 | 4.0 | 2.5 | 2.5 | 5.0 | 7.0 | 7.0 | 7.0 | 10.0 | 9.0 |

The following example illustrates the use of ranking to fuse several similarity measures.

**Example 9.19** *Fusing Similarity Coefficients in a Virtual Screening Program* [13]. In similarity-based virtual screening (Example 9.1) we match a bioactive target structure $A$ against a database of molecules $B_i, i \in \{1, 2, \ldots, N\}$, using any one of several different similarity operators $S_m, m \in \{1, 2, \ldots, M\}$. *A priori* we cannot know which operator will be best for a given target structure or for a given type of activity. Let $x_m(A, B_i)$ denote the corresponding similarity value as measured by the $m$th operator $S_m$. Then, by fusing together the $x_m(A, B_i), m \in \{1, 2, \ldots, M\}$, we may produce a more effective and robust activity measure [13]. Since in general, the $x_m(A, B_i)$ are incommensurate we must normalize them before fusion. This we do by ranking: Let $r_m(A, B_i)$ be the the rank of $x_m(A, B_i)$, where $r_m(A, B_i) = N - h + 1$ if $x_m(A, B_i)$ is the $h$th largest value in $\{x_m(A, B_1), x_m(A, B_2), \ldots, x_m(A, B_N)\}$, then the fused similarity value, $\widetilde{r}(A, B_i)$, is defined as the average value of $r_m(A, B_i), m \in \{1, 2, \ldots, M\}$:

$$\widetilde{r}(A, B_i) = \frac{1}{M} \sum_m r_m(A, B_i) .$$

**Table 9.10** Conversion to probabilities

| Method | Description |
|---|---|
| Platt calibration | Model $f(x)$ using a sigmoid function |
| Histogram binning | Approximate $f(x)$ using a discrete histogram |
| Kernel density | Approximate $f(x)$ using a kernel density function, i.e. a continuous bin histogram |
| Isotonic regression | Model $f(x)$ using an isotonic function, i.e. a function of unknown shape which is non-decreasing |

## 9.8 Conversion to Probabilities

Until now we have only considered normalized techniques in which a sensor measurement $x$, and its corresponding normalized value $y$, have the same physical properties. We now consider a new normalization technique in which this is no longer true. In particular we give $y$ an explicit physical meaning as an a posteriori probability.[3]

We start with the simplest case in which all objects $O$ belong to either a class $C = c_1$ or to a class $C = c_2$. Suppose $x$ is the measured value of $O$. Then our aim is to find two functions $y = f_1(x)$ and $y = f_2(x)$ such that $f_k(x)$ closely approximates the a posteriori probability density function $p(C = c_k|x, I)$:

$$f_k(x) \approx p(C = c_k|x, I) , \quad k \in \{1, 2\} .$$

Since, by definition,

$$p(C = c_1|x, I) = 1 - p(C = c_2|x, I) ,$$

we need only calculate one function $f_2(x)$. We now consider several different approaches to finding $f_2(x)$. For notational convenience we shall, from now on, write $f(x)$ in place of $f_2(x)$. Methods for calculating $f(x)$ are listed in Table 9.10. In the following sections we consider, in turn, the methods of Platt calibration, histogram binning and isotonic regression.

### Platt Calibration

In Platt calibration [22] we convert the sensor measurement $x$ to a posteriori probability $p(C = c_2|x, I)$ using a parametric sigmoid function:

[3] Theoretically, this is the preferred normalization technique since, if successful, we have at our disposal all of the methods of Bayesian analysis (Chap. 8). However, in practice, conversion to a posteriori probabilities is often found to perform badly and for this reason the methods considered previously are often used instead.

$$f(x|\alpha, \beta) = \frac{1}{1 + \exp(\alpha(\beta - x))} ,$$

where $\alpha$ and $\beta$ are two parameters which we optimize using a database $\mathcal{D}$. We suppose that $\mathcal{D}$ contains $N$ objects $O_i, i \in \{1, 2, \ldots, N\}$. Each object has a sensor value $x_i$ and a class label $y_i$, where

$$y_i = \begin{cases} 0 \text{ if } O_i \text{ belongs to class } c_1 \ , \\ 1 \text{ if } O_i \text{ belongs to class } c_2 \ . \end{cases}$$

Then we find the parameters $\alpha$ and $\beta$ by minimizing the negative log likelihood of the data set $\mathcal{D}$:

$$(\alpha, \beta) = \arg\min\Big[-\sum_i \Big(y_i \ln f(y_i|\alpha, \beta) + (1 - y_i) \ln\big(1 - f(x_i|\alpha, \beta)\big)\Big)\Big] .$$

In performing the above optimization procedure we must be careful not to *overfit*. This may happen if the number of objects $O_i$ which belong to $C = c_1$ or to $C = c_2$ is too low. In this case, we may mitigate the effects of overfitting by using modified labels, $\widetilde{y}_i$, in the above minimization procedure, where

$$\widetilde{y}_i = \begin{cases} 1/(N_1 + 2) & \text{if } O_i \text{ belongs to class } C = c_1 \ , \\ (N_2 + 1)/(N_2 + 2) & \text{if } O_i \text{ belongs to class } C = c_2 \ . \end{cases}$$

where $N_k, k \in \{1, 2\}$, is the number of objects which belong to the class $C = c_k$. Platt calibration has been found to work well on calibrating SVM scores.

#### Histogram Binning

In *histogram binning* we do not know the shape of the function $f(x)$ and we cannot, therefore, use Platt calibration. Instead we use a non-parametric method such as frequency histogram or *binning*: Suppose the objects $O_i, i \in \{1, 2, \ldots, N\}$, in the database $\mathcal{D}$ are arranged in order of increasing value, i.e. $x_1 \leq x_2 \leq \ldots \leq x_N$. We divide the $x_i$ into $M$ non-overlapping subsets, or *bins*, $B_m = [b'_m, b''_m), i \in \{1, 2, \ldots, M\}$, where $b'_{m+1} = b''_m$. Then, we define $f(x) = p(C = c_2|x, I), b'_m \leq x \leq b''_m$, as

$$f(x) = \frac{\widetilde{n}_m^{(2)}}{\widetilde{n}_m^{(1)} + \widetilde{n}_m^{(2)}} ,$$

where $\widetilde{n}_m^{(k)} = n_m^{(k)} / \sum_{m=1}^{M} n_m^{(k)}$ and $n_m^{(k)}$ is the number of objects $O_i, i \in \{1, 2, \ldots, N\}$, which fall in the $m$th bin $B_m$ *and* which belong to the class $C = c_k$. *Note*: To avoid overfitting we modify this Eq. as follows:

$$f(x) = \frac{(\widetilde{n}_m^{(2)} + \delta)}{(\widetilde{n}_m^{(1)} + \widetilde{n}_m^{(2)} + 2\delta)} .$$

where $\delta = 1/N$.

### Isotonic Regression

In *isotonic regression* [22] we do not know the shape of the function $y = f(x)$ except that it belongs to the class of *isotonic*, or non-decreasing, functions. Isotonic regression may therefore be regarded as an intermediate approach lying between sigmoid fitting on the one hand and histogram binning on the other.

Isotonic regression has been successfully used to calibrate predictions from SVM's, naive Bayes'classifier (see Sect. 14.3), boosted Naive Bayes'classifier (see Sect. 15.9) and decision trees.

If we assume that the classifier ranks examples correctly, the function $f(x)$ which converts scores to probabilities is non-decreasing. We then use isotonic regression to learn this function.

A commonly used algorithm for computing the isotonic regression is the *pair adjacent violation* (PAV) algorithm. This algorithm finds the stepwise-constant isotonic function that best fits the data according to a mean-squared error criterion. Let $O_i, i \in \{1, 2, \ldots, N\}$, be the training samples arranged such that the measurement values are in order of increasing value, i.e. $x_1 \leq x_2 \leq \ldots \leq x_N$. Let $f^*(x)$ be the sought after isotonic function. Let $f(x_i) = z_i$, then if $f$ is already isotonic, we return $f^* = f$. Otherwise, there must be a subscript $l$ such that $f(x_{l-1}) \leq f(x_l)$. The examples $x_{l-1}$ and $x_l$ are called *pair adjacent violators*, because they violate the isotonic assumption. The values of $f(x_{l-1})$ and $f(x_l)$ are then replaced by their average, so that the $l-1$th and $l$th examples now comply with the isotonic assumption. If this new set of values is isotonic, then $f^*(x_{l-1}) = f^*(x_l) = (f(x_{l-1}) + f(x_l))/2$ otherwise $f^*(x_l) = f(x_l)$. This process is repeated using the new values until an isotonic set of values is obtained.

Isotonic regression has been found to be very effective method for radiometric normalization [14].

### Multi-class Probability Estimates

Thus far we have considered two-class, or binary, probability estimates. We now apply the notion of calibration to *multi-class* probability estimates. Let $c_k, k \in \{1, 2, \ldots, K\}$, denote $K$ classes. Then our aim is to find a multi-dimensional function $f(x)$ such that $f(x)$ closely approximates the a posteriori probability density function $p(\{C = c_k\}|x, I)$:

$$f(x) \approx p(\{C = c_k\}|x, I) .$$

**Table 9.11** Methods for estimating multi-class probabilities $\mathbf{p}$

| Method | Description |
|---|---|
| PKPD | $p_k = -1/(K-2-\sum_{l,l\neq k} r_{kl}^{-1})$; where $r_{kl}$ is the estimate that a given object $O$ belongs to $C=c_k$ assuming it belongs to either $C=c_k$ or $C=c_l$ |
| Hastie-Tibshani | $\mathbf{p} = \arg\min_{\mathbf{p}} \sum_k [\sum_{l\neq k}(r_{kl}/K - p_k/2)]^2$ |
| $WLW_1$ | $\mathbf{p} = \arg\min_{\mathbf{p}} \sum_k [\sum_{l\neq k}(r_{kl}p_l - r_{lk}p_k)]^2$ |
| $WLW_2$ | $\mathbf{p} = \arg\min_{\mathbf{p}} \sum_k \sum_{l\neq k}(r_{kl}p_k - r_{lk}p_k)^2)$ |

The function $f(x)$ represents a multidimensional mapping from one $K$-dimensional space to another $K$-dimensional space. In this case, it is not clear what function shape we should assume for the transformation function. For this reason, the direct calibration of the multi-class probabilities is not recommended. Instead we reduce the multi-class problem into $K(K-1)/2$ binary, or two-class, problems and then combine the binary probability estimates using a *voting* or other similar algorithm.

The following example illustrates the use of a simple voting algorithm to combine the binary probability estimates.

**Example 9.20** *Multi-class Probability Estimates Using a Voting Algorithm* [32, 33]. We consider a multi-class probability estimate involving $K$ classes $c_k, k \in \{1,2,\ldots,K\}$. Let $f_{kl}(x|\alpha_{kl},\beta_{kl})$ denote a probabilistic calibration curve for classes $c_k, c_l, l > k$. The curves $f_{kl}$ may be obtained directly from the classification algorithm or may be estimated using any one of the methods listed in Sect. 9.8. Then the a posteriori probabilities $p(c_k|x,I)$ for an input $x$ are:

$$p(C=c_k|x) = \frac{1}{Z}\sum_{l,l\neq k} f_{kl}(x|\alpha_{kl},\beta_{kl}) ,$$

where $f_{kl}(x|\alpha_{kl},\beta_{kl}) = 1 - f_{lk}(x|\alpha_{lk},\beta_{lk})$ and $Z = \sum_m \sum_n f_{mn}(y|\alpha_{mn},\beta_{mn})$ is a normalization factor which ensures a posteriori probabilities sum to one.

Additional methods [32] for estimating the multi-class probabilities, $\mathbf{p} = (p_1, p_2, \ldots, p_K)^T$, are listed in Table 9.11.

## 9.9 Further Reading

Similarity measures have been intensely investigated in many different fields. Willett and co-workers have published widely on the use of similarity measures in the chemical industry (see e.g. [29–31]). A comprehensive review of these similarity measures is to be found in [2, 8]. Recently [27] investigated the impact of radiometric normalization on classification performance. Binarization is widely used in applications involving image analysis. In these applications an important issue is specification of the threshold $t$. Finally, the advantages of using *calibrated* probability estimates are discussed at length in [9].

## 9.10 Software

**MATLAB**. MATLAB STATISTICS AND MACHINE LEARNING TOOLBOX includes m-files for similarity and histogram matching. m-file for EMD is Earth mover's distance (Author: Yilmaz).

**PYTHON**. SCIKIT-IMAGE, SCIKIT-LEARN and SCIPY include python modules for: histogram matching, data normalization and EMD.

## Problems

**9.1** What is virtual screening? List some of the similarity coefficients which are used in virtual screening.

**9.2** Define binarization and list some of its uses in multi-sensor data fusion.

**9.3** Describe fuzzy normalization. What are the advantages/disadvantages of using a fuzzy normalization scheme.

**9.4** Explain the advantages of using ranking as a normalization procedure.

**9.5** Compare and contrast the following methods for converting measurements to probabilities: Platt calibration, binning, kernels, isotonic regression.

**9.6** Explain Platt calibration. How may we reduce the impact of overfitting on Platt calibration.

**9.7** Describe the process of isotonic regression.

## References

1. Baggerly KA (2001) Probability binning and testing agreement between multivariate immunofluorescence histogram extending the chi-squared test. Cytometry 45:141–150
2. Bender A, Glen RC (2004) Molecular similarity: a key technique in molecular informatics. Org Biomol Chem 2:3204–3218
3. Birge L, Rozenholc Y (2006) How many bins should be put in a regular histogram. Euro Series Appl Indust Math Prob Stat 10:24–45

4. Bruzzone L, Prieto DF (2000) Automatic analysis of the difference image for unsupervised change detection. IEEE Trans Geosci Rem Sens 38:1171–1182
5. Cao X, Li T, Li H, Xia S, Ren F, Sun Y, Xu X (2019) A Robust parameter-free thresholding method for image segmentation. IEEE Access 7:3448–3458
6. Cereto-Massague A, Ojeda MJ, Valls C, Mulero M, Garcia-Vallve S (2014) Molecular fingerprint similarity search in virtual screening. Methods 71:58–63
7. Cha S-H (2008) Taxonomy of nominal type histogram distance measures. Am Conf App Math, pp 325–330
8. Choi S-S, Cha S-H, Tappert CC (2010) A survey of binary similarity and distance measures. J Syst Cyber Inform 8:43–48
9. Cohen I, Goldsmidt M (2004) Properties and benefits of calibrated classifiers. In: Proceedings of the 8th European conference on principles and practice of knowledge discovery in databases, Pisa, Italy, pp 125–136
10. Correa-Morris J, Martinez-Diaz Y, Hernandez N, Mendez-Vazquez H (2015) Novel histograms kernels with structural properties. Patt Recogn Lett 68:146–152
11. Delon J (2004) Midway image equalization. J Math Imag Vis 21:119–134
12. Gilchrist JM, Jerwood D, Ismaiel HS (2005) Comparing and unifying slope estimates across pyschometric function models. Percept Psychophys 67:1289–1303
13. Ginn CMR, Turner DB, Willett P et al (1997) Similarity searching in files of three-dimensional chemical structures: evaluation of the EVA descriptor and combination of rankings using data fusion. J Chem Inform Comp Sci 37:23–37
14. Goswami G, Ratha N, Singh R, et al (2016) Improving classifier fusion via pool adjacent violators normalization. Proc Int Conf Patt Recogn
15. Henning KJ (2003) What is syndromic surveillance? In: Syndromic surveillance: reports from a national conference
16. Ivanov O, Wagner MW, Chapman WW, Olszewski RT (2002) Accuracy of three classifiers of acute gastrointestinal syndrome for syndromic surveillance. Proc AMIA Symp, pp 345–349
17. Jain A, Nandakumar K, Ross A (2005) Score normalization in multimodal biometric systems. Patt Recogn 38:2270–2285
18. Liao J, Wang Y, Yin J et al (2018) Segmentation of rice seedlings using the YCrCb color space and an improved otsu method. Agronomy 8:8110269
19. Ma Y, Gu X, Wang Y (2010) Histogram similarity measure using variable bin size distance. Comp Vis Image Under 114:981–989
20. Madden C, Cheng ED, Piccardi M (2007) Tracking people across disjoint camera views by illumination-tolerant appearance representation. Mach Vis Appl 18:233–247 Tracking people across disjoint camera views by illumination-tolerant appearance representation. Mach Vis Appl 18:233–247
21. Mendolia I, Contino S, Perricone U et al (2020) Convolutional architectures for virtual screening. BMC Bioinform. 21(Suppl 8):310
22. Niculescu-Mizil A, Caruana R (2005) Predicting good probabilities with supervised learning. In: International conference on machine learning
23. O'Conaire C, O'Connor N, Cooke E, Smeaton A (2006) Detection thresholding using mutual information. In: Proceedings of the 1st international conference on computer vision theory and applications, vol 2, pp 408–415
24. Roederer M, Treister A, Moore W et al (2001) Probability binning comparison: a metric for quantitating univariate distribution differences. Cytometry 45:37–46
25. Rubner Y, Tomasi C, Guibas LJ (2000) The earth movers distance as a metric for image retrieval. Int J Comp Vis 40:99–121
26. Sezgin M, Sankur B (2004) Survey over image thresholding techniques and quantitative performance evaluation. J Elect Imag 13:146–165
27. Singh D, Singh B (2020) Investigating the impact of data normalization on classification performance. App Soft Comp 97:105524
28. Suh Y, Ji M, Kim H (2007) Probabilistic class histogram equalization for robust speech recognition. IEEE Sig Proc Lett 14:287–290

29. Willett P (1998) Chemical similarity searching. J Chem Inform Comp Sci 38:983–996
30. Willett P (2000) Textual and chemical information processing: different domains but similar algorithms. Inform Res 5(2)
31. Wilton D, Willet P, Lawson K et al (2003) Comparison of ranking methods for virtual screening in lead-discovery programmes. J Chem Inform Comp Sci 43:469–474
32. Wu T-F, Lin C-J, Weng RC (2004) Probability estimates for multiclass classification by pairwise coupling. J Mach Learn Res 5:975–1005
33. Zadrozny B (1999) Reducing multiclass to binary by coupling probability estimates. In: Neural information processing systems conference, Vancouver
34. Zhang Y, Luo M, Wu P et al (2022) Application of computational biology and artificial intelligence in drug design. Int J Mol Sci 23:13568

# Chapter 10
# Bayesian Inference

**Abstract** In this chapter we give an overview of Bayesian statistics and specifically we consider Bayesian inference methods which are used in data fusion. We first briefly consider the methods of conjugate and non-informative distributions. We then consider in detail the Gaussian Mixture model and the method of expectation-maximization including the problem of missing data. Finally we consider the problem of model selection including Bayes information criterion.

## 10.1 Introduction

In this chapter we give an overview of Bayesian statistics, and in particular, the methods of Bayesian inference as used in data fusion. The basic premise of Bayesian statistics is that all unknowns are treated as random variables and that the knowledge of these quantities is summarized via a probability distribution. The main advantages of using Bayesian statistics are

1. Bayesian statistics is the only known coherent system for quantifying objective and subjective uncertainties.
2. Bayesian statistics provides principled methods for the model estimation and comparison and the classification of new observations.
3. Bayesian statistics provides a natural way to combine different sensor observations.
4. Bayesian statistics provides principle methods for dealing with missing information.

In Table 10.1 we list some of the basic formulas which are used in Bayesian statistics. In a data fusion system, we use Bayesian statistics to represent the multiple sources of information. The corresponding probability distributions act as a powerful *common representational format*.



H. B. Mitchell, *Data Fusion*, https://doi.org/10.1007/978-3-662-71023-4_10

**Table 10.1** Basic formulas used in Bayesian inference

| Name | Formula |
|---|---|
| Pdf (definition) | $p(y\|I)dy = P(y \leq Y \leq y + dy\|I)$ |
| Normalization | $\int p(y\|I)dy = 1$ |
| Expected value | $E(Y) = \int yp(y\|I)dy$ |
| Variance | $\sigma^2 = \int (y - E(Y))^2 p(y\|I)dy$ |
| Product rule | $p(x, y\|I) = p(x\|y, I)p(y\|I)$ |
| Independence | $p(x, y\|I) = p(x\|I)p(y\|I)$ |
| Marginalization | $p(x\|I) = \int p(x, y\|I)dy$ |
| Decomposition | $p(x\|I) = \int p(x\|y, I)p(y\|I)dy$ |
| Bayes' rule | $p(x\|y, I) = p(y\|x, I)p(x\|I)/p(y\|I)$; $p(x\|y, z, I) = p(x\|I)p(y\|x, I)p(z\|x, y, I)/(p(y\|I)p(z\|I))$ |
| Likelihood | $L = p(y\|x, I)$ |

$X$ and $Y$ denote two random variables whose instantiations are $x$ and $y$. For variables with continuous values we do not normally distinguish between the variable and its instantiation. Thus we write $p(y|I)$ and not $p(Y = y|I)$, where $I$ describes the background information. For variables with discrete values this may, however, cause confusion. For these variables we shall, therefore, continue to write $p(Y = y|I)$

## 10.2 Bayesian Analysis

In Bayesian statistics we treat all quantities under consideration as random variables. This includes the observed data, $\mathbf{y}$ and any missing data, $\mathbf{z}$, unknown parameters, $\boldsymbol{\theta}$, and models, $M$. The full process of Bayesian analysis can be described as consisting of three stages:

**Probability Model.** In this stage, we create a joint probability distribution that captures the relationship among all the variables under consideration.

**A Posterior Distribution.** In this stage, we summarize all of our information regarding the different quantities of interest in a set of a posteriori distributions. Typically the a posteriori distribution is a conditional pdf $p(\mathbf{y}|\boldsymbol{\theta}, I)$.

**Model Selection.** In this stage, we evaluate the appropriateness of using a given model. We may then select the best model, or create a new model, or suggest improvements in one of the existing models.

The following example illustrates the role played by Bayesian analysis in the scientific process.

**Example 10.1** *Scientific Theories* [16, 19]. Figure 10.1 (adapted from [19]) illustrates that part of the scientific process in which data is collected and modeled. The heavily-framed boxes denote processes which involve Bayesian inference. In the first box we infer what the model parameters might be given the model and data. In the second box we infer what model is probably closer to the correct model given the data.

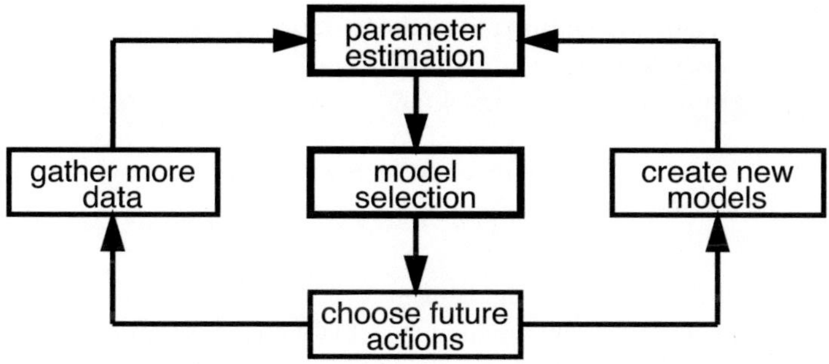


**Fig. 10.1** Shows the role played by Bayesian inference in the scientific process

## 10.3 Probability Model

The standard procedure for setting up a full probability model, or joint probability distribution, is to write down the likelihood (defined as the probability of the observed data given the unknowns) and multiply it by the a priori distribution of the unobserved variables and parameters. If the observed data, $\mathbf{y}$, comes from a parametric pdf $p$ characterized by parameters $\boldsymbol{\theta}$, then the joint pdf can be represented as

$$p(\mathbf{y}, \boldsymbol{\theta}|I) = p(\mathbf{y}|\boldsymbol{\theta}, I)\pi(\boldsymbol{\theta}|\boldsymbol{\lambda}, I) \ , \tag{10.1}$$

where $\boldsymbol{\theta}$ is assumed to come from some a priori distribution, $\pi$, with parameters $\boldsymbol{\lambda}$[1] and $I$ denotes all of our relevant background knowledge.

## 10.4 A Posteriori Distribution

The Bayesian inference is drawn by examining the probability of all possible values of the parameters, $\boldsymbol{\theta}$, after considering the data, $\mathbf{y}$. If the hyperparameters, $\boldsymbol{\lambda}$, are known (or are estimated), then we obtain the a posteriori pdf through the application of Bayes' theorem

$$p(\boldsymbol{\theta}|\mathbf{y}, I) = \frac{\overbrace{p(\mathbf{y}|\boldsymbol{\theta}, I)}^{\text{likelihood}}\,\overbrace{\pi(\boldsymbol{\theta}|\boldsymbol{\lambda}, I)}^{\text{a priori pdf}}}{\underbrace{p(\mathbf{y}|I)}_{\text{evidence}}} = \frac{p(\mathbf{y}|\boldsymbol{\theta}, I)\pi(\boldsymbol{\theta}|\boldsymbol{\lambda}, I)}{\int p(\mathbf{y}|\boldsymbol{\theta}, I)\pi(\boldsymbol{\theta}|\boldsymbol{\lambda}, I)d\boldsymbol{\theta}} \ . \tag{10.2}$$

The denominator $p(\mathbf{y}|I) = \int p(\mathbf{y}|\boldsymbol{\theta}, I)\pi(\boldsymbol{\theta}|\boldsymbol{\lambda}, I)d\boldsymbol{\theta}$ is known as the *evidence* and is a normalizing constant which is required so that $p(\boldsymbol{\theta}|\mathbf{y}, I)$ will integrate to one. The

[1] Hereinafter parameters $\boldsymbol{\lambda}$ are referred to as *hyperparameters* to differentiate them from $\boldsymbol{\theta}$.

evidence is obtained by integrating out all the variables, except for the observed data, $\mathbf{y}$, from the joint distribution.

If the hyperparameters, $\boldsymbol{\lambda}$, are not known, then we may also remove them by integration:

$$p(\boldsymbol{\theta}|\mathbf{y}, I) = \frac{\int p(\mathbf{y}|\boldsymbol{\theta}, I)\pi(\boldsymbol{\theta}|\boldsymbol{\lambda}, I)\pi(\boldsymbol{\lambda}|I)d\boldsymbol{\lambda}}{\int\int p(\mathbf{y}|\boldsymbol{\theta}, I)\pi(\boldsymbol{\theta}|\boldsymbol{\lambda}, I)\pi(\boldsymbol{\lambda}|I)d\boldsymbol{\lambda}d\boldsymbol{\theta}} , \tag{10.3}$$

where $\pi(\boldsymbol{\lambda}|I)$ is the a priori pdf for $\boldsymbol{\lambda}$.[2]

Equation (10.3) tells us how we may systematically update our knowledge of $\boldsymbol{\theta}$ given the data $\mathbf{y}$. For example, if the observations are obtained one at a time (cf. Chap. 13), we can update the a posteriori distribution as follows:

$$\begin{aligned} p(\boldsymbol{\theta}|\mathbf{y}_1, I) &\propto p(\mathbf{y}_1|\boldsymbol{\theta}, I)\pi(\boldsymbol{\theta}|I) , \\ p(\boldsymbol{\theta}|\mathbf{y}_1, \mathbf{y}_2, I) &\propto p(\mathbf{y}_2|\boldsymbol{\theta}, I)p(\boldsymbol{\theta}|\mathbf{y}_1, I) , \\ &\dots \\ p(\boldsymbol{\theta}|\mathbf{y}_1, \mathbf{y}_2, \dots, \mathbf{y}_N, I) &\propto p(\mathbf{y}_N|\boldsymbol{\theta}, I)p(\boldsymbol{\theta}|\mathbf{y}_1, \mathbf{y}_2, \dots, \mathbf{y}_{N-1}, I) . \end{aligned} \tag{10.4}$$

When there is more than one unknown parameter, e.g. $\boldsymbol{\theta} = (\theta_1, \theta_2)$, and we are only interested in one component, say $\boldsymbol{\theta}_1$, those unknown quantities that are not of immediate interest, but are needed by the model, are known as *nuisance parameters*, and are removed by integration:

$$p(\boldsymbol{\theta}_1|\mathbf{y}) = \frac{\int p(\mathbf{y}|\boldsymbol{\theta}_1, \boldsymbol{\theta}_2, I)\pi(\boldsymbol{\theta}_1, \boldsymbol{\theta}_2|I)d\boldsymbol{\theta}_2}{\int p(\mathbf{y}|\boldsymbol{\theta}_1, \boldsymbol{\theta}_2, I)\pi(\boldsymbol{\theta}_1, \boldsymbol{\theta}_2|I)d\boldsymbol{\theta}_1 d\boldsymbol{\theta}_2} . \tag{10.5}$$

Despite the deceptively simple-looking form of Eq. (10.5), the application of Bayesian statistics can be difficult. The most challenging aspects being: (1) Development of a model, $p(\mathbf{y}|\boldsymbol{\theta}, I)\pi(\boldsymbol{\theta}, I)$, which captures the key features of the underlying physical problem. (2) Computation required for deriving the a posteriori distribution.

### *10.4.1 Standard Probability Distribution Functions*

In developing adequate models for the likelihood $p(\boldsymbol{\theta}|\mathbf{y})$ and the a priori and a posteriori distributions, $p(\mathbf{y}|\boldsymbol{\theta}, I)$ and $p(\boldsymbol{\theta}|\mathbf{y}, I)$, we often use standard mathematical distributions. Table 10.2 is a list of some important univariate distributions $f(y)$.

---

[2] Hereinafter $h(\boldsymbol{\lambda})$ is referred to as a *hyperprior* distribution to differentiate it from the a priori distribution $\pi(\boldsymbol{\theta}|I)$.

**Table 10.2** Important univariate distributions f(y)

| Name | Formula |
|---|---|
| Binomial $\mathcal{B}in(y\|n, q)$ | $C(n, q)q^y(1-q)^{n-y}; n \in N$ |
| Chi-Square $\chi^2(y\|n)$ | $\exp -(y/2)y^{(n-2)/2}/(2^{n/2}\Gamma(n/2)); y \geq 0, n \in N$ |
| Exponential $\mathcal{E}(y\|\lambda)$ | $\lambda \exp(-\lambda y); y \geq 0; \lambda > 0$ |
| Gamma $\mathcal{G}a(y\|\alpha, \beta)$ | $y^{\alpha-1} \exp(-y\beta)\beta^\alpha/\Gamma(\alpha); \alpha > 0; \beta > 0$ |
| Inverse Gamma $\mathcal{IG}(y\|\alpha, \beta)$ | $y^{-(\alpha+1)} \exp(-/\beta/y)\beta^\alpha/\Gamma(\alpha); \alpha > 0$ |
| Normal (Gaussian) $\mathcal{N}(y\|\mu, \sigma^2)$ | $(2\pi\sigma^2)^{(-1/2)} \exp -\frac{1}{2}((y-\mu)/\sigma)^2; \sigma > 0$ |
| Student-*t* $\mathcal{S}t(y\|\mu, \sigma, \nu)$ | $\Gamma((\nu+1)/2)/(\sigma\sqrt{\nu\pi}\Gamma(\nu/2))\times$ $[1+(y-\mu)^2/\nu\sigma^2]^{-(\nu+1)/2}; \nu \in N, \nu > 2, \sigma > 0$ |
| Uniform $\mathcal{U}(y\|a, b)$ | $1/(b-a), 0 \leq y \leq 1; a, b > 0$ |

In this table we have explicitly written the variable $y$ for all the distributions. However, for univariate distributions, it is common practice to discard the variable $y$. We shall often adopt this practice e.g. writing $\mathcal{N}(\mu, \sigma^2)$ for the univariate Gaussian distribution

## 10.4.2 Conjugate Priors

An early effort at making the integration required by Eq. (10.2) accessible was the development of the *conjugate priors* [1, 13, 18]. Formally, a conjugate prior is a family of distributions for $\pi(\boldsymbol{\theta}|I)$ that has the same functional form as the likelihood function. As a consequence, when a conjugate prior is used, the functional form of the a posteriori distribution is the same as that of the a priori distribution. Thus although we can choose any functional form for $\pi(\boldsymbol{\theta}|I)$, the conjugate prior enjoys the greatest mathematical and computational advantage. In Table 10.3 we list a few families of conjugate likelihood and a priori distributions [13, 18].

We now illustrate the use of conjugate distributions to calculate the a posteriori pdf for a Gaussian likelihood with different a priori information [23].

**Table 10.3** Conjugate distributions

| Likelihood $p(y_i\|\theta, I)$ | $\pi(\theta\|I)$ | A posteriori $p(\theta\|\{y_i\}, I)$ |
|---|---|---|
| $\mathcal{N}(\theta, \sigma^2)$ [known $\sigma^2$] | $\mathcal{N}(\delta, \tau^2)$ | $\mathcal{N}\big((N\tau^2\bar{y}+\sigma^2\delta)/\Delta, \sigma^2\tau^2/\Delta\big); \Delta = \sigma^2 + N\tau^2$ |
| $\mathcal{N}(\mu, \theta)$ [known $\mu$] | $\mathcal{IG}(\alpha, \beta)$ | $\mathcal{IG}(\alpha + N/2, \beta + \frac{N}{2}W); W = \frac{1}{N}\sum_i(y_i - \mu)^2$ |
| $\mathcal{N}(\mu, 1/\theta)$ [known $\mu$] | $\mathcal{G}a(\alpha, \beta)$ | $\mathcal{G}a(\alpha + N/2, \beta + \frac{N}{2}W); W = \frac{1}{N}\sum_i(y_i - \mu)^2$ |
| $\mathcal{N}(\theta_1, \theta_2)$ | see Example 10.4 | |
| $\mathcal{U}(0, \theta)$ | $\mathcal{P}a(\theta_0, \alpha)$ | $\mathcal{P}a(\max(\theta_0, y), \alpha + 1)$ |
| $\mathcal{G}a(\nu = \frac{1}{2}, 2\theta)$ [known $\nu$] | $\mathcal{IG}(\alpha, \beta)$ | $\mathcal{IG}(\frac{1}{2} + \alpha, 1/(\frac{y}{2} + \beta^{-1}))$ |
| $\mathcal{E}(\theta)$ | $\mathcal{G}a(\alpha, \beta)$ | $\mathcal{G}a(\alpha + 1, \beta + \sum_{i=1}^N y_i)$ |
| $\mathcal{B}in(n, \theta)$ [known $n$] | $\mathcal{B}e(\alpha, \beta)$ | $\mathcal{B}e(\alpha + y, \beta + n - y)$ |

**Example 10.2** *Conjugate Prior for a Gaussian Distribution with Unknown Mean and Known Variance.* We observe $N$ independent samples $y_i, i \in \{1, 2, \ldots, N\}$. If the $y_i$ follow a Gaussian distribution with a known variance $\sigma^2$ and unknown mean $\mu$, the likelihood is $p(\{y_i\}|\mu, I) = \mathcal{N}(\mu, \sigma^2)$. Assuming a Gaussian a priori distribution $p(\mu|\{y_i\}, I) = \mathcal{N}(\mu_0, \sigma_0^2)$, then according to Table 10.3, the a posteriori pdf $p(\mu|\{y_i\}, I)$ is $\mathcal{N}(\mu_1, \sigma_1^2)$, where

$$\mu_1 = \frac{\sigma^2}{N\sigma_0^2 + \sigma^2}\mu_0 + \frac{N\sigma_0^2}{N\sigma_0^2 + \sigma^2}\overline{y} \quad \text{and} \quad \frac{1}{\sigma_1^2} = \frac{1}{\sigma_0^2} + \frac{N}{\sigma^2} \, .$$

Observe that as $N \to \infty$, $\mu_1 \to \overline{y} = \frac{1}{N}\sum_{i=1}^{N} y_i$ and $\sigma_1^2 \to \frac{\sigma^2}{N}$ (cf. Example 1.6).

**Example 10.3** *Conjugate Prior for a Gaussian Distribution with Known Mean and Unknown Variance.* We observe $N$ independent samples $y_i, i \in \{1, 2, \ldots, N\}$. If the $y_i$ follow a Gaussian distribution with a known mean $\mu$ and unknown variance $\sigma^2$, the likelihood is $p(\{y_i\}|\sigma^2, I) = \mathcal{N}(\mu, \sigma^2)$. Assuming the a priori distribution $p(\sigma^2|\{y_i\}, I)$ is an Inverse Gamma distribution $\mathcal{IG}(\alpha_0, \beta_0)$, then according to Table 10.3, the a posteriori pdf $p(\sigma^2|\{y_i\}, I)$ is $\mathcal{IG}(\alpha_1, \beta_1)$, where

$$\alpha_1 = \alpha_0 + \frac{N}{2} \quad \text{and} \quad \beta_1 = \beta_0 + \frac{N}{2}W \, ,$$

and $W = \frac{1}{N}\sum_{i=1}^{N}(y_i - \mu)^2$.

Physically, the parameters $\alpha$ and $\beta$ are related to the expected value $E(\sigma^2)$ and the variance $\mathrm{var}(\sigma^2)$ as follows:

$$\alpha = \frac{\big(E(\sigma^2)\big)^2}{\mathrm{var}(\sigma^2)} + 2 \quad \text{and} \quad \beta = E(\sigma^2)\Big(\frac{\big(E(\sigma^2)\big)^2}{\mathrm{var}(\sigma^2)} + 1\Big) \, .$$

It is often more convenient to work with precision $\lambda$ (defined as $1/\sigma^2$). In this case, the a priori distribution $p(\lambda|\{y_i\}, I)$ is a Gamma distribution $\mathcal{G}a(\alpha_0', \beta_0')$ and the a posteriori pdf $p(\lambda|\{y_i\}, I)$ is $\mathcal{G}a(\alpha_1', \beta_1')$, where

$$\alpha_1' = \alpha_0' + \frac{N}{2} \quad \text{and} \quad \beta_1' = \beta_0' + \frac{N}{2}W \, .$$

**Example 10.4** *Conjugate Prior for a Gaussian Distribution with Unknown Mean and Unknown Precision* [13]. We observe $N$ independent samples $y_i, i \in \{1, 2, \ldots, N\}$. If the $y_i$ follow a Gaussian distribution with unknown mean $\mu$ and unknown precision $\lambda = 1/\sigma^2$, the likelihood is $p(\{y_i\}|\mu, \lambda, I) = \mathcal{N}(\mu, 1/\lambda)$. Assuming the a priori distribution $p(\mu, \lambda|\{y_i\}, I)$ is a Normal-Gamma distribution $\mathcal{NG}(\mu, \lambda|\mu_0, k_0, \alpha_0, \beta_0) = \mathcal{N}(\mu|\mu_0, (k_0\lambda)^{-1})\mathcal{G}a$ $(\lambda|\alpha_0, \beta_0)$, then according to Table 10.3, the a posteriori pdf $p(\mu, \lambda|\{y_i\}, I)$ is $\mathcal{NG}(\mu, \lambda|\mu_1, k_1, \alpha_1, \beta_1)$:

$$\mathcal{NG}(\mu, \lambda|\mu_1, \sigma_1^2, k_1, \alpha_1, \beta_1) = \mathcal{N}(\mu|\mu_1, (k_1\lambda)^{-1})\mathcal{G}a(\lambda|\alpha_1, \beta_1) ,$$

where

$$\begin{aligned}
\mu_1 &= \frac{k_0\mu_0 + N\overline{y}}{k_0 + N} , \\
k_1 &= k_0 + N, , \\
\alpha_1 &= \alpha_0 + \frac{N}{2} , \\
\beta_1 &= \beta_0 + \frac{1}{2}\sum_{i=1}^{N}(y_i - \overline{y})^2 + \frac{k_0 N(\overline{y} - \mu_0)^2}{2(k_0 + N)} .
\end{aligned}$$

By separately integrating out $\mu$ and $\lambda$ we obtain the marginal distributions for $\lambda$ and $\mu$:

$$\begin{aligned}
p(\lambda|\{y_i\}, I) &= \int p(\mu, \lambda|\{y_i\}, I)d\mu = \mathcal{G}a(\lambda|\alpha_1, \beta_1) . \\
p(\mu|\{y_i\}, I) &= \int p(\mu, \lambda|\{y_i\}, I)d\lambda \\
&= \int p(\mu|\lambda, \{y_i\}, I)p(\lambda|\{y_i\})d\lambda .
\end{aligned}$$

### *10.4.3 Non-informative Priors*

Another class of a priori distributions is the *non-informative* or "flat" a priori distributions. These distributions are designed for situations when we have no knowledge of what the true distributions should be. The simplest non-informative a priori pdf for a variable $\theta$ is to assume $\pi(\theta|I)$ is proportional to one: $\pi(\theta|I) \propto 1$. However this choice of pdf is *not* invariant to scale changes.

**Table 10.4** Non-informative priors distributions

| Likelihood $p(y_i\|\theta, I)$ | A priori $\pi(\theta\|I)$ | A posteriori $p(\theta\|\{y_i\}, I)$ |
|---|---|---|
| $\mathcal{N}(\theta, \sigma^2)$ [$\sigma^2$ known] | 1 (U, J, R, M) | $\mathcal{N}(\bar{y}, \sigma^2/N)$; $\bar{y} = \sum_{i=1}^N y_i/N$ |
| $\mathcal{N}(\mu, \theta)$ [$\mu$ known] | 1 (U) | $\mathcal{IG}((N-2), 2/S^2)$; $S^2 = \sum_{i=1}^N (y_i - \bar{y})^2$ |
| | $1/\sigma^2$ (J, R, M) | $\mathcal{IG}(N/2, 2/S^2)$ |
| $\mathcal{N}(\theta_1, \theta_2)$ | 1 (U) | $\theta_1 \sim \mathcal{S}t(\bar{y}, S^2/(N-3), N-3)$ |
| | | $\theta_2 \sim \mathcal{IG}((N-3)/2, 2/S^2)$ |
| | $1/\sigma^4$ (J) | $\theta_1 \sim \mathcal{S}t(\bar{y}, S^2/(N(N+1)), N+1)$ |
| | | $\theta_2 \sim \mathcal{IG}((N+1)/2, 2/S^2)$ |
| | $1/\sigma^2$ (R,M) | $\theta_1 \sim \mathcal{S}t(\bar{y}, S^2/(N(N-1)), N-1)$ |
| | | $\theta_2 \sim \mathcal{IG}((N-1)/2, 2/S^2)$ |

Considerations of this sort led to the concept of Jeffrey's a priori pdfs which are invariant to changes of variables. They are defined as follows:

$$\pi(\theta|I) \propto |I(\theta)|^{1/2} , \tag{10.6}$$

where $I(\theta) = -E(d^2 \log p(\mathbf{y}|\theta)/d\theta^2)$ is the *Fisher information*.

Other non-informative a priori distributions are the reference and the maximal data information prior (MDIP) distributions [17, 32]. In Table 10.4 we list the uniform (U), Jeffrey's (J), reference (R) and the MDIP (M) priors for a Gaussian distribution. The use of non-informative priors is controversal especially for multivariate distributions. *Note*: Not all the priors in Table 10.4 are location and scale independent, although reasonable results are often obtained with a priori distributions which are *not* location or scale independent.

**Example 10.5** *Non-Informative Prior for a Gaussian Distribution with Jointly Unknown Mean and Variance*. We reconsider Example 10.4 but this time use non-informative a priori distributions. We assume no information regarding $\mu$ and $\sigma^2$. In this case, a non-informative distribution, which is invariant to translation and scale changes, is

$$\pi(\mu, \sigma^2|I) = \pi(\mu|\sigma^2, I)\pi(\sigma^2|I) = \frac{1}{\sqrt{2\pi\sigma^2}} \times \frac{1}{\sigma^2} \sim \frac{1}{\sigma^3} .$$

By Bayes' rule, the corresponding a posteriori pdf is

$$p(\mu, \sigma^2|\{y_i\}, I) = \prod_i \frac{\mathcal{N}(y_i|\mu, \sigma^2)\pi(\mu, \sigma^2|I)}{p(y_i|I)} = \frac{1}{\sigma^3\sqrt{2\pi}} \prod_i \frac{\mathcal{N}(y_i|\mu, \sigma^2)}{p(y_i|I)} ,$$

which we may rewrite as:

$$p(\mu, \sigma^2|\{y_i\}, I) = \frac{1}{\Gamma(\frac{N}{2})\sigma^2}\sqrt{\frac{N}{\pi S}}\left(\frac{S}{2\sigma^2}\right)^{(N+1)/2} \exp\left(-\frac{1}{2\sigma^2}(N(\mu - \overline{y})^2 + S)\right),$$

where $\overline{y} = \sum_i y_i/N$ and $S = \sum_i (y_i - \bar{y})^2$.

### 10.4.4 Missing Data

In many problems, it is often fruitful to distinguish between two kinds of unknowns: parameters and hyper-parameters on the one hand and *missing* data on the other hand. Although there is no absolute distinction between the two types of unknowns, the missing data is usually directly related to the individual data and its dimensionality tends to increase as more and more data are observed. On the other hand, the parameters and hyperparameters usually characterize the entire population of observations and are fixed in number.

The following example illustrates a case when data is "missing" because it lies below a given detection limit.

**Example 10.6** *Environmental Data with Below Detection Limit Observations* [29]. The processing of environmental samples containing potentially hazardous chemicals is often complicated by the fact that some of these pollutants are present at trace levels, which cannot be measured reliably and therefore, are reported as unknown values which lie below a certain limit of detection $t$. The corresponding data sets are known as "left-censored", data sets or data sets with "missing" values.

When missing data, $\mathbf{z}$, occurs in a statistical problem, the inference can be achieved by using the "observed-data likelihood", defined as $L = p(\mathbf{y}|\boldsymbol{\theta}, I)$, which can be obtained by integrating out the missing data from the "complete-data likelihood":

$$L = \int p(\mathbf{y}, \mathbf{z}|\boldsymbol{\theta}, I)d\mathbf{z} . \tag{10.7}$$

Bayesian analysis for missing data problems can be achieved coherently through integration:

$$p(\boldsymbol{\theta}|\mathbf{y}, I) = \frac{\int p(\mathbf{y}, \mathbf{z}|\boldsymbol{\theta}, I)\pi(\boldsymbol{\theta}|I)d\mathbf{z}}{\int\int p(\mathbf{y}, \mathbf{z}|\boldsymbol{\theta}, I)\pi(\boldsymbol{\theta}|I)d\mathbf{z}d\boldsymbol{\theta}} . \tag{10.8}$$

Examination of Eq. (10.8) shows that the $\mathbf{z}$ are treated as random variables and that the integration required for eliminating missing data is no different from that required for eliminating nuisance parameters and unknown hyperparameters.

Many applications which involve the use of missing data are solved using the *expectation-maximization* (EM) algorithm.

**Example 10.7** *EM algorithm for left-censored measurements* [29]. Let $y_1, y_2, \ldots, y_N$ denote a random sample of $N$ measurements made from an Gaussian distribution, $\mathcal{N}(\mu, \sigma^2)$, where $K$ measurements lie below the threshold limit $t$. For simplicity we suppose the first $K$ measurements, $y_1, y_2, \ldots, y_K$, are "non-detects", i.e. are below the detection threshold $T$. Then we may estimate $\mu$ and $\sigma^2$ using an iterative "expectation-maximization" (EM) procedure. At each iteration $t$, all non-detects are replaced by the same conditional expected value. Let $\phi$ and $\Phi$ be, respectively the probability and cumulative density functions of the Gaussian distribution. Then, the maximum likelihood estimates of $\mu$ and $\sigma$ at the $(t+1)$th iteration are:

$$\hat{\mu}^{(t+1)} = \frac{1}{N}\Big(\sum_{i=K+1}^{N} y_i + \sum_{i=1}^{K} E_t(y_i|y_i \leq t)\Big) ,$$

$$(\hat{\sigma}^{(t+1)})^2 = \frac{1}{N-1}\Big(\sum_{i=K+1}^{N} (y_i - \hat{\mu}^{(t)})^2 + \sum_{i=1}^{K} E_t\big((y_i - \hat{\mu}^{(t)})^2|y_i \leq t\big)\Big) ,$$

where $z = (T - \hat{\mu}^{(t)})/\hat{\sigma}^{(t)}$ and

$$E_t(y_i|y_i \leq T) = \hat{\mu}^{(t)} - \hat{\sigma}^{(t)}\phi(z)/\Phi(z) ,$$

$$E_t\big((y_i - \hat{\mu}^{(t)})^2|y_i \leq T\big) = (\hat{\sigma}^{(t)})^2\big(1 - z\phi(z)/\Phi(z)\big) .$$

We now consider the application of the EM algorithm to create a new representation consisting of $K$ Gaussian densities.

## 10.5 Gaussian Mixture Model (GMM)

In the Gaussian mixture model (GMM) we model an arbitrary $M$-dimensional pdf as a sum of $K$ Gaussian densities, where $K < M$. The following example explains the main steps in forming the GMM.

**Example 10.8** *Gaussian Mixture Model* (GMM) [20]. In GMM we model a pdf, $\boldsymbol{\theta} = f(\mathbf{y})$, as a sum of $K$ Gaussian distributions $\mathcal{N}(\boldsymbol{\mu}_k, \boldsymbol{\Sigma}_k), k \in \{1, 2, \ldots, K\}$:

$$f(\mathbf{y}) \approx \sum_k \alpha_k \mathcal{N}(\mathbf{y}|\boldsymbol{\mu}_k, \boldsymbol{\Sigma}_k) ,$$

where $(\alpha_k, \boldsymbol{\mu}_k, \boldsymbol{\Sigma}_k), k \in \{1, 2, \ldots, K\}$, are unknown and are to be learnt from a set of training data $(\boldsymbol{\theta}_i, \mathbf{y}_i), i \in \{1, 2, \ldots, N\}$, subject to the constraints

$$\alpha_k > 0 \quad \text{and} \quad \sum_k \alpha_k = 1 .$$

One way of learning the parameters $(\alpha_k, \boldsymbol{\mu}_k, \boldsymbol{\Sigma}_k)$, is to reformulate the mixture model in terms of missing data and then use an iterative EM procedure.

We associate a "hidden" variable $z_i$ with each $\mathbf{y}_i$, where $z_i$ can only take on the values $\{1, 2, \ldots, K\}$. Then we suppose that each value $\boldsymbol{\theta}_i = f(\mathbf{y}_i)$ was generated by randomly choosing $z_i$ from $\{1, 2, \ldots, K\}$ and then drawing $\mathbf{y}_i$ from the Gaussian $\mathcal{N}(\boldsymbol{\mu}_k, \boldsymbol{\Sigma}_k)$, where $k = z_i$. In this case, $\alpha_k$ is given by the proportion of $z_i$ for which $z_i = k$:

$$\alpha_k = \frac{1}{N} \sum_i \delta_{ik} ,$$

where $\delta_{ik}$ is the indicator function:

$$\delta_{ik} = \begin{cases} 1 \text{ if } z_i = k , \\ 0 \text{ otherwise} . \end{cases}$$

Maximum likelihood (ML) estimates of the parameters $(\alpha_k, \boldsymbol{\mu}_k, \boldsymbol{\Sigma}_k), k \in \{1, 2, \ldots, K\}$, are calculated using the EM algorithm (see Algorithm 10).

Although widely used, the EM/GMM algorithm may converge to a local minimum or may fail as a result of singularities, or degeneracies, in the input data. In addition the GMM suffers from the "curse of dimensionality" (Sect. 14.2): As the dimensionality of the input data increases, the number of input vectors which are required to specify the means and covariance matrices, $(\boldsymbol{\mu}_k, \boldsymbol{\Sigma}_k), k \in \{1, 2, \ldots, K\}$, increases exponentially. One way to stop this exponential increase in parameters is to constrain the structure of the covariance matrices.[3]

The following example illustrates using GMM for image segmentation.

**Example 10.9** *Segmentation of SEM Images of Multiphase Materials* [10]. Scanning electron microscopy (SEM) is a powerful technique used to investigate the structural and chemical properties of multiphase materials. One outcome of SEM-based analysis is the calculation of the fractions of the material components which constitute the multiphase material. We assume that each

[3] Traditionally, the covariance matrix is constrained to be diagonal (see Sect. 14.2). In this case the number of parameters required to specify a $M$-dimensional Gaussian pdf is $2M$. The corresponding number of parameters is $M(M+3)/2$ if the covariance matrix is full.

**Algorithm 10:** GMM USING EM ALGORITHM

**Input**: Training data $\boldsymbol{\theta}_i = f(\mathbf{y}_i), i \in \{1, 2, \ldots, N\}$; number of Gaussians $K$; number of iterations: $niter$

**Output**: Gaussian mixture parameters $(\alpha_k, \boldsymbol{\mu}_k, \boldsymbol{\Sigma}_k), k \in \{1, 2, \ldots, K\}$

Randomly initialize $\alpha_k, \boldsymbol{\mu}_k, \boldsymbol{\Sigma}_k, k \in \{1, 2, \ldots, K\}$

**for** $iter \leftarrow 1$ *to* $niter$ **do**

  **E-step:** Calculate the *a posteriori* probability that $z_i = k$ given current parameter values $(\alpha_k, \boldsymbol{\mu}_k, \boldsymbol{\Sigma}_k)$.

  **for** $i \leftarrow 1$ *to* $N$ **do**

    **for** $k \leftarrow 1$ *to* $K$ **do**

      $p(z_i = k|\alpha_k, \boldsymbol{\mu}_k \boldsymbol{\Sigma}_k) = \alpha_k \mathcal{N}(\boldsymbol{y}_i|\boldsymbol{\mu}_k, \boldsymbol{\Sigma}_k) / \sum_{j=1}^{K} \alpha_j \mathcal{N}(\boldsymbol{y}_i|\boldsymbol{\mu}_j, \boldsymbol{\Sigma}_j)$

    **end**

  **end**

  **M-step:** Update parameters $(\alpha_k, \boldsymbol{\mu}_k, \boldsymbol{\Sigma}_k)$ using probability $p(z_i = k|\alpha_k, \boldsymbol{\mu}_k \boldsymbol{\Sigma}_k)$.

  **for** $k \leftarrow 1$ *to* $K$ **do**

    $\alpha_k = \sum_{i=1}^{N} w_{ik}/N, \quad \text{where } w_{ik} = p(z_i = k|\alpha_k, \boldsymbol{\mu}_k, \boldsymbol{\Sigma}_k)$

    $\boldsymbol{\mu}_k = \frac{1}{\alpha_k N} \sum_{i=1}^{N} w_{ik} \mathbf{y}_i, \quad \text{where } \alpha_k N = \sum_{i=1}^{N} w_{ik}$

    $\boldsymbol{\Sigma}_k = \frac{1}{\alpha_k N} \sum_{i=1}^{N} w_{ik} (\mathbf{y}_i - \boldsymbol{\mu}_k)(\mathbf{y}_i - \boldsymbol{\mu}_k)^T$

  **end**

**end**

**return** $(\alpha_k, \boldsymbol{\mu}_k, \boldsymbol{\Sigma}_k), k \in \{1, 2, \ldots, K\}$

material component contributes a *Gaussian distribution* to the overall distribution of pixel gray-levels in the SEM image. Then, disregarding all pixel interactions, we use a GMM to segment the SEM image.

Let $\mathbf{X}$ denote the gray-level SEM image consisting of $N$ pixels $\{x_1, x_2, \ldots, x_N\}$, where $x_i$ is the gray-level of the $i$th pixel. Disregarding pixel interactions, we model the gray-levels $x_i, i \in \{1, 2, \ldots, N\}$, as independent random values drawn from a mixture of $K$ Gaussian distributions:

$$p(x_i|\boldsymbol{\Theta}) \approx \sum_k \pi_k \mathcal{N}(x_i|\mu_k, \Sigma_k) , \tag{10.9}$$

where $K$ is the number of material components and $\boldsymbol{\Theta} = (\pi_k, \mu_k, \Sigma_k), k \in \{1, 2, \ldots, K\}$, denotes the (unknown) mixing coefficients, means and variances of the GMM. The log-likelihood of $\boldsymbol{\Theta}$, given the image $\mathbf{X}$, is

$$\log L(\boldsymbol{\Theta}|\mathbf{X}) = \log \prod_i \sum_k \pi_k \mathcal{N}(\mathbf{x}_i|\mu_k, \Sigma_k) = \sum_i \log \sum_k \pi_k \mathcal{N}(x_i|\mu_k, \Sigma_k) .$$

whose maximum likelihood (ML) solution is found using the EM algorithm: At iteration $(t + 1)$ the a posteriori pdf and ML parameter values are:

$$\text{E-step} \qquad p(k|\mathbf{x}_i, \boldsymbol{\Theta}^{(t)}) = \frac{\pi_k^{(t)} \mathcal{N}(x_i|\mu_k^{(t)}, \Sigma_k^{(t)})}{\sum_{h=1}^{K} \pi_h^{(t)} \mathcal{N}\big(x_i|\mu_h^{(t)}, \Sigma_h^{(t)}\big)} .$$

M-step
$$\pi_k^{(t+1)} = \frac{1}{N}\sum_i p\big(k|\mathbf{x}_i, \mathbf{\Theta}^{(t)}\big) ,$$
$$\mu_k^{(t+1)} = \frac{\sum_i x_i p\big(k|x_i, \mathbf{\Theta}^{(t)}\big)}{\sum_i p\big(k|x_i, \mathbf{\Theta}^{(t)}\big)} ,$$
$$\Sigma_k^{(t+1)} = \frac{\sum_i \big(x_i - \mu_k^{(t+1)}\big)\big(x_i - \mu_k^{(t+1)}\big)^T p(k|x_i, \mathbf{\Theta}^{(t)})}{\sum_i p(k|x_i, \mathbf{\Theta}^{(t)})} .$$

If $p(k|x_i, \mathbf{\Theta})$ and $\big(\pi_k, \mu_k, \Sigma_k\big), k \in \{1, 2, \ldots, K\}$, denote the final pdf and parameter values (at convergence), then

$$\pi_k = \text{Fraction of the material corresponding to } k\text{th component} .$$

We may furthermore segment $\mathbf{X}$ by giving each pixel $x_i$ a label $y_i = k^*$, where

$$y_i = k^* \quad \text{if} \quad k^* = \arg\max_k p(k|x_i, \mathbf{\Theta}) .$$

### Distance Between GMM's

In comparing two GMM's we often use a *symmetric* Kullback-Leibler (KL) distance. Given two distributions $F$ and $G$ the symmetric KL distance is defined as

$$d_{\text{symKL}}(F, G) = \frac{1}{2}\big(d_{\text{KL}}(F|G) + d_{\text{KL}}(G|F)\big) ,$$

where $d_{\text{KL}}(F|G) = \int F(x)\log\big(F(x)/G(x)\big)dx$ is the corresponding (asymmetric) KL distance.

For two GMM's: $F = \sum_{k=1}^K \alpha_k \mathcal{N}(\mathbf{m}_k, \mathbf{S}_k)$ and $G = \sum_{k=1}^K \beta_k \mathcal{N}(\boldsymbol{\mu}_k, \mathbf{\Sigma}_k)$, there is no closed-form expression for $d_{\text{KL}}(F|G)$ except when $K = 1$. In this case, if $f$ and $g$ denote the corresponding $M$-dimensional Gaussian pdf's, then

$$d_{\text{KL}}(f|g) = \frac{1}{2}\Big(\log\Big(\frac{\det(\mathbf{\Sigma})}{\det(\mathbf{S})}\Big) + \text{trace}\big(\mathbf{\Sigma}^{-1}\mathbf{S}\big) - M + (\mathbf{m} - \boldsymbol{\mu})\mathbf{\Sigma}^{-1}(\mathbf{m} - \boldsymbol{\mu})\Big) .$$

If $K > 1$, we may approximate $d_{\text{KL}}(F|G)$ as follows. Let $\widetilde{f} = \mathcal{N}(\widetilde{\mathbf{m}}, \widetilde{\mathbf{S}})$ and $\widetilde{g} = \mathcal{N}(\widetilde{\boldsymbol{\mu}}, \widetilde{\mathbf{\Sigma}})$ be two Gaussian pdf's whose means and covariance matrices match, respectively, that of $F$ and $G$:

$$\widetilde{\mathbf{m}} = \sum_k \alpha_k \mathbf{m}_k \quad \text{and} \quad \widetilde{\mathbf{S}} = \sum_k \alpha_k\big(\mathbf{S}_k + (\mathbf{m}_k - \widetilde{\mathbf{m}})(\mathbf{m}_k - \widetilde{\mathbf{m}})^T\big) ,$$
$$\widetilde{\boldsymbol{\mu}} = \sum_k \beta_k \boldsymbol{\mu}_k \quad \text{and} \quad \widetilde{\mathbf{\Sigma}} = \sum_k \beta_k\big(\mathbf{\Sigma}_k + (\boldsymbol{\mu}_k - \widetilde{\boldsymbol{\mu}})(\boldsymbol{\mu}_k - \widetilde{\boldsymbol{\mu}})^T\big) .$$

Then, $d_{\mathrm{KL}}(F|G) \approx d_{\mathrm{KL}}(\hat{f}|\hat{g})$, where

$$d_{\mathrm{KL}}(f|g) = \frac{1}{2}\Big(\log\Big(\frac{\det(\hat{\mathbf{\Sigma}})}{\det(\hat{\mathbf{S}})}\Big) + \mathrm{trace}\big(\hat{\mathbf{\Sigma}}^{-1}\hat{\mathbf{S}}\big) - M + (\hat{\mathbf{m}} - \hat{\boldsymbol{\mu}})\hat{\mathbf{\Sigma}}^{-1}(\hat{\mathbf{m}} - \hat{\boldsymbol{\mu}})\Big) .$$

### GMM with Spatial Relationships*

In Example 10.9 we used the GMM to segment a SEM input image $\mathbf{X} = \{x_1, x_2, \ldots, x_N\}$ into $K$ classes by *disregarding* all pixel interactions. This is possible for SEM images where the pixel interactions are weak [10]. However, in electro-optical images, the pixel interactions are usually much stronger and *cannot* be disregarded. Nevertheless, by using a modified GMM which includes pixel interactions, we may segment a electro-optical image.

In the modified GMM, we let the conditional probability $p(x_i|\mathbf{\Theta})$ in Eq. (10.9) depend also on the neighbors of $x_i$: For each pixel $x_i$ we erect a small window $W_i$ centered on $x_i$. If $\overline{x_i}$ is the mean pixel gray-level in $W_i$, then we incorporate $\overline{x_i}$ into the calculation of $p(x_i|\mathbf{\Theta})$:

$$p(x_i|\mathbf{\Theta}) \approx (1-\beta)\sum_k \pi_k \mathcal{N}(x_i|\mu_k, \Sigma_k) + \beta \sum_k \pi_k \mathcal{N}(\overline{x_i}|\mu_k, \Sigma_k) , \qquad (10.10)$$

where $\beta$ determines the relative weight of the pixel interactions [4]. The log-likelihood of Eq. (10.10) is

$$\log L(\mathbf{\Theta}|\mathbf{X}) = \sum_i \log \sum_k \pi_k \Big((1-\beta)\mathcal{N}(x_i|\mu_k, \Sigma_k) + \beta \mathcal{N}(\overline{x_i}|\mu_k, \Sigma_k)\Big) ,$$

whose maximum likelihood solution is found using the iterative EM algorithm:

$$\text{E-step}: \qquad p(k|x_i, \mathbf{\Theta}^{(t)}) = \frac{\pi_k^{(t)}\Big[(1-\beta)\mathcal{N}(x_i|\mu_k, \Sigma_k) + \beta\mathcal{N}(\overline{x_i}|\mu_k, \Sigma_k)\Big]}{\sum_{j=1}^{K}\pi_j^{(t)}\Big[(1-\beta)\mathcal{N}(x_i|\mu_j, \Sigma_j) + \beta\mathcal{N}(\overline{x_i}|\mu_j, \Sigma_j)\Big]} ,$$

$$\text{M-step} \qquad \pi_k^{(t+1)} = \frac{1}{N}\sum_{i=1}^{N} p\big(k|x_i, \mathbf{\Theta}^{(t)}\big) ,$$

$$\mu_k^{(t+1)} = \frac{\sum_i \Big[(1-\beta)x_i + \beta\overline{x_i}\Big] p\big(k|x_i, \mathbf{\Theta}^{(t)}\big)}{\sum_i p\big(k|x_i, \mathbf{\Theta}^{(t)}\big)} ,$$

$$\Sigma_k^{(t+1)} = \frac{p(k|x_i, \mathbf{\Theta}^{(t)})\Big[d\big(x_i, \mu_k^{(t)}\big) + \beta d\big(\overline{x_i}, \mu_k^{(t)}\big)\Big]}{\sum_i p\big(k|x_i, \mathbf{\Theta}^{(t)}\big)} ,$$

where $d(x, \mu) = (x-\mu)(x-\mu)^T$. After convergence, we give each pixel $x_i$ a label $y_i = k^*$, where

$$y_i = k^* \quad \text{if} \quad k^* = \arg\max_k p(k|x_i, \mathbf{\Theta}) .$$

Experimental results [4, 27] show that using this simple modified GMM model can improve the performance of image segmentation especially for noisy images. Ref. [4] recommends setting $\beta$ equal to the noise variance in the image.

A slightly more sophisticated model was proposed by [33] who modeled the pixel interactions using the weighted arithmetic mean of the gray-levels $x_j,\ j \in W_i$. In this case, Eq. (10.9) becomes

$$p(x_i|\boldsymbol{\Theta}) \approx \sum_k \pi_{ik} \sum_{j \in W_i} \frac{w_j(i)}{R_i} \mathcal{N}(x_j|\mu_j, \Sigma_j) , \tag{10.11}$$

where the strength of the weights $w_j(i)$ decrease as the distance between $x_i$ and $x_j$ increases and $R_i = \sum_{j \in W_i} w_j(i)$.

The corresponding a posteriori pdf $p(k|x_i, \boldsymbol{\Theta})$ and the maximum likelihood parameter values are found using the iterative EM algorithm:

$$\begin{aligned}
p(k|x_i, \boldsymbol{\Theta}^{(t)}) &= \frac{\pi_{ik}^{(t)} \sum_{j \in W_i} \frac{w_j(i)}{R_i} \mathcal{N}\big(x_j|\mu_k^{(t)}, \Sigma_k\big)}{\sum_{h=1}^{K} \pi_{ih}^{(t)} \sum_{j \in W_i} \frac{w_j(i)}{R_i} \mathcal{N}\big(\mathbf{x}_j|\mu_h^{(t)}, \Sigma_h\big)} , \\
\pi_{ik}^{(t+1)} &= \frac{\sum_{j \in W_i} \frac{w_j(i)}{R_i} p(k|x_j, \boldsymbol{\Theta}^{(t)})}{\sum_{h=1}^{K} \sum_{j \in W_i} \frac{w_j(i)}{R_i} p(h|x_j, \boldsymbol{\Theta}^{(t)})} , \\
\mu_k^{(t+1)} &= \frac{\sum_{i=1}^{N} \sum_{j \in W_i} \frac{w_j(i)}{R_i} p(k|x_i, \boldsymbol{\Theta}^{(t)}) x_j}{\sum_{i=1}^{N} p(k|x_i, \boldsymbol{\Theta}^{(t)})} , \\
\Sigma_k^{(t+1)} &= \frac{\sum_{i=1}^{N} \sum_{j \in W_i} \frac{w_j(i)}{R_i} p(k|x_i, \boldsymbol{\Theta}^{(t)}) \big(x_j - \mu_k^{(t)}\big)\big(x_j - \mu_k^{(t)}\big)^T}{\sum_{i=1}^{N} p(k|x_i, \boldsymbol{\Theta}^{(t)})} ,
\end{aligned}$$

**Example 10.10** *Image Segmentation Using GMM with Saliency Map* [7]. Recently state-of-the-art image segmentation was obtained using a saliency map $S$ for the weights $w_j(i)$ in Eq. (10.11).

A saliency map $S$ is derived from the human visual system (HVS) and varies according to the amount of information perceived by the HVS. Given an input image $\mathbf{X}$, $S$ is calculated from the residual spectrum $R(f)$:

$$R(f) = L(f) - A(f) ,$$

where $A(f)$ is the average spectrum of $I$ and $L(f)$ is image log spectrum.

Let $W_i$ be a $3 \times 3$ window centered on pixel $x_i$ (coordinates $(r, c)$). Then for each window $W_i$, there are $J = 9$ neighboring pixels $x_j \in W_i$, whose saliency values are $\{S(r-1, c-1), S(r, c-1), \ldots, S(r+1, c+1)\}$. By setting the weights $w_j(i)$ in Eq. (10.11) equal to the corresponding saliency values:

$$
\begin{aligned}
w_1(i) =&\ S(r-1, c-1)\ , \\
w_2(i) =&\ S(r, c-1)\ , \\
&\ \vdots \\
w_9(i) =&\ S(r+1, c+1)\ .
\end{aligned}
$$

the authors obtain state-of-the-art segmentation performance.

### 10.5.1 Examples

GMM has been widely used in a very wide variety of different data fusion applications. To illustrate the power and versatility of GMM we give two examples.

#### Speaker Recognition

One of the first and still popular uses of GMM is in speaker recognition. The following example describes GMM-based text-independent speaker *verification* system.

**Example 10.11** *GMM Text-independent Speaker Verification* [6, 8, 25]. Speaker recognition is a biometric technology wherein the speakers voice signal is analyzed to uniquely recognize the speaker. The reason why we may use speech for this purpose is that speech is a product of the speaker's individual (unique) anatomy (vocal track geometry) and his linguistic background.

Given a segment of speech $\mathbf{Y}$ and a hyothesized speaker $S$ the task of speaker verification is to determine whether, or not, $\mathbf{Y}$ was spoken by $S$. The hypothesized speaker is assumed to come from a list of $K$ speakers $\{S_1, S_2, \ldots, S_K\}$. Mathematically, the task may be formulated as a test between the hypothesis $\mathcal{H}_0$ that $\mathbf{Y}$ is from the hypothesized speaker $S$ and the hypothesis $\mathcal{H}_1$ that $\mathbf{Y}$ is *not* from the hypothesized speaker $S$.

To decide between the two hypotheses we use the likelihood ratio test (LRT) given by

$$
\text{Decision} = \begin{cases} \mathcal{H}_0 & \text{if } \ LRT \geq T\ , \\ \mathcal{H}_1 & \text{otherwise}\ . \end{cases}
$$

where $p(\mathbf{Y}|\mathcal{H}_i)$ is the pdf for hypothesis $\mathcal{H}_i, i \in \{0, 1\}$ evaluated on the observed speech segment $\mathbf{Y}$, $T$ is a decision threshold and $LRT = p(\mathbf{Y}|\mathcal{H}_0)/p(\mathbf{Y}|\mathcal{H}_1)$. We often use the log likelihood ratio test. In this case,

$$\text{Decision} = \begin{cases} \mathcal{H}_0 & \text{if } \ln p(\mathbf{Y}|\mathcal{H}_0) - \ln p(\mathbf{Y}|\mathcal{H}_1) \geq thr \ , \\ \mathcal{H}_1 & \text{otherwise} \ . \end{cases}$$

where $thr$ is the appropriate log threshold.

To calculate the pdf's $p(\mathbf{Y}|\mathcal{H}_0)$ and $p(\mathbf{Y}|\mathcal{H}_1)$ we extract features from the speech signal. Then we use the GMM to create a set of statistical models for each speaker.

### Point Set Registration Using the Gaussian Mixture Model*

Coherent point drift (CPD) [24] is a probabilistic approach for the spatial alignment of two point data sets and works for both both rigid and non-rigid transforms. In CPD the problem of spatial alignment is formulated as a problem in pdf estimation: One set of points represents the centroids of the GMM and the other set of points represents the data points at which the pdf is estimated. Spatial alignment is performed by fitting the GMM centroids to the data points by maximizing the pdf.

**Example 10.12** *Coherent Point Drift using GMM* [24, 28, 34]. Let $\mathbf{A} = \{\mathbf{a}_1, \mathbf{a}_2, \ldots, \mathbf{a}_M\}$ and $\mathbf{B} = \{\mathbf{b}_1, \mathbf{b}_2, \ldots, \mathbf{b}_N\}$ denote two sets of 2D $(x, y)$ or 3D $(x, y, z)$ data points. The points $\mathbf{A}$ are assumed to be GMM centroids and the points $\mathbf{B}$ are assumed to be generated by the GMM. The goal is to find the optimal GMM centroids to fit $\mathbf{B}$. Due to the presence of noise and outliers, we use a *robust* GMM in which, in addition to the Gaussian components, we include a single uniform component with a weight $\omega$ (see Sect. 12.3.4). Assuming all the Gaussian components have the same a priori probability $\pi_m = 1/M$ and are independently distributed with the same isotropic variance $\sigma^2$, the mixture model then takes the form:

$$p(\mathbf{B}) = \prod_{i=1}^{N} p(\mathbf{b}_i) = \prod_{i=1}^{N} \Big(\omega \frac{1}{N} + (1-\omega) \sum_{m=1}^{M} \pi_m p(\mathbf{b}_i|m)\Big) \ ,$$

where $\pi_m = 1/M, 0 \leq \omega \leq 1$ and

$$p(\mathbf{b}_i|m) = \frac{1}{(2\pi\sigma^2)^{D/2}} \exp -\frac{1}{2\sigma^2} \|\mathbf{b}_i - \mathcal{T}(\mathbf{a}_m, \boldsymbol{\theta})\|^2 \ .$$

In this equation $\mathcal{T}(\mathbf{a}_m, \boldsymbol{\theta})$ denotes the transformed GMM centroids, where $\mathcal{T}$ has parameters $\boldsymbol{\theta}$. These parameters, along with the variance $\sigma^2$, are optimized by maximizing the probability $p(\mathbf{B})$ using an iterative EM algorithm. At each iteration, the steps in the EM algorithm are:

**E-step**. Use "old" parameter values (calculated in the previous iteration) to compute the a posteriori pdf:

$$p_{\text{old}}(m|\mathbf{b}_i) = \frac{\exp - \|\mathbf{b}_i - \mathcal{T}(\mathbf{a}_m, \boldsymbol{\theta}_{\text{old}})\|^2 / 2\sigma^2_{\text{old}}}{\sum_{m=1}^{M} \exp - \|\mathbf{b}_i - \mathcal{T}(\mathbf{a}_m, \boldsymbol{\theta}_{\text{old}})\|^2 / 2\sigma^2_{\text{old}} + c}$$

where $c = (2\pi\sigma^2_{\text{old}})^{D/2} \frac{\omega}{1-\omega} \frac{M}{N}$.

**M-step**. Calculate "new" transformation parameters $\boldsymbol{\theta}_{\text{new}}$ and "new" variance $\sigma^2_{\text{new}}$ by maximizing $p(\mathbf{B})$, or equivalently minimizing the negative log-likelihood function. Leaving out terms which are constant wrt $\boldsymbol{\theta}$ and $\sigma^2$, the function to be minimized is

$$Q(\boldsymbol{\theta}, \sigma^2) = \frac{1}{2\sigma^2} \sum_{i=1}^{N} \sum_{m=1}^{M} p_{\text{old}}(m|\mathbf{b}_i) \times \|\mathbf{b}_i - \mathcal{T}(\mathbf{a}_m, \boldsymbol{\theta})\|^2 + \frac{N_p D}{2} \ln(\sigma^2)$$

where $N_p = \sum_{i=1}^{N} \sum_{m=1}^{M} p_{\text{old}}(\mathbf{b}_i|m) \leq N$ is the estimated number of matching points.

In the above algorithm the value of $\omega$ is set manually. The optimum value of $\omega$ may, however, be found using an optimization algorithm given in [31].

## 10.6 Model Selection

At times we may have more than one model, and then our interest may focus on assessing the fitness of each model and conducting model selection. To illustrate Bayesian model selection, we focus on the comparison between a "null" model, $M = 0$, and an alternative model, $M = 1$. In this case, the corresponding joint pdf is:

$$p(\mathbf{y}, \boldsymbol{\theta}, M = m|I) = p(\mathbf{y}|\boldsymbol{\theta}, M = m, I)\pi(\boldsymbol{\theta}, M = m|I) \ . \tag{10.12}$$

If we assume that $\mathbf{y}$ depends on the models through their parameters, then Eq. (10.12) becomes:

$$p(\mathbf{y}, \boldsymbol{\theta}, M = m|I) = p(\mathbf{y}|\boldsymbol{\theta}_m)\pi(\boldsymbol{\theta}_m|M = m, I)\pi(M = m|I) \ , \tag{10.13}$$

where $\pi(\boldsymbol{\theta}_m|M = m, I)$ is the a priori distribution for the parameters, $\boldsymbol{\theta}_m$, given the model $M = m$ and $\pi(M = m|I)$ is the a priori probability of the model $M = m$. The a posteriori probability for the model $M = m$ is proportional to the evidence,

$$p(M = m|\mathbf{y}, I) \propto \int p(\mathbf{y}|\boldsymbol{\theta}_m, I)\pi(\boldsymbol{\theta}_m|M = m, I)\pi(M = m|I)d\boldsymbol{\theta}_m \ , \tag{10.14}$$

and the optimal model is

$$m_{\text{OPT}} = \arg\max_m \big(p(M = m|\mathbf{y}, I)\big) . \tag{10.15}$$

The choice of the a priori distribution, $\pi(M = m|I)$, is clearly dependent on the application. If both models appear to be equally likely and contain the same number of parameters, then $\pi(M = 0|I) = \frac{1}{2} = \pi(M = 1|I)$.

In comparing different models we often use a simple approximation to the a posteriori pdf, $p(\mathbf{y}|\boldsymbol{\theta}, M = m, I)$, which is known as the *Laplace approximation.*

## 10.6.1 Laplace Approximation

The Laplace approximation assumes that the a posteriori distribution, $p(\boldsymbol{\theta}|\mathbf{y}, M = m, I)$, has only one mode, which we approximate with a Gaussian function. We proceed as follows. Let

$$\begin{aligned} t(\boldsymbol{\theta}) &= \ln p(\boldsymbol{\theta}|\mathbf{y}, M = m, I) , \\ &= \ln p(\mathbf{y}|\boldsymbol{\theta}, M = m, I) + \ln \pi(\boldsymbol{\theta}|M = m, I) + \text{const} . \end{aligned} \tag{10.16}$$

Then we expand $t(\boldsymbol{\theta})$ about $\hat{\boldsymbol{\theta}}$, its maximum a posteriori (MAP) value, i.e. around the value of $\boldsymbol{\theta}$ that maximizes $p(\boldsymbol{\theta}|\mathbf{y}, M = m, I)$:

$$\begin{aligned} t(\boldsymbol{\theta}) &= t(\hat{\boldsymbol{\theta}}) + (\boldsymbol{\theta} - \hat{\boldsymbol{\theta}}) \left.\frac{\partial t(\boldsymbol{\theta})}{\partial \boldsymbol{\theta}}\right|_{\boldsymbol{\theta}} = \hat{\boldsymbol{\theta}} + \frac{1}{2}(\boldsymbol{\theta} - \hat{\boldsymbol{\theta}})^T \left.\frac{\partial^2 t(\boldsymbol{\theta})}{\partial \boldsymbol{\theta} \partial \boldsymbol{\theta}^T}\right|_{\boldsymbol{\theta}} = \hat{\boldsymbol{\theta}}(\boldsymbol{\theta} - \hat{\boldsymbol{\theta}}) + \ldots \\ &\approx t(\hat{\boldsymbol{\theta}}) + (\boldsymbol{\theta} - \hat{\boldsymbol{\theta}}) H(\boldsymbol{\theta}) (\boldsymbol{\theta} - \hat{\boldsymbol{\theta}})^T , \end{aligned} \tag{10.17}$$

where

$$H(\hat{\boldsymbol{\theta}}) = \left.\frac{\partial^2 \ln p(\boldsymbol{\theta}|\mathbf{y}, M = m, I)}{\partial \boldsymbol{\theta} \partial \boldsymbol{\theta}^T}\right|_{\boldsymbol{\theta} = \hat{\boldsymbol{\theta}}} . \tag{10.18}$$

Substituting Eq. (10.17) into the logarithm of Eq. (10.12) and integrating over $\boldsymbol{\theta}$, yields

$$\begin{aligned} \ln p(\mathbf{y}|M = m, I) &= \ln \int \exp\big(t(\boldsymbol{\theta})\big) d\boldsymbol{\theta} , \\ &\approx t(\hat{\boldsymbol{\theta}}) + \frac{1}{2} \ln |2\pi H^{-1}| , \\ &\approx \ln p(\mathbf{y}|\hat{\boldsymbol{\theta}}, M = m, I) + \ln \pi(\hat{\boldsymbol{\theta}}|M = m, I) + \frac{d}{2} \ln 2\pi - \frac{1}{2} \ln |H| , \end{aligned} \tag{10.19}$$

**Table 10.5** Approximations to logarithm of bayesian evidence $\ln p(\mathbf{y}|M=m,I)$

| Name | Approximation to $\ln p(\mathbf{y}|M=m,I)$ |
|---|---|
| Laplace | $\ln p(\mathbf{y}|\hat{\boldsymbol{\theta}},M=m,I)+\ln\pi(\hat{\boldsymbol{\theta}}|M=m,I)+\frac{d}{2}\ln(2\pi)-\frac{1}{2}\ln|H|$, where $\hat{\boldsymbol{\theta}}=\arg\max(p(\boldsymbol{\theta}|\mathbf{y},M=m,I))$; $d$ is dimension of parameter space $\boldsymbol{\theta}$ |
| BIC | $\ln p(\mathbf{y}|\hat{\boldsymbol{\theta}},M=m,I)-\frac{d}{2}\ln N$ |
| AIC | $\ln p(\mathbf{y}|\hat{\boldsymbol{\theta}},M=m,I)-d$ |
| KIC | $\ln p(\mathbf{y}|\hat{\boldsymbol{\theta}},M=m,I)-3d/2$ |
| Cheeseman-Stutz | $\ln p(\mathbf{y},\hat{\mathbf{z}}|M=m,I)+\ln p(\mathbf{y}|\hat{\boldsymbol{\theta}},M=m,I)-\ln p(\mathbf{y},\hat{\mathbf{z}}|\hat{\boldsymbol{\theta}},M=m,I)$, where $\hat{\mathbf{z}}=\arg\max p(\mathbf{z}|\mathbf{y},\hat{\boldsymbol{\theta}},M=m,I)$ |

BIC, AIC and KIC denote, respectively, Bayesian, Akaike and Kullback Information Criterion

where $d$ is the dimensionality of the parameter space $\boldsymbol{\theta}$. The corresponding approximate expression for $p(\mathbf{y}|M=m,I)$ is

$$p(\mathbf{y}|M=m,I)\approx\sqrt{(2\pi)^d|H^{-1}|}p(\mathbf{y}|\hat{\boldsymbol{\theta}},M=m,I)\pi(\hat{\boldsymbol{\theta}}|M=m,I)\ . \qquad (10.20)$$

In Table 10.5 we list some approximations to the natural logarithm of the Bayesian evidence which are commonly used in model selection. The following example illustrates Bayesian model selection to select the optimum number of Gaussians in GMM [20].

**Example 10.13** *Selecting the Number of Gaussians in a GMM* [12, 26]. In a GMM we approximate a function $f(y)$ using a sum of $K$ Gaussian pdf's:

$$f(y)\approx\sum_k\alpha_k\mathcal{N}(y|\mu_k,\sigma_k^2)\ ,$$

subject to the constraints

$$\alpha_k>0\ \text{ and }\ \sum_k\alpha_k=1\ .$$

Traditionally the user supplies the value of $K$. However, if the user is unable to supply a value for $K$, it may be determined by selecting the best GMM from a set of candidate models, each model having a different number of Gaussians. After optimally fitting the parameters of the different models (using the EM algorithm) we find the optimum value for $K$ (i.e. the optimum model) by choosing the model with the largest Bayesian evidence.

The following example illustrates the use of Bayesian model selection to estimate the intrinsic dimensionality of a given data set $\{y_i\}$.

**Example 10.14** *Intrinsic Data Dimensionality* [21]. A central issue in pattern recognition is to estimate the intrinsic dimensionality $d$ of the input data $\{\mathbf{y}_i\}$. One approach to estimating $d$ is let $d$ equal the optimum number principal components we need to retain in a principal component analysis (PCA) of the $\{\mathbf{y}_i\}$ (see Sect. 4.7.1). By interpreting PCA within a Bayesian framework, we are able to derive the following BIC approximation for the evidence:

$$p(\{\mathbf{y}_i\}|L) = \Big(\prod_{j=1}^{L} \lambda_j\Big)^{-N/2} \sigma^{-N(D-L)} N^{-(m+L)/2} ,$$

where $D$ is the dimension of the input data and $m = DL - L(L+1)/2$. The intrinsic dimensionality of the input data is then defined as

$$d = L_{\max} = \arg\max p(\{\mathbf{y}_i\}|L) .$$

Apart from the model selection methods listed in Table 10.5 there are a large number of non-Bayesian model selection methods available. The more popular of these methods are listed in Table 14.6. Recently [9] questioned the use of the Bayesian model selection approximations given in Table 10.5 for applications involving the classification of an object $O$. They suggest a new criterion, known as *Bayesian Entropy Criterion*, which is defined as follows. Suppose each measurement $\mathbf{y}_i$ has a label $z_i$ associated with it. Then,

$$\ln p(\mathbf{z}|\mathbf{y}, M = m, I) \approx \ln p(\mathbf{y}, \mathbf{z}|\hat{\boldsymbol{\theta}}, M = m, I) - \ln p(\mathbf{y}|\widetilde{\boldsymbol{\theta}}, M = m, I) , \quad (10.21)$$

where $\widetilde{\boldsymbol{\theta}} = \arg\max_{\boldsymbol{\theta}} \ln p(\mathbf{y}|\boldsymbol{\theta}, M = m, I)$.

## 10.6.2 Bayesian Model Averaging (BMA)

An alternative to model selection is Bayesian Model Averaging (BMA) [15]. In BMA we do not select a single "optimal" model. Instead we take the uncertainty involved in selecting an optimal model into consideration by averaging over the a posteriori distribution of multiple models. In a BMA framework, the a posteriori pdf, $p(\boldsymbol{\theta}|\mathbf{y}, I)$, is given by the following formula:

$$\begin{aligned} p(\boldsymbol{\theta}|\mathbf{y}, I) &= \sum_m p(\boldsymbol{\theta}|\mathbf{y}, M = m, I) p(M = m|\mathbf{y}, I) , \\ &= \frac{\sum_m p(\boldsymbol{\theta}|\mathbf{y}, M = m, I)\pi(M = m|I) p(\mathbf{y}|M = m, I)}{\sum_m \pi(M = m|I) p(\mathbf{y}|M = m, I)} , \end{aligned} \quad (10.22)$$

**Table 10.6** Algorithms used in Bayesian inference

| Name | Description |
|---|---|
| Conjugate Priors | This is an *exact* method for Bayesian Inference |
| Laplace Approx. | Approximation often used in point parameter estimation and model selection |
| Variational Algs. | Uses set of variational parameters which are iteratively updated to minimize cross-entropy between approximate and true pdf |
| Belief Propagation | Involves propagating pdf's using local message passing algorithm |
| EM | Optimization algorithm. Convenient when problem involves missing data |
| MCMC | All-purpose approximate algorithm for Bayesian inference |

EM: Expectation-Maximization algorithm, MCMC: Markov Chain Monte Carlo

where $\pi(M = m|I)$ is the a priori probability of the model $M = m$. In general, BMA has a better performance than any single model that could reasonably have been selected. In practice, we limit the summation in Eq. (10.22) to a small number of models.[4] In Sect. 15.2 we illustrate the use of the BMA as a general framework for combining classifiers in a multiple classifier system (MCS).

## 10.7 Computation

In many practical problems, the main obstacle to using the Bayesian method is required computation. However, the introduction of iterative simulation methods, such as data augmentation and the more general Markov chain Monte Carlo (MCMC), have opened up the Bayesian method to mainstream applications.

Table 10.6 lists some of the optimization and integration algorithms which are used in Bayesian inference.

### Markov Chain Monte Carlo (MCMC)

Although the integrals in Bayesian inference are often intractable, sampling methods can be used to numerically evaluate them. There are a wide range of practical sampling techniques which go under the generic name "Monte Carlo". Given an a posteriori pdf, $p(\mathbf{y}|I)$, we approximate the integral $\int f(\mathbf{y})p(\mathbf{y}|I)d\mathbf{y}$ by using the average of $N$ independent samples, $\mathbf{y}_1, \mathbf{y}_2, \ldots, \mathbf{y}_N$, which are drawn from $p(\mathbf{y}|I)$. Mathematically, we have

$$\int f(\mathbf{y})p(\mathbf{y}|I)d\mathbf{y} \approx \frac{1}{N}\sum_i f(\mathbf{y}_i) . \qquad (10.23)$$

[4] In practice we often discard all models whose a priori probabilities $\pi(M = m|I)$ are less than 5% of the a priori probability of the optimum model.

Unfortunately, in many practical applications, $p(\mathbf{y}|I)$ is only non-zero in an extremely small region of the parameter, or $\mathbf{y}$, space. The problem, then, is how to efficiently generate the samples $\mathbf{y}_i$. The MCMC method works by constructing a random walk (Markov chain) in the parameter space such that the probability for being in a region of the space is proportional to the posterior density for that region. We start with some value of $\mathbf{y}$, calculate the a posteriori density there, take a small step and then recalculate the density. The step is accepted or rejected according to some rule and the process is repeated. The resulting output is a series of points in the sample space. The various methods differ according to the rules used to make the moves in the parameter space and the rules determining whether or not a step is accepted. It turns out to be relatively simple to invent such rules that are guaranteed to produce a set of steps that correctly sample the a posteriori distribution.

## 10.8 Software

**MATLAB**. MATLAB IMAGE PROCESSING and STATISTICS AND MACHINE LEARNING TOOLBOX includes m-files for different pdf's, dimension reduction, feature extraction, binarization (Otsu) and GMM (including image segmentation and coherent point drift).

**PYTHON**. SCIKIT-LEARN includes Python modules for dimension reduction, feature selection and extraction, and GMM.

## 10.9 Further Reading

Modern textbooks on Bayesian inference are: [1, 13, 19, 30]. In addition there is a wide choice of specialist review articles including [2, 11]. Useful books and review articles on approximate methods for Bayesian inference are: [5, 22]. Specific reference on MCMC are: [14]; on the EM algorithm are: [3, 20, 26]. Full-length discussions on model selection are given in [19]. For a discussion of the philosophy which underlies the Bayesian paradigm we recommend [16].

## Problems

**10.1** Define the main premise of Bayesian inference. What are its relative advantages and disadvantages?

**10.2** Compare and contrast fuzzy logic and Bayesian inference as a framework for multi-sensor data fusion.

**10.3** What are (i) conjugate priors? (ii) Non-informative priors? Give some examples.

**10.4** Explain the EM algorithm. Write down the main steps in the EM algorithm for modeling a given function as a mixture of Gaussians (Example 10.8).

**10.5** Describe Bayesian model averaging? Compare and contrast it with other model selection procedures.

## References

1. D'Agostini G (2003) Bayesian reasoning in data analysis. World Scientific, Singapore
2. D'Agostini G (2003) Bayesian inference in processing experimental data: principles and basic applications. Rep Prog Phys 66:1383–1419
3. Archambeau C (2005) Probabilistic models in noisy environments. PhD thesis, Universite Catholque de Louvain, Belgium
4. Balafar MA (2011) Spatial based expectation maximizing (EM). Diagn Path 6:103
5. Beal MJ (2003) Variational algorithms for approximate Bayesian inference. PhD thesis, University of London
6. Barai B, Chakraborty T, Das N et al (2022) Closed-set speaker identification using VQ and GMM based models. Int J Speech Tech 25:173–196
7. Bi H, Tang H, Yang G et al (2018) Accurate image segmentation using Gaussian mixture model with saliency map. Patt Analy App 21:869–878
8. Bimbot F, Bonastre J-F, Fredouille C et al (2004) A tutorial on text-independent speaker verification. Eurasip J Appl Sig Proc 4:430–451
9. Bouchard G, Celeux G (2006) Selection of generative models in classification. IEEE Trans Patt Anal Mach Intell 28:544–554
10. Chatzigeorgiou M, Vrigkas M, Beazi-Katsioti M et al (2023) Segmentation of SEM images of multiphase materials: when Gaussian mixture models are accurate? J Micr 289:58–70
11. Dose V (2003) Bayesian inference in physics: case studies. Rep Prog Phys 66:1421–1461
12. Figueiredo M, Jain AK (2002) Unsupervised learning of finite mixture models. IEEE Trans Patt Anal Mach Intell 24:381–396
13. Gelman A, Larlin JS, Stern HS, Rubin DR (2003) Bayesian data analysis. Published by Chapman and Hall
14. Gilks WR, Richardson S, Spiegelhater DJ (1996) Markov Chain Monte Carlo in practice. Published by Chapman and Hall
15. Hoeting JA, Madigan D, Raftery AE et al (1999) Bayesian model averaging: a tutorial. Stat Sci 14:382–417
16. Howson C, Urbach P (1996) Scientific reasoning: the Bayesian approach, 2nd edn. Published by Open Court Publishing Co., Illinois, USA
17. Kass RE, Wasserman L (1994) Formal rules for selecting prior distributions: a review and annotated bibliography. Technical Report 583, Department of Statistics, Carnegie Mellon University
18. Lambert B (2018) A student's guide to Bayesian statistics. Published by Sage Publications
19. MacKay DJC (2003) Information theory, inference and learning algorithms. Published by Cambridge University Press
20. McLachlan GJ, Krishnan T (1997) The EM algorithm and extensions. Published by John Wiley and Sons
21. Minka TP (2000) Automatic choice of dimensionality for PCA. NIPS
22. Minka TP (2001) A family of algorithms for approximate Bayesian inference. Phd thesis, Masschusetts Institute of Technology, Cambridge, Massachusetts, USA

23. Murphy KP (2007) Conjugate Bayesian analysis of the Gaussian distribution. Notes available on internet
24. Myronenko A, Song X (2010) Point-set registration: coherent point drift. IEEE Trans Patt Analy Mach Intell 32:2262–2275
25. Nainan S, Kulkarni V (2021) Enhancement in speaker recognition for optimized speech features using GMM, SVM and 1-D CNN. Int J Speech Tech 24:809–822
26. Paalanen P, Kristian J, Kamarainen J-K et al (2006) Feature representation and discrimination based on Gaussian mixture model probability densities - practices and algorithms. Patt Recogn 39:1346–1358
27. Raghavendro U, Gudigar A, Paul A et al (2023) Brain tumor detection and screening using artificial intelligence techniques: current trends and future perspectives. Comp Biol Med 163:107063
28. Shen L, Zhu J, Fan C, Huang X, et al (2019) Novel co-variant feature point matching based on Gaussian mixture model. arXiv:1910.11981
29. Singh A, Nocerino J (2002) Robust estimation of mean and variance using environmental data sets with below detection limit observations. Chemometrics Intell Lab Syst 60:69–86
30. Sivia DS, Skilling J (2006) Data analysis: a Bayesian tutorial, 2nd edn. Published by Oxford University Press
31. Wang P, Wang P, Qu Z-G et al (2011) A refined coherent point drift (CPD) algorithm for point set registration. Sci China 54:2639–2646
32. Yang R, Berger JO (1997) A catalog of noninformative priors. Plann Infer 79:223–235. Discussion Paper 97-42, ISDS, Duke University, Durham, NC, USA
33. Zhang H, Wu QMJ, Nguyen TM (2013) Incorporating mean template into finite mixture model for image segmentation. IEEE Trans Neural Netw Learn Syst 24:328–335
34. Zang Y, Lindenbergh R (2019) An improved coherent point drift method for TLS point cloud registration of complex scenes. Int Archives Photogramm Remote Sens Spatial Inform Sci 1169–1175

# Chapter 11
# Parameter Estimation

**Abstract** In this chapter we consider parameter estimation which we regard as an application of Bayesian statistics to the problem of data fusion. We first consider curve fitting which we regard as a problem in fusing multiple sensor observations. Maximum a posteriori, maximum likelihood and minimum mean square error solutions are compared and we determine under what conditions these solutions are equal. Finally we consider the linear Gaussian model and the problem of change point detection.

## 11.1 Introduction

In this chapter we consider *parameter estimation*, which is our first application of Bayesian statistics to the problem of data fusion. Let $\mathbf{y} = \left(\mathbf{y}_i^T, \mathbf{y}_2^T, \ldots, \mathbf{y}_N^T\right)^T$ denote a set of $N$ multi-dimensional sensor observations $\mathbf{y}_i, i \in \{1, 2, \ldots, N\}$, where $\mathbf{y}_i = (y_i^{(1)}, y_i^{(2)}, \ldots, y_i^{(M)})^T$. We suppose the $\mathbf{y}_i$ were generated by a parametric distribution $p(\mathbf{y}|\boldsymbol{\theta}, I)$, where $\boldsymbol{\theta} = (\theta_1, \theta_2, \ldots, \theta_K)^T$ are the parameters and $I$ is any background information we may have. Then parameter estimation involves estimating $\boldsymbol{\theta}$ from $\mathbf{y}$.

In the Bayesian framework we assume an a priori distribution $\pi(\boldsymbol{\theta}|I)$ for $\boldsymbol{\theta}$. The *a posteriori* distribution, $p(\boldsymbol{\theta}|\mathbf{y}, I)$, represents the pdf of $\boldsymbol{\theta}$ after observing $\mathbf{y}$, where the relation between the a priori and the *a posteriori* distributions is given by Bayes' Theorem:

$$p(\boldsymbol{\theta}|\mathbf{y}, I) = \frac{p(\mathbf{y}|\boldsymbol{\theta}, I)\pi(\boldsymbol{\theta}|I)}{p(\mathbf{y}|I)} . \qquad (11.1)$$

In many applications, parameter estimation represents the major fusion algorithm in a data fusion system. In Table 11.1 we list some of these applications together with their classification.



H. B. Mitchell, *Data Fusion*, https://doi.org/10.1007/978-3-662-71023-4_11

**Table 11.1** Applications in which parameter estimation is primary fusion algorithm

| Class | | Application |
|---|---|---|
| Early-DaO | Example 2.5 | Time-of-Flight ultrasonic sensor |
| | Example 3.2 | Image smoothing |
| | Example 3.9 | Heart rate estimation using a Kalman filter |
| | Example 4.19 | Debiased cartesian coordinates |
| | Example 9.16 | Psychometric functions |
| | Example 11.4 | Kriging: fusing digital elevation model & Rainfall measurements |
| Early-FeO | Example 4.2 | Bayesian triangulation of direction-bearing measurements |
| | Example 4.20 | PCA for vehicle sound signature application |
| | Example 4.22 | Recursive LDA |
| | Example 4.21 | Principal discriminant analysis (PDA) for small sample size problems |
| | Example 11.5 | Fitting a finite-length straight-line segment in 2D |
| | Example 11.6 | Non-intrusive speech quality estimation using GMM's |
| | Example 12.14 | Target track initialization using the Hough transform |
| Inter-FeO | Example 9.19 | Fusing Similarity Coefficients in a Virtual Screening Programme |
| | Example 14.9 | Feature selection: drug discovery |

## 11.2 Parameter Estimation

Once the *a posterior* distribution is available we can use it to estimate the parameters $\boldsymbol{\theta}$. In what follows we shall suppose $\boldsymbol{\theta}$ is a scalar parameter, $\theta$. We define a *loss function*, $L = L(t, \theta)$, which links $\theta$ to its estimated value $t$. By minimizing the expected value of $L$, we obtain the *optimum* estimated value $\hat{\theta}$, where

$$\hat{\theta} = \min_t \int_{-\infty}^{\infty} L(t, \theta) p(\theta | \mathbf{y}, I) d\theta \ . \tag{11.2}$$

Clearly the estimates, $\hat{\theta}$, will vary with the choice of loss function. Three commonly used loss functions are:

**Quadratic Loss Function**. This is defined as $L(t, \theta) = \Delta^2$, where $\Delta = t - \theta$. In this case, the corresponding MMSE (minimum mean square error) estimate, $\hat{\theta}_{\text{MMSE}}$, is equal to the expected value of $\theta$:

$$\hat{\theta}_{\text{MMSE}} = E(\theta) \ . \tag{11.3}$$

**Absolute Loss Function**. This is defined as $L(t, \theta) = |\Delta|$. In this case, the corresponding MMAE (minimum mean absolute error) estimate, $\hat{\theta}_{\text{MMAE}}$, is equal to the median of the distribution of $\theta$:

$$\hat{\theta}_{\text{MMAE}} = \text{med}\big(p(\theta | \mathbf{y}, I)\big) \ . \tag{11.4}$$

**Zero-One Loss Function.** This is defined as:

$$L(t,\theta) = \begin{cases} 0 & \text{if } |\Delta| \le b \ , \\ 1 & \text{if } |\Delta| > b \ . \end{cases} \tag{11.5}$$

In this case, the corresponding estimate is the center of the interval of width $2b$ which has the highest probability. If $b$ goes to zero, then the zero-one estimate converges to the maximum *a posteriori* (MAP) estimate:

$$\hat{\theta}_{\text{MAP}} = \arg\max_{\theta}\big(p(\theta|\mathbf{y}, I)\big) \ . \tag{11.6}$$

Assuming $p(\theta|\mathbf{y}, I)$ is differentiable, $\hat{\theta}_{\text{MAP}}$ is given by: $\left.\frac{\partial \ln p(\theta|\mathbf{y},I)}{\partial\theta}\right|_{\theta=\hat{\theta}_{\text{MAP}}} = 0.$

Although the above loss functions are widely used, in some circumstances it may be better to use a different loss function. The following example illustrates the use of a *robust* version of the quadratic loss function.[1]

**Example 11.1** *Color Constancy and the Minimum Local Mass (MLM) Loss Function* [6]. Color constancy is defined as the ability of a visual system to maintain object colour appearance across variations in factors extrinsic to the object. Human vision exhibits approximate color constancy. Let $p(\theta|\mathbf{y}, I)$ denote the *a posteriori* distribution for the illuminants and the surface colors in a given scene. By definition, the best estimate of the surface color estimate is

$$\hat{\theta} = \min_t \int_{-\infty}^{\infty} L(t,\theta)p(\theta|\mathbf{y}, I)d\theta \ ,$$

where $L$ is an appropriate loss function. For this application, Ref. [6], found that none of the above loss functions were appropriate. Instead they used a "robust" quadratic loss function:

$$L(t,\theta) = \begin{cases} \Delta^2 & \text{if } |\Delta| < \alpha \ , \\ \alpha^2 & \text{otherwise} \ , \end{cases}$$

where $\Delta = t - \theta$ and $\alpha$ is a fixed constant.

Table 11.2 lists some the common loss functions along with their Bayesian estimators. The following example illustrates the calculation of several estimates for the exponential distribution.

[1] The concept of a *robust* loss function is explained in more detail in Chap. 12, where we consider the subject of *robust* statistics.

**Table 11.2** Loss functions $L(t,\theta)$ and their Bayesian Estimators $\hat{\theta}$

| Name | $L(t,\theta)$ | $\hat{\theta} \equiv \min_t \int_{-\infty}^{\infty} L(t,\theta)p(\theta\vert\mathbf{y},I)d\theta$ |
|---|---|---|
| Quadratic | $\Delta^2$ | $\hat{\theta} = E(p(\theta\vert\mathbf{y},I))$, i. e. $\hat{\theta}$ is expectation of $p(\theta\vert\mathbf{y})$ |
| Absolute | $\vert\Delta\vert$ | $\hat{\theta}$ is median of $p(\theta\vert\mathbf{y},I)$, i. e. $\int_{-\infty}^{\hat{\theta}} p(\theta\vert\mathbf{y},I)d\theta = \frac{1}{2} = \int_{\hat{\theta}}^{\infty} p(\theta\vert\mathbf{y},I)d\theta$ |
| Zero-One | 1 if $(\vert\Delta\vert - b) \geq 0$, otherwise 0 | $\hat{\theta}$ is center of interval of width $2b$ which has highest probability. If $b \to 0$, then $\hat{\theta} = \arg\max\big(p(\theta\vert\mathbf{y},I)\big)$ |
| Local mass | $\min(\Delta^2,\alpha^2)$ | $\hat{\theta} = \min_t \int L(t,\theta)p(\theta\vert\mathbf{y},I)d\theta$ |
| Hubert-$t$ | $\min(\vert\Delta\vert,\alpha)$ | $\hat{\theta} = \min_t \int L(t,\theta)p(\theta\vert\mathbf{y},I)d\theta$ |

**Example 11.2** *Exponential Distribution.* Assume an exponential likelihood:

$$p(y|\theta,I) = \begin{cases} \theta\exp(-\theta y) & \text{if } y > 0\,, \\ 0 & \text{otherwise}\,. \end{cases}$$

In this case, the corresponding prior conjugate distribution is also exponential:

$$\pi(\theta|I) = \begin{cases} \alpha\exp(-\alpha\theta) & \text{if } \theta > 0\,, \\ 0 & \text{otherwise}\,. \end{cases}$$

By Bayes' theorem, the corresponding *a posteriori* pdf is

$$\begin{aligned} p(\theta|y,I) &= \frac{p(y|\theta,I)\pi(\theta|\alpha,I)}{\int p(y|\theta,I)\pi(\theta|\alpha,I)d\theta} = \frac{\alpha\theta\exp\big(-(\alpha+y)\theta\big)}{\int \alpha\theta\exp\big(-(\alpha+y)\theta\big)d\theta}\,, \\ &= (\alpha+y)^2\theta\exp\big(-(\alpha+y)\theta\big)\,. \end{aligned}$$

where $\int \alpha\theta\exp\big(-(\alpha+y)\theta\big)d\theta = \alpha/(\alpha+y)^2$.

The corresponding maximum likelihood (ML) and maximum *a posteriori* (MAP) estimates are given, respectively, by the location of the peaks in $p(y|\theta,I)$ and $p(\theta|y,I)$:

$$\hat{\theta}_{\mathrm{ML}} = \frac{1}{y} \quad \text{and} \quad \hat{\theta}_{\mathrm{MAP}} = \frac{1}{\alpha+y}\,,$$

the minimum mean square error (MMSE) estimate is given by

$$\begin{aligned} \hat{\theta}_{\mathrm{MMSE}}(y) &= \int \theta p(\theta|y,I)d\theta\,, \\ &= \int (\alpha+y)^2\theta^2\exp\big(-(\alpha+y)\theta\big)d\theta = \frac{2}{\alpha+y}\,, \end{aligned}$$

and the minimum mean absolute error (MMAE) estimate is given by

$$\hat{\theta}_{\text{MMAE}}(y) = \frac{T}{\alpha + y} ,$$

where $T \approx 1.68$ is the solution of $(1 + T)e^{-T} = \frac{1}{2}$.

## 11.3 Bayesian Curve Fitting*

Curve fitting is an important technique in data fusion where it is commonly used to fuse together multiple sensor observations. We may regard curve fitting as a problem in parameter estimation: Given a set of pairs of input measurements $(y_i, u_i), i \in \{1, 2, \ldots, N\}$, we estimate the parameters of an algebraic curve which best fits these data. A classic example of curve fitting is straight-line regression, where we fit a straight-line to several sensor measurements $y_i, i \in \{1, 2, \ldots, N\}$.

The following example illustrates the use of multiple straight-line regression in a multi-target tracking algorithm.

**Example 11.3** *Track Initialization Using Straight-Line Regression* [4, 5]. In surveillance applications we often use a Kalman filter (KF) (see Sect. 13.3) to track all objects which move in a given surveillance area. The KF is a sequential algorithm in which the tracks are updated as new sensor observations are received. To initiate the KF we must, however, create an initial set of tracks, one track for each object. This is often performed using a Hough transform (see Example 12.14).

A curve fitting technique which is widely used in data fusion is Kriging (see Sect. 4.3). The following example illustrates the fusion of a digital elevation model (DEM) and rain gauge measurements using a Kriging estimator.

**Example 11.4** *Kriging: Fusing DEM and Rainfall Measurements* [16]. High resolution estimates of the spatial variability of rainfall are important for the identification of locally intense storms which can lead to floods and especially to flash floods. Given a limited number of rainfall gauges we may generate an estimate of the rainfall at any point $\mathbf{u}_0$ by ordinary Kriging, or curve-fitting the $N$ nearby rain gauge measurements $y(\mathbf{u}_i), i \in \{1, 2, \ldots, N\}$ (see Sect. 4.3). We now consider the situation where the rain gauge measurements are supplemented by elevation data which we assume is correlated with the rainfall and which is available for all $\mathbf{u}$.

In Kriging with an external trend (KED) we estimate the rainfall at the point $\mathbf{u}_0$ using a Kriging estimator (Eq. 4.5):

$$\hat{y}(\mathbf{u}_0) = \mu(\mathbf{u}_0) + \sum_i \lambda_i (y_i - \mu(\mathbf{u}_0)) ,$$

in which we model the mean value $\mu(\mathbf{u}_0)$ as a linear function of $z$. The unknown mean, $\mu_0 = \mu(\mathbf{u}_0)$, and the weights, $\lambda_i, i \in \{1, 2, \ldots, N\}$, are given by the following matrix equation:

$$\begin{pmatrix} \lambda_1 \\ \vdots \\ \lambda_N \\ \mu_0 \end{pmatrix} = \begin{pmatrix} \Sigma(\mathbf{u}_1, \mathbf{u}_1) & \ldots & \Sigma(\mathbf{u}_1, \mathbf{u}_N) & 1 & z(\mathbf{u}_1) \\ \vdots & \ddots & \vdots & \vdots & \vdots \\ \Sigma(\mathbf{u}_N, \mathbf{u}_1) & \ldots & \Sigma(\mathbf{u}_N, \mathbf{u}_N) & 1 & z(\mathbf{u}_N) \\ 1 & \ldots & 1 & 0 & 0 \\ z(\mathbf{u}_1) & \ldots & z(\mathbf{u}_N) & 0 & 0 \end{pmatrix}^{-1} \begin{pmatrix} \Sigma(\mathbf{u}_0, \mathbf{u}_1) \\ \vdots \\ \Sigma(\mathbf{u}_0, \mathbf{u}_N) \\ 1 \\ z(\mathbf{u}_0) \end{pmatrix} ,$$

where $\Sigma(\mathbf{u}_i, \mathbf{u}_j)$ is the spatial covariance of the rainfall $y$. In KED the weights $\lambda_i$ sum to one. In this case, we can write $\hat{y}(\mathbf{u}_0)$ more simply as

$$\hat{y}(\mathbf{u}_0) = \sum_i \lambda_i y_i .$$

The following example graphically illustrates the power of *Bayesian* curve fitting. We consider the problem of optimally fitting a straight-line *segment* to a set of 2D sensor measurements $\mathbf{y}_i, i \in \{1, 2, \ldots, N\}$. The algorithm calculates both the slope and intercept of the straight-line *and* the end points of the straight-line segment.

**Example 11.5** *Fitting a Straight-Line Segment in 2D* [25]. We consider the problem of fitting a straight-line *segment* $L \equiv L(\boldsymbol{\theta})$, to a set of 2D points, $\mathbf{z}_i = (x_i, y_i), i \in \{1, 2, \ldots, N\}$, where $\boldsymbol{\theta} = \big((x_A, y_A), (x_B, y_B)\big)$ denotes the Cartesian coordinates of the end points, $A$ and $B$ (Fig. 11.1). Let $\mathbf{z} = \{\mathbf{z}_1, \mathbf{z}_2, \ldots, \mathbf{z}_N\}$, then our problem reduces to calculating the *a posteriori* pdf $p(\boldsymbol{\theta}|\mathbf{z}, I)$. Using Bayes' theorem and introducing the dummy variable, $\mathbf{P}$, gives

$$p(\boldsymbol{\theta}|\mathbf{z}, \mathbf{P}, I) = \frac{p(\mathbf{z}|\boldsymbol{\theta}, I)\pi(\boldsymbol{\theta}|I)}{p(\mathbf{z}|I)} = \frac{p(\mathbf{z}|\mathbf{P}, I)\pi(\mathbf{P}|\boldsymbol{\theta}, I)\pi(\boldsymbol{\theta}|I)}{p(\mathbf{z}|I)} ,$$

where $\mathbf{P} = (x, y)$ is any point on the straight-line $L(\boldsymbol{\theta})$. We treat $\mathbf{P}$ as a nuisance parameter which we eliminate by integration. If the $\mathbf{z}_i$ are independent, and assuming a uniform distribution for $\pi(\boldsymbol{\theta}|I)$, we obtain

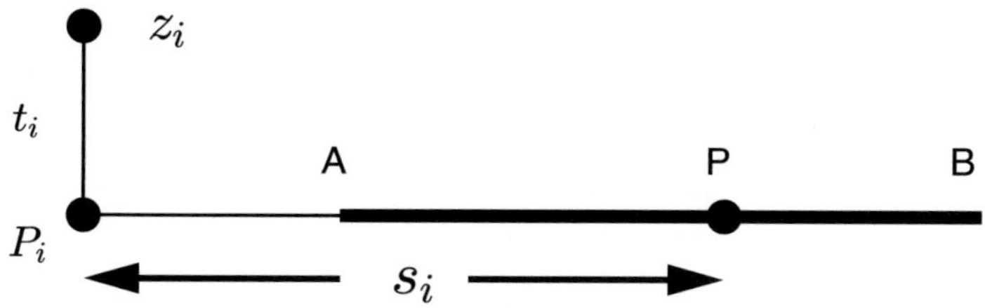


**Fig. 11.1** Shows the straight-line segment $AB$. $\mathbf{P}$ is any point which lies on $AB$, while $\mathbf{z}_i$ is any point which lies beside $AB$. $\mathbf{P}_i$ is the point closest to $\mathbf{z}_i$ which lies on $AB$ or on its extension. $s_i$ and $t_i$ are, respectively, distances $|\mathbf{P}_i - \mathbf{P}|$ and $|\mathbf{z}_i - \mathbf{P}_i|$

$$p(\boldsymbol{\theta}|\mathbf{z}, I) \propto \prod_i \int_{L(\boldsymbol{\theta})} p(\mathbf{z}_i|\mathbf{P}, I)\pi(\mathbf{P}|\boldsymbol{\theta}, I)d\mathbf{P} .$$

Assuming a Gaussian noise model with zero-mean and variance $\sigma^2$, and a uniform distribution for $\pi(\mathbf{P}|\boldsymbol{\theta}, I)$, gives

$$p(\boldsymbol{\theta}|\mathbf{z}, I) \propto \prod_i \int_{L(\boldsymbol{\theta})} \frac{1}{\sqrt{2\pi\sigma^2}} \exp -\frac{1}{2}\left(\frac{\mathbf{z}_i - \mathbf{P}}{\sigma}\right)^2 d\mathbf{P} .$$

Let $\mathbf{P}_i$ denote the point on $L(\boldsymbol{\theta})$ which is nearest to $\mathbf{P}$, then $|\mathbf{z}_i - \mathbf{P}_i| = \sqrt{s_i^2 + t_i^2}$, where $s_i = |\mathbf{P} - \mathbf{P}_i|$ and $t_i = |\mathbf{z}_i - \mathbf{P}_i|$. We obtain

$$p(\boldsymbol{\theta}|\mathbf{z}, I) \propto \prod_i \int_{s_A}^{s_B} \frac{1}{\sqrt{2\pi\sigma^2}} \exp -\left(\frac{s_i^2 + t_i^2}{2\sigma^2}\right) ds_i ,$$

which may be solved numerically. An important special case occurs when we let the line extend to $\pm\infty$. In this case the *a posteriori* probability is

$$p(\boldsymbol{\theta}|\mathbf{z}, I) = \prod_i \frac{1}{\sqrt{2\pi\sigma^2}} \exp -\left(\frac{t_i^2}{2\sigma^2}\right) ,$$

and the corresponding maximum *a posteriori* (MAP) value for $\theta$ is

$$\hat{\boldsymbol{\theta}}_{\mathrm{MAP}} = \arg\max \prod_i \frac{1}{\sqrt{2\pi\sigma^2}} \exp -\left(\frac{t_i^2}{2\sigma^2}\right) = \arg\min \sum_i t_i^2 .$$

In this case, the maximum *a posteriori* straight-line, $L(\hat{\boldsymbol{\theta}}_{\mathrm{MAP}})$, is identical to the straight-line found using the standard least square approach (see Sect. 11.5).

## 11.4 Maximum Likelihood

In some circumstances we do not assume any a priori information or, equivalently, we assume a uniform a priori distribution $\pi(\boldsymbol{\theta}|I)$. In this case, the *a posteriori* pdf $p(\boldsymbol{\theta}|\mathbf{y}, I)$ is proportional to the likelihood function $p(\mathbf{y}|\boldsymbol{\theta}, I)$:

$$\begin{aligned} p(\boldsymbol{\theta}|\mathbf{y}, I) &= p(\mathbf{y}|\boldsymbol{\theta}, I)\pi(\boldsymbol{\theta}|I)/p(\mathbf{y}|I) , \\ &\propto p(\mathbf{y}|\boldsymbol{\theta}, I) , \end{aligned} \tag{11.7}$$

and the maximum *a posteriori* (MAP) and maximum likelihood (ML) solutions are identical:

$$\hat{\boldsymbol{\theta}}_{\mathrm{MAP}} \equiv \hat{\boldsymbol{\theta}}_{\mathrm{ML}} , \tag{11.8}$$

where $\hat{\boldsymbol{\theta}}_{\mathrm{MAP}} = \arg\max_{\boldsymbol{\theta}} p(\boldsymbol{\theta}|\mathbf{y}, I)$ and $\hat{\boldsymbol{\theta}}_{\mathrm{ML}} = \arg\max_{\boldsymbol{\theta}} p(\mathbf{y}|\boldsymbol{\theta}, I)$. Assuming a differentiable function, $\hat{\boldsymbol{\theta}}_{\mathrm{ML}}$ is found from $\left.\frac{\partial \ln p(\mathbf{y}|\boldsymbol{\theta}, I)}{\partial \boldsymbol{\theta}}\right|_{\boldsymbol{\theta}=\hat{\boldsymbol{\theta}}_{\mathrm{ML}}} = \mathbf{0}$. If multiple solutions exist, then $\hat{\boldsymbol{\theta}}_{\mathrm{ML}}$ is the one that maximizes the likelihood function.

The ML estimator is perhaps the most widely used estimation method and is often used in combination with other optimization/estimation procedures.

**Example 11.6** *Non-intrusive Speech Quality Estimation Using GMM's* [11]. The evaluation and assurance of speech quality is critically important for telephone service providers. A nonintrusive speech quality measurement algorithm is shown in Fig. 11.2. Perceptual features are extracted from test speech. The labeling module labels the feature vectors of each time frame as belonging to one of three possible classes: voiced ($V$), unvoiced ($U$) or inactive ($I$). Offline, high-quality undistorted speech signals are used to produce a reference model of the behaviour of clean speech features. This is accomplished by modeling the pdf of the features for each class $\Lambda \in \{U, V, I\}$ with a mixture model GMM($\Lambda$). Features extracted from the test signal which belong to a class $\Lambda$ are assessed by calculating a "consistency" measure with respect to GMM($\Lambda$). The consistency measure is defined as

$$c(\Lambda) = \frac{1}{N} \sum_i \ln p(y_i|\Lambda) ,$$

where $p(y_i|\Lambda)$ is the likelihood function of the $i$th test signal, $y_i$, given the model GMM($\Lambda$). The three consistency values $c(\Lambda)$, $\Lambda \in \{U, V, I\}$ are fused together and then converted to a speech quality $Q$ by passing through an appropriate transfer function (see Example 9.16).

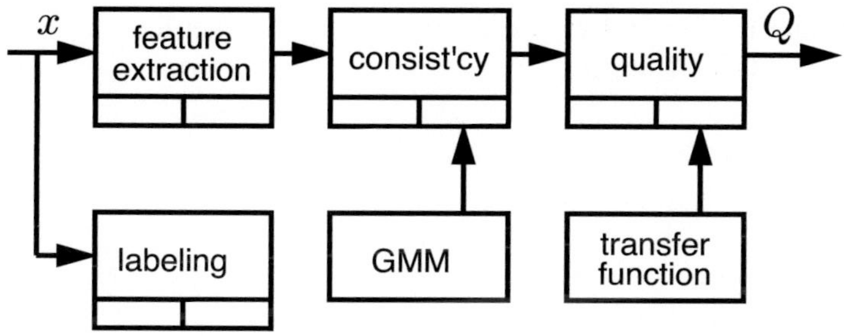


**Fig. 11.2** Shows the non-intrusive speech quality estimation algorithm. The principal modules in the algorithm are: (1) Labeling. This labels each time frame of the speech signal as voiced ($V$), unvoiced ($U$) or inactive ($I$). (2) Feature extraction. This extracts perceptual features from each time frame. (3) Consistency. This calculates a consistency measure $c(\Lambda)$ and (4) Quality. This calculates the speech quality $Q$

Although the ML procedure is widely used in many applications, a certain degree of caution should be exercised when using it. The reason is that the ML has no mechanism for regulating its solution. As a consequence, the solution may prove to be completely wrong, or as in the following example, wildly optimistic.

**Example 11.7** *Completion of a Doctoral Thesis: ML versus MAP* [7]. A doctoral student gives his supervisor a *precise* time, $T$, for the completion of his thesis. If the supervisor belongs to the "maximum likelihood" school then he does not use any a priori information and will uncritically accept the students estimate. If, on the other hand, the supervisor belongs to the "Bayesian" school, he will use all the a priori information which is available. E.g. he may use his knowledge of human psychology and conjecture that a student working under pressure will probably underestimate the time required to complete his PhD thesis. The result is that although the likelihood function is only high for $t \sim T$, the *a priori* probability that $T$ is a reasonable date, given the student is working under pressure, is sufficiently low to make the estimate $t \approx T$ highly unlikely.

## 11.5 Least Squares

For some applications we cannot define an *a posteriori* pdf $p(\boldsymbol{\theta}|\mathbf{y}, I)$, where $\mathbf{y} = (\mathbf{y}_1^T, \mathbf{y}_2^T, \ldots, \mathbf{y}_N^T)^T$. In this case we may optimally determine the parameters $\boldsymbol{\theta}$ by using a deterministic, i.e. *non-probabilistic*, method such as the method of least squares. For simplicity, we shall restrict ourselves to 1D scalar measurements $\mathbf{y} = (y_1, y_2, \ldots, y_N)^T$, which are noisy samples of an "ideal" function $f(\boldsymbol{\theta})$. If the $y_i$ are independent, then the least square estimate $\hat{\boldsymbol{\theta}}_{\text{LS}} = (\hat{\theta}_1, \hat{\theta}_2, \ldots, \hat{\theta}_K)^T$, is defined as the parameter values which minimize a weighted sum of the square differences $(y_i - f(\boldsymbol{\theta}))^2$. Mathematically,

$$\hat{\boldsymbol{\theta}}_{\text{LS}} = \arg\min_{\boldsymbol{\theta}} \sum_i \left( \frac{y_i - f(\boldsymbol{\theta})}{\sigma_i} \right)^2 , \tag{11.9}$$

where $\sigma_i^2$ is the noise variance in $y_i$. If we assume $\big(y_i - f(\boldsymbol{\theta})\big) \sim \mathcal{N}(0, \sigma_i^2)$, then the corresponding likelihood function is

$$p(\mathbf{y}|\boldsymbol{\theta}, I) = \prod_i \mathcal{N}(y_i | f(\boldsymbol{\theta}), \sigma_i^2) . \tag{11.10}$$

The ML solution $\hat{\boldsymbol{\theta}}_{\text{ML}}$ is numerically identical to the LS solution $\hat{\boldsymbol{\theta}}_{\text{LS}}$, where $\hat{\boldsymbol{\theta}}_{\text{ML}} = \arg\max_{\boldsymbol{\theta}} \prod_i \mathcal{N}\big(y_i | f(\boldsymbol{\theta}), \sigma_i^2\big)$ and $\hat{\boldsymbol{\theta}}_{\text{LS}} = \arg\max_{\boldsymbol{\theta}} \sum_i \big((y_i - f(\boldsymbol{\theta}))/\sigma_i\big)^2$.

Additionally, if we assume a uniform *a priori* distribution $\pi(\boldsymbol{\theta}|I)$ for $\boldsymbol{\theta}$, then the maximum *a posteriori* solution $\hat{\boldsymbol{\theta}}_{\text{MAP}}$ becomes numerically equal to $\hat{\boldsymbol{\theta}}_{\text{ML}}$ and $\hat{\boldsymbol{\theta}}_{\text{LS}}$.

In general, the MAP, ML and LS solutions are numerically equal, if the *a priori* distribution $\pi(\boldsymbol{\theta}|I)$ is uniform and the input measurement vector $\mathbf{y}$ follows a multivariate Gaussian distribution: $\mathbf{y} \sim \mathcal{N}(\mathbf{y}|\mathbf{f}(\boldsymbol{\theta}), \boldsymbol{\Sigma})$. Mathematically,

$$\hat{\boldsymbol{\theta}}_{\text{MAP}} = \hat{\boldsymbol{\theta}}_{\text{ML}} = \hat{\boldsymbol{\theta}}_{\text{LS}}$$

where $\hat{\boldsymbol{\theta}}_{\text{MAP}} = \arg\max_{\boldsymbol{\theta}} \big(\pi(\boldsymbol{\theta}|I)\mathcal{N}\big(\mathbf{y}|\mathbf{f}(\boldsymbol{\theta}), \boldsymbol{\Sigma}\big)\big), \hat{\boldsymbol{\theta}}_{\text{ML}} = \arg\max_{\boldsymbol{\theta}} \big(\mathcal{N}\big(\mathbf{y}|\mathbf{f}(\boldsymbol{\theta}), \boldsymbol{\Sigma}\big)\big)$ and $\hat{\boldsymbol{\theta}}_{\text{LS}} = \arg\min_{\boldsymbol{\theta}} \big((\mathbf{y} - \mathbf{f}(\boldsymbol{\theta}))^T \boldsymbol{\Sigma}^{-1} (\mathbf{y} - \mathbf{f}(\boldsymbol{\theta}))\big)$.

The following example illustrates the LS procedure by fitting a set of experimental covariance measurements with a parametric covariance model. This procedure is widely used in geostatistical applications including geographical information systems (GIS) (see Sect. 4.3).

**Example 11.8** *Fitting Covariance Measurements to a Spatial Covariance Model* [8]. In Sect. 4.3 we described the construction of a GIS using the Kriging estimators. The calculation of these estimators require a positive-definite spatial covariance matrix $\boldsymbol{\Sigma}(\mathbf{u}_i, \mathbf{u}_j)$. One way of ensuring this is to fit the experimentally measured $\boldsymbol{\Sigma}(\mathbf{u}_i, \mathbf{u}_j)$ values to a parametric model $\boldsymbol{\Sigma}(\mathbf{u}_i, \mathbf{u}_j|\boldsymbol{\theta})$ which has been designed to be positive definite. Traditionally, we determine the parameter vector $\boldsymbol{\theta}$ using the method of least squares in which we suppose all $\mathbf{y}_i$ are equally important. In this case the calculation of $\boldsymbol{\theta}$ reduces to

$$\hat{\boldsymbol{\theta}}_{\text{LS}} = \arg\min_{\boldsymbol{\theta}} \Big( \big(\boldsymbol{\Sigma}(\mathbf{u}_i, \mathbf{u}_j) - \boldsymbol{\Sigma}(\mathbf{u}_i, \mathbf{u}_j|\boldsymbol{\theta})\big)^T \big(\boldsymbol{\Sigma}(\mathbf{u}_i, \mathbf{u}_j) - \boldsymbol{\Sigma}(\mathbf{u}_i, \mathbf{u}_j|\boldsymbol{\theta})\big) \Big) .$$

This is known as the *ordinary* least square procedure.

## 11.6 Linear Gaussian Model

The *Linear Gaussian Model* describes a linear model between two vectors $\mathbf{y}$ and $\boldsymbol{\theta}$ with Gaussian errors. Mathematically, the model is

$$\mathbf{y} = \mathbf{H}\boldsymbol{\theta} + \mathbf{w} , \tag{11.11}$$

where $\mathbf{y} = (y_1, y_2, \ldots, y_N)^T$ is a $N \times 1$ vector of scalar measurements, $\mathbf{H}$ is a known $N \times K$ matrix, $\boldsymbol{\theta} = (\theta_1, \theta_2, \ldots, \theta_N)^T$ is a $K \times 1$ random vector of linear coefficients with *a priori* pdf, $\mathcal{N}(\boldsymbol{\mu}, \boldsymbol{\Sigma})$, and $\mathbf{w} = (w_1, w_2, \ldots, w_N)^T$ is a $N \times 1$ noise vector with pdf, $\mathcal{N}(\mathbf{0}, \sigma^2\mathbf{I})$, and independent of $\boldsymbol{\theta}$.

The linear Gaussian model forms the basis for much statistical modelling which is used in data fusion. In Table 11.3 we list some common statistical models which may be written in the form of a linear Gaussian model. The following example shows how we may re-write the polynomial model as a linear Gaussian model.

**Example 11.9** *Polynomial Model* [20]. The polynomial model is defined as

$$y_i = \sum_k a_k u_i^{k-1} + w_i ,$$

or, in matrix form, as

$$\begin{pmatrix} y_0 \\ y_1 \\ \vdots \\ y_{N-1} \end{pmatrix} = \begin{pmatrix} 1 & u_0 & u_0^2 & \ldots & u_0^{K-1} \\ 1 & u_1 & u_1^2 & \ldots & u_1^{K-1} \\ \vdots & \vdots & \vdots & \ddots & \vdots \\ 1 & u_{N-1} & u_{N-1}^2 & \ldots & u_{N-1}^{K-1} \end{pmatrix} \begin{pmatrix} a_1 \\ a_2 \\ \vdots \\ a_K \end{pmatrix} + \begin{pmatrix} w_0 \\ w_1 \\ \vdots \\ w_{N-1} \end{pmatrix} .$$

By comparing this equation with (11.11) we make the following identification:

**Table 11.3** Representations which use the Linear Gaussian Model

| Name | Equivalent linear gaussian model |
|---|---|
| Sinusoidal | $y_i = \sum_{k=1}^{K} [a_k \sin(\omega_k t_i) + b_k \cos(\omega_k t_i)] + w_i$ |
| Autoregressive (AR) | $y_i = \sum_{k=1}^{K} a_k y_{i-k} + w_i$ |
| AR with External Input (ARX) | $y_i = \sum_{k=1}^{K} a_k y_{i-k} + \sum_{l=0}^{L} b_k u_{i-l} + w_i$ |
| Nonlinear Autoregressive (NAR) | $y_i = \sum_{k=1}^{K} a_k y_{i-k} + \sum_{l=1}^{L} \sum_{m=1}^{M} b_{lm} y_{i-l} y_{i-m} + \ldots + w_i$ |
| Polynomial | $y_i = \sum_{k=1}^{K} a_k u_i^{k-1}$ |

$$\mathbf{H} = \begin{pmatrix} 1 & u_0 & u_0^2 & \dots u_0^{K-1} \\ 1 & u_1 & u_1^2 & \dots u_1^{K-1} \\ \vdots & \vdots & \vdots & \ddots & \vdots \\ 1 & u_{N-1} & u_{N-1}^2 & \dots u_{N-1}^{K-1} \end{pmatrix} \quad \text{and} \quad \boldsymbol{\theta} = (a_1, a_2, \dots, a_N)^T .$$

By Bayes' theorem, the *a posteriori* pdf of the linear Gaussian model parameters is

$$p(\mathbf{H}, \sigma, \boldsymbol{\theta}|\mathbf{y}, I) = \frac{p(\mathbf{y}|\mathbf{H}, \boldsymbol{\theta}, \sigma, I)\pi(\mathbf{H}, \boldsymbol{\theta}, \sigma|I}{p(\mathbf{y}|I)} , \tag{11.12}$$

where

$$p(\mathbf{y}|\mathbf{H}, \sigma, \boldsymbol{\theta}, I) = (2\pi\sigma^2)^{-N/2} \exp\left[-\frac{(\mathbf{y} - \mathbf{H}\boldsymbol{\theta})(\mathbf{y} - \mathbf{H}\boldsymbol{\theta})^T}{\sigma^2}\right] . \tag{11.13}$$

Assuming that $\mathbf{H}$, $\boldsymbol{\theta}$ and $\sigma$ are statistically independent, we have

$$p(\mathbf{H}, \sigma, \boldsymbol{\theta}|\mathbf{y}, I) = \frac{p(\mathbf{y}|\mathbf{H}, \boldsymbol{\theta}, \sigma, I)\pi(\mathbf{H}|I)\pi(\boldsymbol{\theta}|I)\pi(\sigma|I)}{p(\mathbf{y}|I)} , \tag{11.14}$$

In the case where there is no *a priori* information concerning $\mathbf{H}$, $\boldsymbol{\theta}$ and $\sigma$ we may use uniform priors for the elements of $\mathbf{H}$ and $\boldsymbol{\theta}$ and Jeffrey's prior for $\sigma$. In this case, the joint *a posteriori* pdf for $\mathbf{H}$, $\boldsymbol{\theta}$ and $\sigma$ becomes

$$\begin{aligned} p(\mathbf{H}, \boldsymbol{\theta}, \sigma|\mathbf{y}, I) &\propto \frac{p(\mathbf{y}|\mathbf{H}, \boldsymbol{\theta}, \sigma, I)}{p(\mathbf{y}|I)} \times \frac{1}{\sigma} , \\ &\propto \frac{1}{\sigma^{(N+1)}} \exp\left[-\left(\frac{(\mathbf{y} - \mathbf{H}\boldsymbol{\theta})^T(\mathbf{y} - \mathbf{H}\boldsymbol{\theta})}{2\sigma^2}\right)\right] . \end{aligned} \tag{11.15}$$

Until now we assumed the noise components $w_i$ are iid, where $w_i \sim \mathcal{N}(0, \sigma^2)$. However, in some applications the noise components $w_i$ are not independent. In this case $\mathbf{w} \sim \mathcal{N}(\mathbf{0}, \boldsymbol{\Sigma})$, where the off-diagonal components of $\boldsymbol{\Sigma}$ are not zero. However, we may always convert the measurements so that $w_i \sim \mathcal{N}(0, \sigma^2)$ by applying a *whitening transformation* $\mathbf{D}$.

**Example 11.10** *Whitening Transformation* [17]. We assume $\boldsymbol{\Sigma}$ is positive definite. In this case it can be factored as $\boldsymbol{\Sigma}^{-1} = \mathbf{D}^T\mathbf{D}$, where $\mathbf{D}$ is an $N \times N$ invertible matrix. The matrix $\mathbf{D}$ acts as a *whitening* transformation when applied to $\mathbf{w}$, since

$$E\big(\mathbf{D}\mathbf{w}(\mathbf{D}\mathbf{w})^T\big) = \mathbf{D}\boldsymbol{\Sigma}\mathbf{D}^T = \mathbf{D}\mathbf{D}^{-1}(\mathbf{D}^T)^{-1}\mathbf{D}^T = \mathbf{I} ,$$

where $\mathbf{I} = \text{diag}\,\overbrace{(1, 1, \ldots, 1)}^{N}$. If we now apply $\mathbf{D}$ to $\mathbf{y} = \mathbf{H}\boldsymbol{\theta} + \mathbf{w}$ we obtain a linear Gaussian model:

$$\mathbf{y}' = \mathbf{D}\mathbf{y} = \mathbf{D}\mathbf{H}\boldsymbol{\theta} + \mathbf{D}\mathbf{w} = \mathbf{H}'\boldsymbol{\theta} + \mathbf{w}' ,$$

where $\mathbf{w}' = \mathbf{D}\mathbf{w} \sim \mathcal{N}(\mathbf{0}, \mathbf{I})$.

The linear Gaussian model and the *a posteriori* pdf $p(\mathbf{H}, \boldsymbol{\theta}, \sigma|\mathbf{y}, I)$ are widely used in data fusion. By way of an example, we give three different types of application.

### 11.6.1 Line Fitting

In curve fitting the matrix $\mathbf{H}$ is completely specified. In this case, our objective is to estimate the vector $\boldsymbol{\theta}$ assuming $\mathbf{H}$ is given but having no *a priori* knowledge regarding $\boldsymbol{\theta}$ or $\sigma$. In this case $\mathbf{H}$ does not appear in (11.15). We eliminate $\sigma$ by marginalization. Then the resulting *a posteriori* pdf is

$$p(\boldsymbol{\theta}|\mathbf{y}, I) = \frac{1}{2}\Gamma\left(\frac{N-1}{2}\right)\left(\frac{(\mathbf{y} - \mathbf{H}\boldsymbol{\theta})^T(\mathbf{y} - \mathbf{H}\boldsymbol{\theta})}{2}\right)^{-((N-1)/2)} . \tag{11.16}$$

Maximizing this expression gives us the MAP estimate for the parameter vector $\boldsymbol{\theta}$:

$$\boldsymbol{\theta}_{\text{MAP}} = (\mathbf{H}^T\mathbf{H})^{-1}\mathbf{H}^T\mathbf{y} . \tag{11.17}$$

The following example illustrates the fitting of a straight-line to a set of points.

**Example 11.11** *Straight-Line Regression.* We consider a set of points $(u_i, y_i), i \in \{1, 2, \ldots, N\}$, which we assume obey a straight-line equation:

$$y_i = mu_i + c + w_i ,$$

where $w_i \sim \mathcal{N}(0, \sigma^2)$ are independent random noise. The corresponding matrix equation is

$$\begin{pmatrix} y_1 \\ \vdots \\ y_N \end{pmatrix} = \begin{pmatrix} 1 & u_1 \\ \vdots & \vdots \\ 1 & u_N \end{pmatrix} \begin{pmatrix} c \\ m \end{pmatrix} + \begin{pmatrix} w_1 \\ \vdots \\ w_N \end{pmatrix} .$$

It follows that

$$\mathbf{H}^T\mathbf{H} = \begin{pmatrix} N & \sum_i u_i \\ \sum_i u_i & \sum_i u_i^2 \end{pmatrix} ,$$

$$(\mathbf{H}^T\mathbf{H})^{-1} = \frac{1}{N\sum_i u_i^2 - (\sum_i u_i)^2} \begin{pmatrix} \sum_i u_i^2 & -\sum_i u_i \\ -\sum_i u_i & N \end{pmatrix} ,$$

$$\mathbf{H}^T\mathbf{y} = \begin{pmatrix} \sum_i y_i \\ \sum_i u_i y_i \end{pmatrix} .$$

Substituting these terms in Eq. (11.17) gives the following expression for the MAP estimate of $(c, m)^T$:

$$\begin{pmatrix} c \\ m \end{pmatrix} = \frac{1}{\sum_i u_i^2 - N\overline{u}^2} \begin{pmatrix} \overline{y}\sum_i u_i^2 - \overline{u}\sum_i u_i y_i \\ \sum_i u_i y_i - \overline{u}\overline{y} \end{pmatrix} ,$$

where $\overline{u} = \sum_{i=1}^N u_i/N$ and $\overline{y} = \sum_{i=1}^N y_i/N$. This solution is identical to the solution obtained using the method of ML or LS. This is not unexpected since we assumed non-informative priors for $\boldsymbol{\theta}$ and $\sigma$.

The following example illustrates multi-metric fusion using linear regression.

**Example 11.12** *Multi-Metric Fusion for Visual Quality Assessment* [18]. Various image quality metrics have been developed to reflect human visual quality. However, so far, there is no single quality metric that is suitable for all image inputs. In [18] the authors construct a multi-metric fused score (MMF) by non-linearly fusing multiple image quality metrics. Given $N$ training images, let $q_i^{(m)}, m \in \{1, 2, \ldots, M\}$, denote the quality score obtained with the $m$th quality metric on the $i$th training image. Then we define the MMF score as

$$MMF_i = \mathbf{W}^T\mathbf{Q}_i + b ,$$

where $\mathbf{Q}_i = (q_i^{(1)}, q_i^{(2)}, \ldots, q_i^{(M)})^T$ and $\mathbf{W} = (w^{(1)}, w^{(2)}, \ldots, w^{(M)})^T$. We determine $\mathbf{W}$ and the bias $b$ by minimizing the difference between $MMF_i$ and the mean opinion score ($MOS_i$) obtained by human observers on the $i$th training image.

Experimentally, MMF is found to outperform all of the individual quality measures $q^{(m)}, m \in \{1, 2, \ldots, M\}$.

### 11.6.2 Change Point Detection

The central concept which underlies a linear change point detector is that an observed time sequence is modeled by different models at different points in time. The models are known but the time instants at which the models change are not known. In this case, our objective is to estimate $\mathbf{H}$. We eliminate $\boldsymbol{\theta}$ and $\sigma$ from (11.15) by marginalization [21]. The resulting *a posteriori* pdf is

$$p(\mathbf{H}|\mathbf{y}, I) \propto \frac{\left(\mathbf{y}^T\mathbf{y} - \mathbf{y}\mathbf{H}(\mathbf{H}^T\mathbf{H})^{-1}\mathbf{H}^T\mathbf{y}\right)^{((N-K)/2)}}{\sqrt{|\mathbf{H}^T\mathbf{H}|}} . \tag{11.18}$$

The following example illustrates Eq. (11.18) given the input measurements, $\mathbf{y} = (y_1, y_2, \ldots, y_N)^T$, and assuming there is a single (unknown) change point.

**Example 11.13** *Single Step in Piecewise Constant Time Series* [13]. We consider a time series $y_i$:

$$y_i = \begin{cases} \mu_1 + w_i & \text{if } i \leq m , \\ \mu_2 + w_i & \text{if } i > m , \end{cases}$$

where $m$ is an unknown integer $(1 \leq m \leq M)$ and $w_i \sim \mathcal{N}(0, \sigma^2)$. Then

$$\begin{pmatrix} y_1 \\ \vdots \\ y_m \\ y_{m+1} \\ \vdots \\ y_N \end{pmatrix} = \begin{pmatrix} 1 & 0 \\ \vdots & \vdots \\ 1 & 0 \\ 0 & 1 \\ \vdots & \vdots \\ 0 & 1 \end{pmatrix} \begin{pmatrix} \mu_1 \\ \mu_2 \end{pmatrix} + \begin{pmatrix} w_1 \\ \vdots \\ w_m \\ w_{m+1} \\ \vdots \\ w_N \end{pmatrix} .$$

It follows that

$$\mathbf{H}^T\mathbf{H} = \begin{pmatrix} m & 0 \\ 0 & N-m \end{pmatrix} ,$$

$$(\mathbf{H}^T\mathbf{H})^{-1} = \frac{1}{m(N-m)} \begin{pmatrix} N-m & 0 \\ 0 & m \end{pmatrix} ,$$

$$\mathbf{y}^T\mathbf{H} = (\mathbf{H}^T\mathbf{y})^T = \Big(\sum_{i=1}^{m} y_i \quad \sum_{i=m+1}^{N} y_i\Big) ,$$

$$\mathbf{y}^T\mathbf{y} = \sum_{i=1}^{N} y_i^2 .$$

Substituting all these terms in Eq. (11.18) gives the *a posteriori* pdf $p(m|\mathbf{y}, I)$:

$$p(m\mathbf{y}, i) \propto \frac{[\mathbf{y}^T\mathbf{y} - \mathbf{y}^T\mathbf{H}(\mathbf{H}^T\mathbf{H})^{-1}\mathbf{H}^T\mathbf{y}]^{-(N-m)/2}}{\sqrt{\det(\mathbf{H}^T\mathbf{H})}} ,$$

$$= \frac{1}{\sqrt{m(N-m)}}\Big[\sum_{i=1}^{N} y_i^2 - \frac{1}{m}\Big(\sum_{i=1}^{m} y_i\Big)^2 - \frac{1}{N-m}\Big(\sum_{i=m+1}^{N} y_i\Big)^2\Big]^{-(N-2)/2} .$$

The corresponding MAP estimate of $m$ is

$$m_{\mathrm{MAP}} = \arg\max_m p(m|\mathbf{y}, I) ,$$

and the associated $\mu_1$ and $\mu_2$ values are:

$$\mu_1 = \frac{1}{m_{\mathrm{MAP}}}\sum_{i=1}^{m_{\mathrm{MAP}}} y_i \quad \text{and} \quad \mu_2 = \frac{1}{N - m_{\mathrm{MAP}} + 1}\sum_{i=m_{\mathrm{MAP}}+1}^{N} y_i .$$

Single change point detection with more complicated curves are also easily handled as shown in the following two examples.

**Example 11.14** *Single Ramp in a Time Series* [13]. We consider a time series $y_i$:

$$y_i = \begin{cases} \mu + w_i & \text{if } i \le m , \\ \mu + r(i-m) + w_i & \text{if } i > m , \end{cases}$$

where $w_i \sim \mathcal{N}(0, \sigma^2)$ are independent random noise. The corresponding matrix equation is

$$\begin{pmatrix} y_1 \\ y_2 \\ \vdots \\ y_m \\ y_{m+1} \\ y_{m+2} \\ \vdots \\ y_N \end{pmatrix} = \begin{pmatrix} 1 & 0 \\ 1 & 0 \\ \vdots & \vdots \\ 1 & 0 \\ 1 & 1 \\ 1 & 2 \\ \vdots & \vdots \\ 1 & N-m \end{pmatrix} \begin{pmatrix} \mu \\ r \end{pmatrix} + \begin{pmatrix} w_1 \\ w_2 \\ \vdots \\ w_m \\ w_{m+1} \\ w_{m+2} \\ \vdots \\ w_N \end{pmatrix} .$$

**Example 11.15** *Mixed Two and Three-Order AR Process* [13]. We consider a time series $y_i$:

$$y_i = \begin{cases} \sum_{j=1}^{2} \alpha_j y_{i-j} + w_i \text{ if } i \leq m \ , \\ \sum_{j=1}^{3} \beta_j y_{i-j} + w_i \text{ if } i > m \ , \end{cases}$$

where $w_i \sim \mathcal{N}(0, \sigma^2)$ are independent random noise variables. The corresponding matrix equation is

$$\begin{pmatrix} y_1 \\ \vdots \\ y_{m-1} \\ y_m \\ \vdots \\ y_N \end{pmatrix} = \begin{pmatrix} y_0 & y_{-1} & 0 & 0 & 0 \\ \vdots & & \ddots & & \vdots \\ y_{m-1} & y_{m-2} & 0 & 0 & 0 \\ 0 & 0 & y_m & y_{m-1} & y_{m-2} \\ \vdots & & & & \\ 0 & 0 & y_{N-1} & y_{N-2} & y_{N-3} \end{pmatrix} \begin{pmatrix} \alpha_1 \\ \alpha_2 \\ \beta_1 \\ \beta_2 \\ \beta_3 \end{pmatrix} + \begin{pmatrix} w_1 \\ \vdots \\ w_{m-1} \\ w_m \\ \vdots \\ w_N \end{pmatrix} .$$

In the most general case the number of change points is unknown. In each segment the $y_i$ are modeled as independent polynomial functions of unknown order. Efficient algorithms for solving the more general case have been developed, including an exact curve fitting algorithm [12].

## 11.7 Generalized Millman Formula *

The *generalized Millman's formula* [23] deals with the optimal linear combination of an arbitrary number of correlated estimates. Suppose we have $K$ local estimates, $\boldsymbol{\mu}_1, \boldsymbol{\mu}_2, \ldots, \boldsymbol{\mu}_K$, of an unknown $L$-dimensional random vector $\boldsymbol{\theta}$. Then the optimal *linear* estimate of $\boldsymbol{\theta}$ is

$$\widetilde{\boldsymbol{\mu}} = \sum_k \mathbf{W}_k \boldsymbol{\mu}_k \ , \tag{11.19}$$

where

$$\sum_k \mathbf{W}_k = \text{diag}(\overbrace{1, 1, \ldots, 1}^{L}) \ , \tag{11.20}$$

and the $\mathbf{W}_k$ are $L \times L$ constant weighting matrices determined from the least square error criterion

$$(\mathbf{W}_1, \mathbf{W}_2, \ldots, \mathbf{W}_K) = \min_{\mathbf{W}_k} \Big( |\boldsymbol{\theta} - \sum_k \mathbf{W}_k \boldsymbol{\mu}_k|^2 \Big) \ . \tag{11.21}$$

The solution to this problem is given by the following set of $(K-1)$ linear equations

$$\begin{aligned}
&\sum_{k=1}^{K-1} \mathbf{W}_k(\boldsymbol{\Sigma}_{k,1} - \boldsymbol{\Sigma}_{k,K}) + \mathbf{W}_K(\boldsymbol{\Sigma}_{K,1} - \boldsymbol{\Sigma}_{K,K}) = 0 \;, \\
&\sum_{k=1}^{K-1} \mathbf{W}_k(\boldsymbol{\Sigma}_{k,2} - \boldsymbol{\Sigma}_{k,K}) + \mathbf{W}_K(\boldsymbol{\Sigma}_{K,2} - \boldsymbol{\Sigma}_{K,K}) = 0 \;, \\
&\qquad \vdots \\
&\sum_{k=1}^{K-1} \mathbf{W}_k(\boldsymbol{\Sigma}_{k,K-1} - \boldsymbol{\Sigma}_{k,K}) + \mathbf{W}_K(\boldsymbol{\Sigma}_{K,K-1} - \boldsymbol{\Sigma}_{K,K}) = 0 \;,
\end{aligned} \tag{11.22}$$

subject to Eq. (11.19), where $\boldsymbol{\Sigma}_{k,k} \equiv \boldsymbol{\Sigma}_k$ is the $L \times L$ covariance matrix $\boldsymbol{\Sigma}_{k,k} = E\big((\boldsymbol{\theta} - \boldsymbol{\mu}_k)(\boldsymbol{\theta} - \boldsymbol{\mu}_k)^T\big)$ and $\boldsymbol{\Sigma}_{k,l}$, $k \neq l$ is the $L \times L$ cross-covariance matrix $\boldsymbol{\Sigma}_{k,l} = E\big((\boldsymbol{\theta} - \boldsymbol{\mu}_k)(\boldsymbol{\theta} - \boldsymbol{\mu}_l)^T\big)$. In many applications, only the covariance matrices, $\boldsymbol{\Sigma}_{k,k}$, $k \in \{1, 2, \ldots, K\}$, are known, while the cross-covariances, $\boldsymbol{\Sigma}_{k,l}$, $k \neq l$, are unknown. In this case, we often assume $\boldsymbol{\Sigma}_{k,l} = 0$, $k \neq l$. This may, however, cause problems, especially in decentralized systems (see Sect. 3.4.2).

**Example 11.16** *Millman's Formula* [14]. Given two *uncorrelated* estimates $\boldsymbol{\mu}_1$ and $\boldsymbol{\mu}_2$, the generalized Millman formula reduces to

$$\widetilde{\boldsymbol{\mu}} = \boldsymbol{\Sigma}_{22}(\boldsymbol{\Sigma}_{11} + \boldsymbol{\Sigma}_{22})^{-1}\boldsymbol{\mu}_1 + \boldsymbol{\Sigma}_{11}(\boldsymbol{\Sigma}_{11} + \boldsymbol{\Sigma}_{22})^{-1}\boldsymbol{\mu}_2 \;,$$

where $\widetilde{\boldsymbol{\mu}}$ is the optimum linear estimate.

**Example 11.17** *Bar-Shalom-Campo Formula* [3]. Given two *correlated* estimates $\boldsymbol{\mu}_1$ and $\boldsymbol{\mu}_2$, the generalized Millman formula reduces to the Bar-Shalom-Campo formula:

$$\begin{aligned}
\widetilde{\boldsymbol{\mu}} = {} & (\boldsymbol{\Sigma}_{22} - \boldsymbol{\Sigma}_{21})(\boldsymbol{\Sigma}_{11} + \boldsymbol{\Sigma}_{22} - \boldsymbol{\Sigma}_{12} - \boldsymbol{\Sigma}_{21})^{-1}\boldsymbol{\mu}_1 \\
& + (\boldsymbol{\Sigma}_{11} - \boldsymbol{\Sigma}_{12})(\boldsymbol{\Sigma}_{11} + \boldsymbol{\Sigma}_{22} - \boldsymbol{\Sigma}_{12} - \boldsymbol{\Sigma}_{21})^{-1}\boldsymbol{\mu}_2 \;,
\end{aligned}$$

where $\widetilde{\boldsymbol{\mu}}$ is the optimum linear estimate.

## 11.8 Further Reading

Bayesian parameter estimation is discussed in many books and monographs. Good accounts are to be found in [1, 2, 10, 15, 17, 19, 24]. The linear Gaussian models are discussed at length in [22]. A full description of Bayesian curve fitting is given in [9].

## 11.9 Software

**MATLAB**. MATLAB CURVE FITTING and STATISTICS AND MACHINE LEARNING toolboxes includes m-files on:maximum likelihood estimates, Gaussian models, line fitting. BEAST (Bayesian estimator of abrupt change, seasonality and trend) a matlab toolbox for change detection [26]. It is also available in python.

**PYTHON**. SCIPY and SCIKIT-LEARN include python modules for: curve fitting and Gaussian models. BEAST is a set of python libraries for change detection [26].

## Problems

**11.1** Describe how we may use Bayesian analysis to find a finite length straight-line segment to a set of noisy points.

**11.2** Define the method of maximum likelihood. When should it reduce to MAP (maximum a posteriori)?

**11.3** Define the method of least squares. When does it reduce to MAP?

**11.4** Define the linear Gaussian model. Describe straight-line regression using the linear Gaussian model.

**11.5** Describe the detection of a single step in a piece-wise constant time using the linear Gaussian model.

## References

1. D'Agostini G (2003) Bayesian reasoning in data analysis. World Scientific, Singapore
2. D'Agostini G (2003) Bayesian inference in processing experimental data: principles and basic applications. Rep Prog Phys 66:1383–1419
3. Bar-Shalom Y, Campo L (1986) The effect of the common process noise on the two sensor fused track covariance. IEEE Trans Aero Elect Syst 22:803–805

4. Bar-Shalom Y, Li X (1995) Multitarget-Multisensor tracking: principles and techniques. YBS Publishing, Storrs, CT
5. Blackman SS, Popoli RF (1999) Design and analysis of modern tracking Systems. Published by Artech House, Norwood, MA
6. Brainard DH, Freeman WT (1997) Bayesian color constancy. J Opt Soc Am 14A:1393–1411
7. Choudrey RA (2002) Variational methods for Bayesian independent component analysis. PhD thesis, University of Oxford
8. Cressie NAC (1993) Statistics for spatial data. Published by Wiley
9. Denison DGT, Holmes C, Mallick BK et al (2002) Bayesian methods for nonlinear classification and regression. Published by Wiley
10. Dose V (2003) Bayesian inference in physics: case studies. Rep Prog Phys 66:1421–1461
11. Falk TH, Chan W-Y (2006) Nonintrusive speech quality estimation using Gaussian mixture models. IEEE Sig Proc Lett 13:108–111
12. Fearnhead P (2005) Exact Bayesian curve fitting and signal segmentation. IEEE Trans Sig Proc 53:2160–2166
13. Fitzgerald WJ, Godsill SJ, Kokaram AC et al (1999) Bayesian methods in signal and image processing. In: Bayesian statistics, vol 6. Published by Oxford University Press
14. Gelb A (1974) Applied optimal estimation. Published by MIT Press, MA, USA
15. Gelman A, Larlin JS, Stern HS, Rubin DR (2003) Bayesian data analysis. Published by Chapman and Hall
16. Goovaerts P (2000) Geostatistical approaches for incorporating elevation into the spatial interpolation of rainfall. J Hydrol 228:113–129
17. Kay SM (1993) Fundamentals of statistical signal processing: estimation theory. Published by Prentice-Hall
18. Liu T-J, Lin W, Kuo C-C Jay (2011) A multi-metric fusion approach to visual quality assessment. International workshop quality multimedia experience, pp 72–77
19. MacKay DJC (2003) Information theory. Inference and learning algorithms. Published by Cambridge University Press
20. Punska O (1999) Bayesian approaches to multi-sensor data fusion. MPhil thesis, Signal Processing and Communications Laboratory, Department of Engineering, University of Cambridge
21. Ruanaidh JJK, Fitzgerald WJ (1996) Numerical Bayesian methods applied to signal processing. Published by Springer
22. Roweis S, Ghahramani Z (1999) A unifying review of linear Gaussian models. Neural Comput 11:305–345
23. Shin V, Lee Y, Choi T-S (2006) Generalized Millman's formula and its application for estimation problems. Sig Proc 86:257–266
24. Sivia DS, Skilling J (2006) Data analysis: a bayesian tutoral, 2nd ed. Oxford University Press
25. Werman M, Keren D (2001) A Bayesian method for fitting parametric and nonparametric models to noisy data. IEEE Trans Patt Anal Mach Intell 23:528–534
26. Zhao K, Wulder MA, Hu T et al (2019) Detecting change-point, trend, and seaonality in satellite time series data to track abrupt changes ad nonlinear dynamics: a Bayesian ensemble algorithm. Rem Sens Env 232:111181

# Chapter 12
# Robust Statistics

**Abstract** In this chapter we consider the subject of robust statistics and, in particular, robust parameter estimation. Robust statistics is defined as statistical methods which are relatively insensitive to the presence of outliers. We start by discussing robust techniques which reduce the influence of outliers without actually identifying the outliers. We show how we may robustify the Bayesian paradigm by using likelihood functions with enhanced tails. As an example we consider the Student-t mixture model which is a robust Gaussian mixture model. Finally we consider robust statistics in computer vision including robust subspace methods.

## 12.1 Introduction

In this chapter we shall consider the subject of *robust statistics* and, in particular, *robust parameter estimation*. Robust statistics is defined as the study of statistical methods which are relatively insensitive to the presence of outliers, i.e. input data which is "strange" or "incompatible" with the remaining input data. It might be thought that outliers are a rare event. This is not, however, true. In fact, a common experience in all scientific experiments is that repeated measurements of supposedly one and the same quantity result, occasionally, in values which are in striking disagreement with all the others.

Increasingly, the need to use robust statistics in data fusion is being recognized. The reason for this is the devastating effect outliers may have on standard statistical procedures as illustrated in the following example.

**Example 12.1** *Fitting a Straight-Line in the Presence of Outliers.* We reconsider Example 11.5 except this time we suppose that in addition to the $N$ inlier points $(x_i, y_i)$, $i \in \{1, 2, \ldots, N\}$, we have a single outlier $(x_{N+1}, y_{N+1})$. If the outlier lies sufficiently far from the inliers, then by inspection the optimal straight-line (i.e. the line with the minimum sum of square deviations) will pass through the center of mass $(\bar{x}, \bar{y})$ and through the outlier $(x_{N+1}, y_{N+1})$ (line BB in Fig. 12.1). This is in spite of the fact that all the inliers lie on the line AA.


H. B. Mitchell, *Data Fusion*, https://doi.org/10.1007/978-3-662-71023-4_12

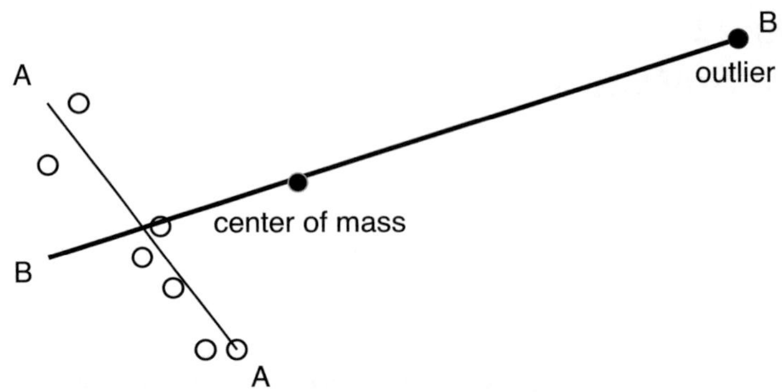


**Fig. 12.1** Shows a set of 2D points (including an outlier). Without the outlier the optimal straight-line is $AA$ while with the outlier the optimal straight-line is $BB$

**Table 12.1** Applications in which robust estimation is primary fusion algorithm

| Class | | Application |
|---|---|---|
| Early-DaO | Example 2.9 | Fusion of experimental neutron fission cross-section measurements |
| | Example 11.1 | Color constancy and the minimum local mass (MLM) Loss Function |
| | Example 12.2 | Gating sensor observations in target tracking |
| | Example 12.3 | Calibration of the SI Unit of the Volt |
| | Example 12.8 | "Good-and-Bad" Fusion of neutron fission cross section measurements |
| | Example 12.9 | Hubble's constant using combinatorial likelihood |
| Early-FeO | Example 4.21 | Principal discriminant analysis (PDA) for Small-Sample Problems |
| | Example 12.7 | Student mixture model |
| | Example 12.12 | Robust LDA-L1 algorithm |
| | Example 12.14 | Target tracking initialization using a hough transform |

In many applications, robust parameter estimation represents the major fusion algorithm in a data fusion system. In Table 12.1, we list some of these applications together with their classification.

Apart from the applications listed in Table 12.1, robust parameter estimation techniques are widely used in forming a common representational format (Chaps. 4–9).

## 12.2 Outliers

The simplest approach to dealing with outliers is to eliminate them before parameter estimation takes place. One area in which outliers are routinely eliminated before any parameter estimation takes place is in target tracking when it is known as *gating* and is illustrated in the following example.

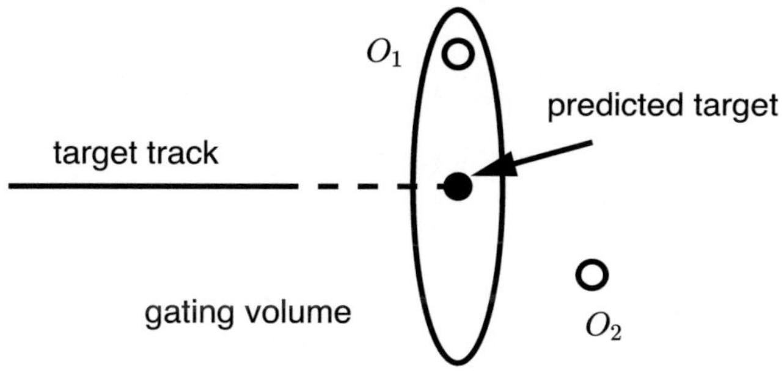


**Fig. 12.2** Shows gating of two sensor observations $O_1$ and $O_2$ made at time step $k$, where $O_1$ falls outside the gate while $O_2$ falls inside the gate. The gate is the ellipse centered on the expected (predicted) sensor observation $\hat{\mathbf{y}}$ for time step $k$

**Example 12.2** *Gating Sensor Observations in Target Tracking* [5]. Recursive filters (see Chap. 13) are routinely used to track a moving target. A necessary part of a tracking filter is to eliminate spurious sensor observations which are not associated with the target. In gating this is done by drawing a validation volume, or *gate*, around the predicted sensor observation (Fig. 12.2). The gate is usually formed in such a way that the probability that a true observation (i.e. an observation which is associated with the target) falls within the gate (assuming it is detected) is given by a gating probability $P_G$.

Let $\boldsymbol{\theta}_k$ denote the state of the target at time step $k$ and let $\mathbf{y}_k$ denote the corresponding sensor measurement, where

$$\mathbf{y}_k = \mathbf{H}\boldsymbol{\theta}_k + \mathbf{w}_k \ ,$$

and $\mathbf{w}_k \sim \mathcal{N}(\mathbf{0}, \mathbf{R})$. If a Kalman filter is used to track the target, then the predicted state of the target at time step $k$ is

$$p(\boldsymbol{\theta}_k | \mathbf{y}_{1:k-1}, I) = \mathcal{N}(\boldsymbol{\mu}_{k|k-1}, \boldsymbol{\Sigma}_{k|k-1}) \ ,$$

where $\mathbf{y}_{1:k-1} = (\mathbf{y}_1^T, \mathbf{y}_2^T, \ldots, \mathbf{y}_{k-1}^T)^T$ is the sequence of measurements made upto, and including, the time step $k-1$. In this case, the gating region $G$ is defined as the ellipsoid

$$G = \{\mathbf{y} | (\mathbf{y} - \hat{\mathbf{y}}_{k|k-1})^T \hat{\mathbf{S}}_{k|k-1}^{-1} (\mathbf{y} - \hat{\mathbf{y}}_{k|k-1}) \leq \gamma\} \ ,$$

where $\hat{\mathbf{y}}_{k|k-1} = \mathbf{H}_k \boldsymbol{\mu}_{k|k-1}$ is the predicted sensor measurement and $\hat{\mathbf{S}}_{k|k-1} = \mathbf{H}_k \boldsymbol{\Sigma}_{k|k-1} \mathbf{H}_k^T + \mathbf{R}_k$ is the predicted sensor covariance matrix. The threshold $\gamma$ maybe chosen to give a specified gating probability $P_G$. Alternatively, we may choose a threshold such that a true measurement falling within the gate is more likely than a false alarm. In 2D the corresponding threshold is given by

$$\gamma = 2\ln(P_D) - 2\ln\big(2\pi\lambda(1-P_D)|\hat{\mathbf{S}}_{k|k-1}|^{1/2}\big) ,$$

where $P_D$ is the probability of detection of the target and $\lambda$ is the spatial density of false alarms.

A popular procedure for detecting outliers is the *Hampel identifier*. Given $N$ measurements $x_i$, $i \in \{1, 2, \ldots, N\}$, this procedure outputs a outlier identifier for each $x_i$, where

$$I_i = \begin{cases} 1 \quad \text{(outlier)} & \text{if } |x_i - m| > 3S , \\ 0 & \text{otherwise} . \end{cases}$$

where $m$ and $S$ are, respectively, the median and median absolute deviation (MAD) of the $x_i$:

$$\begin{aligned} m &= \text{median}\Big(x_1, x_2, \ldots, x_N\Big) , \\ S &= 1.4826 \times \text{median}\Big(|x_1 - m|, |x_2 - m|, \ldots, |x_N - m|\Big) . \end{aligned} \qquad (12.1)$$

Eliminating spurious observations may, however, cause unforeseen problems as the following example shows.

**Example 12.3** *Calibration of the SI Unit of the Volt* [8]. In 1973 and again in 1986 a committee of the International Council of Scientific Unions performed a least squares adjustment of several independent measurements concerning $K_V$ (a constant involved in the calibration of the SI unit of the Volt). In performing the least square adjustment the committee eliminated measurements which were deemed incompatible with the remaining measurements. A large change was found between $K_V(1973)$ and $K_V(1986)$ although as the committee explain, the large change in $K_V$ could have been avoided if two measurements which seemed to be discrepant with the remaining measurements had *not* been deleted in the 1973 adjustment.

In order to avoid problems of this sort we often prefer to use "robust" techniques which reduce the influence of the outliers without actually identifying (and subsequently eliminating) the outliers. Given $N$ measurements $x_i$, $i \in \{1, 2, \ldots, N\}$, this often [11] reduces to simply replacing the sample mean $\bar{x}$ by the sample median $m$:

$$\bar{x} = \frac{1}{N}\sum_i x_i \rightarrow m = \text{median}\big(x_1, x_2, \ldots, x_N\big) , \qquad (12.2)$$

and replacing the sample standard deviation $\sigma$ by the sample *median absolute deviation* (MAD) $S$:

$$\sigma = \left(\frac{1}{N-1}\sum_i (x_i - \bar{x})^2\right)^{1/2} \rightarrow S = 1.4826 \times \text{median}\big(|x_1 - m|, \ldots, |x_N - m|\big) . \tag{12.3}$$

An alternative robust scale estimator which does not require a location estimate of the data [1, 17] is $S_{RS}$:

$$S_{RS} = 1.1926 \times \text{median}_i\big(\text{median}(|x_i - x_1|, |x_i - x_2|, \ldots, |x_i - x_N|)\big) , \tag{12.4}$$

This is computed as follows: For each $i$ compute the median of $\{|x_i - x_j|; j = 1, 2, \ldots, N\}$. This yields $N$ numbers. The median of these numbers is $S_{RS}$.

The following two example illustrate the robustification using Eqs. (12.2-12.3). See also the $K$-median cluster algorithm (Sect. 8.4.2).

**Example 12.4** *Median-based Otsu Thresholding* [26]. Given an input image $I$, a widely used global thresholding algorithm is method of Otsu (see Sect. 9.6). Here we use a threshold $T$ to divide the image pixels into: "low-intensity" pixels whose gray-levels are below $T$ and "high-intensity" pixels whose gray-levels are above $T$. The optimum threshold $T^*$ maximizes the between-class variance (see Eq. 9.3):

$$T^* = \arg\max_T \big(p_0(T)(\mu_0(T) - \mu)^2 + p_1(T)(\mu_1(T) - \mu)^2\big) ,$$

where $p_0(T)$ and $p_1(T)$ are the relative number of pixels in each group and $\mu$ is the mean gray-level of the image $I$.

The Otsu algorithm may be robustified [26] by simply replacing $\mu_0$, $\mu_1$ and $\mu$ with the medians $m_0$, $m_1$ and $m$. The robust threshold is then

$$T_{\text{robust}} = \arg\max_T \big(p_0(T)(m_0(T) - m)^2 + p_1(T)(m_1(T) - m)^2\big) .$$

**Example 12.5** *Robust Saccade Detection* [22]. The analysis of gaze behavior is important for understanding visual attention. An important component of gaze behaviour is *saccades*: ballistic eye movements that rapidly re-orient the eye and in turn re-orient the image impinging on the retina. Saccades typically last between 20 and 100 msec. Because of their ballistic nature, saccades can be differentiated from other gaze events by examining the eye velocity. Traditionally, a saccade velocity threshold $T$ is calculated as:

$$T = \mu + 3\sigma \ ,$$

where $\mu$ and $\sigma$ is the mean and standard deviation of the eye velocity samples. This equation is robustified by replacing $\mu$ with median $m$ and $\sigma$ with MAD $S$:

$$T_{\text{robust}} = m + 3S \ .$$

A powerful generalization of the median filter is the Hampel filter. This is illustrated in the following example:

**Example 12.6** *Hampel Filters* [16]. Given a noisy signal $x_i$, $i \in \{-\infty, -\infty + 1, \ldots, \infty\}$, we smooth the $x_i$ using a sliding-window Hampel filter:

$$y_i = \begin{cases} x_i & \text{if } |x_i - m_i| \leq \alpha S_i \ , \\ m_i & \text{otherwise} \ . \end{cases}$$

where $y_i$ denotes the $i$th smoothed value and $m_i$ is the median of a *moving window* of $(2K+1)$ $x_i$ values:

$$m_i = \text{median}(x_{i-K}, x_{i-K+1}, \ldots, x_{i+K}) \ ,$$

and $S_i$ is the corresponding *mean* absolute deviation of the moving window:

$$S_i = \frac{1}{2K+1} \sum_{k=-K}^{K} |x_{i+k} - m_i| \ .$$

In Example 6.18 we use Hempel filter to smooth the input signals before optical flow estimation.

## 12.3 Robust Parameter Estimation

The aim of robust parameter estimation techniques is to reduce the influence of the outliers without actually identifying and eliminating the outliers. The basic idea is to use a likelihood function, $p(\mathbf{y}|\boldsymbol{\theta}, I)$, whose tail size varies in proportion to the number of outliers which are present. Our aim is that as we move away from $\hat{\boldsymbol{\theta}}$, the mean value of $\boldsymbol{\theta}$, the tail will decrease but not too quickly nor too slowly.[1]

[1] In this context, the Gaussian distribution, $\mathcal{N}(\boldsymbol{\mu}, \boldsymbol{\Sigma})$, is *not* robust since its likelihood drops off very quickly as we move more than a few standard deviations away from $\boldsymbol{\mu}$.

**Table 12.2** Robust likelihood functions

| Name | Formula |
|---|---|
| Student-*t* | $\Gamma((\nu+1)/2)/(\sqrt{\nu\pi}\Gamma(\nu/2)\sigma) \times (1+((y-\mu)/\sigma\sqrt{\nu})^2)^{-(\nu+1)/2}$. The parameter $\nu$ controls the thickness of the tails, the thickness decreasing with an increase in $\nu$ |
| "Good-and-Bad" | $\alpha\mathcal{N}(y\|\mu,\sigma^2)+(1-\alpha)\mathcal{N}(y\|\mu,\beta^2\sigma^2)$, where $\mathcal{N}(y\|\mu,\sigma^2)$ is the "Good", or "normal", likelihood function which is used to represent the inliers and $\mathcal{N}(y\|\mu,\beta^2\sigma^2)$ is the "Bad", or "aberrant", likelihood function which is used to represent the outliers, where $\beta >> 1$. The parameter $\alpha$ represents the expected proportion of inliers in the input data |
| Gaussian + Step | $\mathcal{N}(y\|\mu,\sigma^2)+F$, where $F$ is a constant which ensures the probability of an outlier never falls too low |

In Table 12.2 we list some standard functions which may be used to model a robust one-dimensional likelihood function.

### 12.3.1 Student-t Function

The multivariate Student-*t* likelihood function is

$$p(\boldsymbol{\theta}|\mathbf{y},I) = \mathcal{S}t(\mathbf{y}|\boldsymbol{\mu},\boldsymbol{\Sigma},\nu)\ , \tag{12.5}$$

where the Student-*t* function, $\mathcal{S}t(\boldsymbol{\mu},\boldsymbol{\Sigma},\nu)$, is a unimodal function with adjustable tails. In one-dimension it is defined as

$$\mathcal{S}t(y|\mu,\sigma^2,\nu) = \frac{\Gamma((\nu+1)/2)}{\sqrt{\nu\pi}\times\Gamma((\nu/2))\sigma}\left[1+\frac{(y-\mu)^2}{\nu\sigma^2}\right]^{-(\nu+1)/2}\ , \tag{12.6}$$

where the parameter, $\nu$, controls the thickness of the tails, the thickness decreasing with an increase in $\nu$. When $\nu > 30$, there is virtually no difference between $\mathcal{S}t(y|\mu,\sigma,\nu)$ and $\mathcal{N}(y|\mu,\sigma^2)$.

The following example illustrates the use of $K$ Student-*t* likelihood functions in a *robust* finite mixture model.

**Example 12.7** *Student-t Mixture Model (SMM)* [3]. We reconsider Example 10.8. This time we use a robust mixture of $K$ Student-*t* functions, instead of a mixture of $K$ Gaussian functions. Our aim is to model a 1D function $\theta = f(y)$:

$$f(y) \simeq \sum_k \alpha_k \mathcal{S}t(y|\mu_k,\sigma_k^2,\nu_k)\ ,$$

where $\alpha_k, \mu_k, \sigma_k, \nu_k, k \in \{1, 2, \ldots, K\}$, are unknown and are to be learnt from a set of training data $(y_i, \theta_i), i \in \{1, 2, \ldots, N\}$, subject to the constraints

$$\alpha_k > 0 \quad \text{and} \quad \sum_k \alpha_k = 1 \; .$$

The ML parameter values can be estimated using the expectation-maximization (EM) algorithm (Sect. 10.8). Theoretically the degrees of freedom, $\nu_k$, can also be estimated by the EM algorithm. However, in practice [3] it is better to consider the $\nu_k$ as regular hyperparameters which are learnt by an exhaustive search.

We associate a "hidden" variable $z_i$ with each $y_i$, where $z_i$ can only take on the values $\{1, 2, \ldots, K\}$. Then, the EM algorithm for this problem consists of the following two steps, which are repeated until convergence:

E-step
$$p(z_i = k|y_i, \phi_k, \nu_k, I) = \frac{p(y_i|z_i = k, \phi_k, \nu_k, I)p(\phi_k|z_i = k, I)}{p(y_i|I)} \; ,$$
$$Q(z_i = k|y_i, \phi_k, \nu_k) = \frac{\nu_k + 1}{\nu_k + \big((y_i - \mu_k)^2/(\nu_k \sigma_k^2)\big)} \; ,$$

M-step
$$\alpha_k = \frac{1}{N} \sum_i w_{ik} \; ,$$
$$\mu_k = \frac{\sum_i w_{ik} Q(z_i = k|y_i, \phi_k, \nu_k) y_i}{\sum_i w_{ik} Q(z_i = k|y_i, \phi_k, \nu_k)} \; ,$$
$$\sigma_k^2 = \frac{\sum_i w_{ik} Q(z_i = k|y_i, \phi_k, \nu_k)(y_i - \mu_j)^2}{\sum_i w_{ik} Q(z_i = k|y_i, \phi_k, \nu_k)} \; ,$$

where $\phi_k = (\alpha_k, \mu_k, \sigma_k)$ and $w_{ik} = p(z_i = k|y_i, \phi_k, \nu_k, I)$.

Subspace techniques (Sect. 4.7) such as principal component analysis (PCA), linear discriminant analysis (LDA) and canonical correlation analysis (CCA), are sensitive to outliers and atypical observations. For this reason robust version of the different subspace techniques have been developed (see Sect. 12.5).

### *12.3.2 "Good-and-Bad" Likelihood Function*

The "Good-and-Bad" likelihood function uses a weighted sum of two Gaussian functions:

$$p(\mathbf{y}|\boldsymbol{\theta}, I) = \alpha \mathcal{N}(\mathbf{y}|\boldsymbol{\mu}, \boldsymbol{\sigma}^2) + (1 - \alpha)\mathcal{N}(\mathbf{y}|\boldsymbol{\mu}, \beta^2 \boldsymbol{\sigma}^2) \; , \tag{12.7}$$

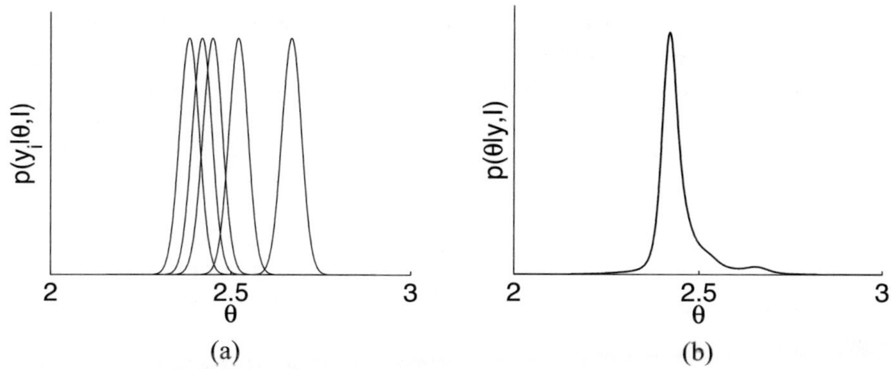


**Fig. 12.3** (**a**) Shows the Gaussian likelihood functions $p(y_i|\theta, I) = \mathcal{N}(y_i, \sigma_i^2)$ for the experimental measurements $y_i$ and error bars $\sigma_i$. (**b**) Shows the *a posteriori* pdf $p(\theta|\mathbf{y}, I)$ calculated using a "Good-and-Bad" likelihood function with $\beta = 10$

where $\mathcal{N}(\mathbf{y}|\boldsymbol{\mu}, \boldsymbol{\sigma}^2)$ is a "Good" likelihood function and is used to represent the inliers, $\mathcal{N}(\mathbf{y}|\boldsymbol{\mu}, \beta^2\boldsymbol{\sigma}^2)$ is a "Bad" likelihood function and is used to represent the outliers, and $\beta >> 1$. We weight the two Gaussian functions according to the expected proportion of inliers and outliers in the input data.

The "Good-and-Bad" likelihood function is often used to fuse together discordant measurements. We revisit Example 2.9 and use the "Good-and-Bad" likelihood function to fuse five independent neutron fission cross-sections.

**Example 12.8** *"Good-and-Bad" Fusion of Neutron Fission Cross-Section Measurements.* Table 2.1 lists $N = 5$ experimental measurements $\mathbf{y} = (y_1, y_2, \ldots, y_N)^T$ and error bars $\boldsymbol{\sigma} = (\sigma_1, \sigma_2, \ldots, \sigma_N)^T$, obtained in a neutron fission cross-section experiment. The corresponding likelihoods, $p(y_i|\theta, I) \sim \mathcal{N}(y_i, \sigma_i^2)$, are shown in Fig. 12.3.

Data point #5 is clearly an outlier being eight standard deviations away from the mean of the other four points. Assuming a flat prior on the cross-section $\theta$, the *a posteriori* pdf $p(\theta|\mathbf{y}, I)$ is proportional to the product of the individual likelihoods:

$$p(\theta|\mathbf{y}, I) \propto \prod_i p(y_i|\theta, I) .$$

Using "Good" and "Bad" Gaussian likelihood functions, $\mathcal{N}(y_i, \sigma_i^2)$ and $\mathcal{N}(y_i, \beta^2\sigma_i^2)$, the *a posteriori* pdf becomes

$$p(\theta|\mathbf{y}, I) \propto \int \prod_i \Big(\alpha\mathcal{N}(\theta|y_i, \sigma_i^2) + (1-\alpha)\mathcal{N}(\theta|y_i, \beta^2\sigma_i^2)\Big)\pi(\alpha|I)d\alpha ,$$

where $\pi(\alpha|I)$ is the *a priori* pdf of $\alpha$. Assuming a flat distribution for $\pi(\alpha|I)$, we obtain the *a posteriori* pdf shown in Fig. 12.3(b). This result is close to the product of $\mathcal{N}(\theta|y_i, \sigma_i^2)$ for $i \in \{1, 3, 4\}$, which is what we might expect if we believe that the points #2 and #5 are outliers that should be discounted.

Different variants of the "Good-and-Bad" likelihood have been developed and are widely used for dealing with spurious data. When the number of measurements is not too large ($N \leq 10$) the combinatorial "Good-and-Bad" likelihood is a good choice.

### Combinatorial "Good-and-Bad" Likelihood

In this variant we modify the "Good-and-Bad" likelihood function by including the probability that each input value belongs to the inlier distribution. We revisit Example 2.10 and use the combinatorial "Good-and-Bad" likelihood to fuse together a set of thirteen experimental measurements.

**Example 12.9** *Hubble's Constant Using Combinatorial Likelihood* [2, 19]. Table 2.2 lists $N = 13$ experimental measurements $\mathbf{y} = (y_1, y_2, \ldots, y_N)^T$ and error bars $\boldsymbol{\sigma} = (\sigma_1, \sigma_2, \ldots, \sigma_N)^T$ of Hubble's constant $\theta$. The measurements include a wide variety of different techniques and are assumed to be independent. We divide the measurements $\{y_i\}$ into "good" and "bad" measurements. Altogether there are $2^N$ different ways, or combinations, of doing this. We use a function $\delta_{ki}, i \in \{1, 2, \ldots, N\}$, to indicate which measurement is "Good" and which measurement is "Bad" in a given combination. Mathematically,

$$\delta_{ki} = \begin{cases} 1 \text{ if } i\text{th measurement in } k\text{th combination is "Good"} , \\ 0 \text{ if } i\text{th measurement in } k\text{th combination is "Bad"} . \end{cases}$$

Since the measurements are independent, the joint likelihood function for the $k$th combination is

$$p_k(\mathbf{y}|\theta, I) = \prod_i \left(\delta_{ki}\mathcal{N}(y_i|\theta, \sigma_i^2) + (1 - \delta_{ki})\mathcal{N}(y_i|\theta, \beta^2\sigma_i^2)\right) ,$$

and the corresponding *a posteriori* pdf $p(\theta|\mathbf{y}, I)$ is proportional to

$$\sum_{k=1}^{2^N} \pi(k|I)\ p_k(\mathbf{y}|\theta, I) ,$$

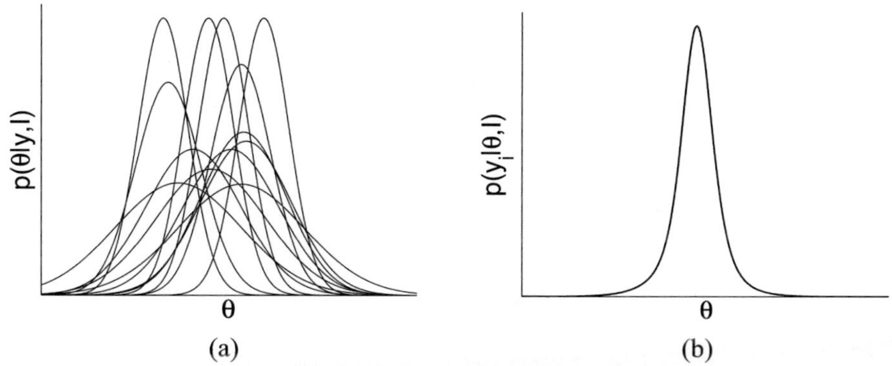


**Fig. 12.4** (**a**) Shows the Gaussian likelihoods $p(y_i|\theta, I) = \mathcal{N}(y_i, \sigma_i^2)$ for cross-sections $y_i$ and error bars $\sigma_i$. (**b**) Shows the *a posteriori* pdf $p(\theta|\mathbf{y}, I)$ calculated using a "Good-and-Bad" likelihood function with $\beta = 10$. *Note*: The curve was calculated assuming that $\pi(k|I)$ is inversely proportional to the total number of combinations which have the same number of "Good" measurements

where $\pi(k|I)$ is the *a priori* probability of the $k$th combination. By carrying out the summation over all $2^N$ combinations, we obtain the *a posteriori* pdf shown in Fig. 12.4. The pdf, $p(\theta|\mathbf{y}, I)$, bears a superficial resemblance to a Gaussian function, which is an *outcome* of the calculation and *not* an assumed input.

### 12.3.3 Kumar-Garg-Zachery (KGZ) Model

The Kumar-Garg-Zachery model is described in Sect. 2.8. In this model, the likelihood $p(y_i|\theta, I)$ that $y_i$ is a "good" measurement (i.e. an inlier) is modeled using a modified Gaussian distribution:

$$p(y_i|\theta, I) == \frac{1}{\sigma_i\sqrt{2\pi}} \exp - \frac{(\theta - y_i)^2}{\alpha_i^2}$$

where the parameter $\alpha_i$ *jointly* depends on the intrinsic variance $\sigma_i^2$ of the $i$th measurement *and* the distance between $y_i$ and the other measurements $y_k, k \neq i, k \in \{1, 2, \ldots, N\}$. The idea is that as the probability that $y_i$ is *not* a "good" measurement increases so the variance $\alpha_i$ should also increase. Mathematically, we define $\alpha_i^2$ as

$$\alpha_i^2 = \frac{2\sigma_i^2 m'}{\prod_{\substack{k=1\\k\neq i}}^{N} \left(\max(|y_i - y_k|, \sigma_i)\right)^2} .$$

where $m' = \max_i \left( \prod_{\substack{k=1\\k\neq i}}^{N} \left(\max(|y_i - y_k|, \sigma_i)\right)^2 \right)$.

Using the KSG model, the *a posteriori* pdf $p(\theta|\mathbf{y}, I)$ is given by

$$p(\theta|\mathbf{y}, I) = \frac{\pi(\theta|I)}{p(\mathbf{y}|I)} \prod_{i=1}^{N} \frac{1}{\sigma_i\sqrt{2\pi}} \times \exp -\left[\frac{(\theta - y_i)^2}{2\sigma_i^2}\Big(1 - \frac{1}{m'}\prod_{\substack{k=1\\k\neq i}}^{N}\big(\max(|y_i - y_k|, \sigma_i))^2\big)\Big)\right] . \tag{12.8}$$

Examples 2.9 and 2.10 illustrate using Eq. (12.8) to fuse a set of experimental measurements.

### 12.3.4 Gaussian Plus Constant

The Gaussian plus constant likelihood function is defined as

$$p(\mathbf{y}|\boldsymbol{\theta}, I) = \gamma \mathcal{N}(\mathbf{y}|\boldsymbol{\mu}, \boldsymbol{\sigma}^2) + (1-\gamma)\frac{1}{v} , \tag{12.9}$$

where $\gamma$ is a mixing parameter, $0 \leq \gamma \leq 1$, and $v$ is just a constant over which it is assumed the outlier distribution is uniform. We often write $(1-\gamma)/v$ as a constant $F$ (see Example 2.11).

Equation (12.9) is often used to robustify the Gaussian mixture model:

$$p(\mathbf{y}|\boldsymbol{\theta}, I) = \gamma \sum_{m=1}^{M} \pi_m \mathcal{N}(\mathbf{y}|\boldsymbol{\mu}_m, \sigma_m^2) + (1-\gamma)\frac{1}{v} . \tag{12.10}$$

Example 10.12 illustrates the use of Eq. (12.10).

## 12.4 Classical Robust Estimators

One popular classical robust technique is the $M$ estimators which are a generalization of the maximum likelihood and least squares estimators. The $M$-estimate of $a$ is defined as

$$\hat{a} = \arg\min_a \sum_i \rho\Big(\frac{y_i - a}{\sigma_i}\Big) , \tag{12.11}$$

where $\rho(u)$ is a robust loss function that grows subquadratically and is monotonically nondecreasing with increasing $|u|$ and $\sigma_i^2$ is the variance, or scale, associated with the residual errors $\epsilon_i = y_i - a$. The many $M$-estimators that have been proposed differ in shape of the loss functions $\rho(u)$. Some common loss functions are listed in Table 12.3.

**Table 12.3** Robust loss functions $L(t, \theta)$

| Name | $L(t, \theta)$ |
|---|---|
| $L_1$ | $\|y\|$ |
| $L_1 - L_2$ | $2(\sqrt{1 + y^2/2} - 1)$ |
| $L_p$ | $\|y\|^p/p$ |
| Beaton and Tukey | $c^2(1 - (1 - (u/c)^2)^3)/6$ if $\|u\| \leq c$; otherwise $c^2/6$ |
| Cauchy | $c^2 \log(1 + (u/c)^2)/2$ |
| Huber | $u^2/2$ if $\|u\| \leq k$, otherwise $k(2\|u\| - k/2)$ |

Inherent in the different $M$-estimators is the need to estimate the standard deviation $\sigma$ of the residual errors $\epsilon_i$. A robust standard deviation which may be used for this purpose is

$$\sigma = 1.4826\left(1 + \frac{5}{N-1}\right)\text{med}_i |y_i - a| \, . \tag{12.12}$$

**Example 12.10** *Robust Savitsky-Golay (SG) Filters* [15]. SG filters are widely used to smooth noisy data points in a time series (see Example 7.4). They work as follows. Let $x$ denote the unknown clean signal which is corrupted by random noise $w$. Then the observed noisy version of $x$ at the $k$th data point is $y_k = x_k + w_k$. Given the noisy data $y$, we choose a symmetric window about each point $y_k$, perform polynomial approximation over that window, and assign the value of the polynomial function at the central index of the window to the selected data point. The process is repeated for all $k$.

Consider the data points $k \in \{-K, -K+1, \ldots, K\}$ over which $y$ is approximated by an $N$th order polynomial function whose value at index $k$ is $p_k = \sum_{n=0}^{N} a_n k^n$, where $a_n$, $n \in \{0, 1, \ldots, N\}$, are the coefficients of the polynomial. In a robust SG the optimum coefficients $a_n^*$ are found by minimizing the $L_1$ error:

$$\mathbf{a}^* = \{a_0^*, a_1^*, \ldots, a_N^*\} = \arg\min \sum_{k=-K}^{K} |y_k - p_k| \, .$$

Unfortunately no analytical expressions are available for $\mathbf{a}^*$ although the coefficients may be found using an iterative procedure.

### 12.4.1 Least Median of Squares

A widely used alternative to the $M$ estimators is the *least median of squares* (LMedS) estimator in which we calculate the unknown parameter $a$ by solving the following nonlinear minimization problem:

$$\hat{a} = \arg\Bigl(\min \operatorname{med}_i \bigl(y_i - a\bigr)^2\Bigr) . \tag{12.13}$$

Increasingly, algorithms based on a least squares (LS) fit are being replaced with the LMedS algorithm.

**Example 12.11** *Using Least Median of Squares for Structural Superposition of Flexible Proteins* [12]. Structural comparison of two different conformations of the same protein are often based on a LS fit, which is sensitive to outliers, i.e. to large displacements. By replacing the LS algorithm with the LMedS algorithm we obtain substantial improvement in robustness (against both large displacements and other outliers) [12].

## 12.5 Robust Subspace Techniques*

Subspace techniques (see Sect. 4.7), such as principal component analysis (PCA), linear discriminant analysis (LDA) and canonical correlation analysis (CCA) are sensitive to outliers and to atypical observations. In creating robust subspace techniques, the basic idea is to replace non-robust estimators with robust versions. Unfortunately, the corresponding robust PCA, LDA and CCA algorithms [6, 10, 24] involve difficult optimization problems which must be solved numerically.

The following example describes the method used to robustify LDA.

**Example 12.12** *Robust LDA-L1 Algorithm* [6, 28]. Linear discriminant analysis (LDA) is a *supervised* technique in which we suppose each input measurement $\mathbf{y}_n$ is associated with a given class $C = c_k, k \in \{1, 2, \ldots, K\}$. The aim of LDA is to find a $L$-dimensional subspace $\mathbf{U} = (\mathbf{u}_1, \mathbf{u}_2, \ldots, \mathbf{u}_L)$ in which the different classes are maximally separated. If $\boldsymbol{\theta}_n = \bigl(\theta_n^{(1)}, \theta_n^{(2)}, \ldots, \theta_n^{(L)}\bigr)^T$ denotes the $L$-dimensional LDA representation of $\mathbf{y}_n$, then

$$\boldsymbol{\theta}_n = \mathbf{U}^T \mathbf{y}_n ,$$

where $\mathbf{U}$ is found by solving the following optimization problem:

$$\mathbf{U} = \arg\max_{W} \frac{\sum_{k=1}^{K} n_k \|\mathbf{U}^T(\boldsymbol{\mu}_k - \boldsymbol{\mu}_G)\|_2^2}{\sum_{k=1}^{K}\sum_{j\in c_k} \|\mathbf{U}^T(\mathbf{x}_j - \boldsymbol{\mu}_k)\|_2^2} \quad (12.14)$$

where $n_k$ is the number of input measurements which belong to class $c_k$, $\boldsymbol{\mu}_k$ is the mean vector for the $k$th class, $\boldsymbol{\mu}_G = \frac{1}{N}\sum_k n_k \boldsymbol{\mu}_k$ is the global mean vector and $\|\cdot\|_2$ is the $L_2$-norm. Analytical solutions of Eq. (12.14) are given in Eq. (4.8). To robustify Eq. (12.14) we simply replace the $L_2$-norm with a $L_1$-norm, giving:

$$\mathbf{U} = \arg\max_{W} \frac{\sum_{k=1}^{K} n_k \|\mathbf{U}^T(\boldsymbol{\mu}_k - \boldsymbol{\mu}_G)\|_1}{\sum_{k=1}^{K}\sum_{j\in C_k} \|\mathbf{U}^T(\mathbf{x}_j - \boldsymbol{\mu}_k)\|_1} . \quad (12.15)$$

Equation (12.15) is solved numerically subject to the constraint $\mathbf{U}^T\mathbf{U} = \mathbf{I}$ which ensures orthonormality.

## 12.6 Robust Statistics in Computer Vision

Special purpose robust statistical techniques are required for computer vision and image analysis applications. The input data in these applications are characterized by multiple structures and models, in which each model accounts for a relatively small percentage of the input data [23]. In these circumstances the classical $M$ and LMedS estimators, which require at least 50% inliers, are no longer reliable. In Table 12.4 we list some of the special robust techniques which have been developed for computer vision and image analysis applications [23]. The following example illustrates the RANSAC technique.

**Example 12.13** *RANSAC in Computer Vision* [13, 27]. The RANSAC algorithm works (see also Sect. 6.5) as follows. First we generate a model hypothesis from the input data $D$ by randomly selecting a subset of $p$ points and fitting a given parametric model of dimension $p$. The quality of the model is evaluated by finding the number of input data which are consistent with the estimated model. Mathematically, the quality is given by

$$Q = \sum_i \rho(e_i^2) ,$$

where $\rho(e_i^2) = 0$ if $e^2 < T^2$, otherwise 1. The random selection of a subset of $p$ points is repeated until a suitable termination criterion is fulfilled. If several

**Table 12.4** Robust techniques for computer vision and image analysis

| Name | Description |
|---|---|
| Hough | Hough Transform is a voting technique in which we count the number of data features that are mapped into each cell in quantized parameter space. Its main drawbacks are: excessive storage space, computational complexity and limited parameter precision |
| RANSAC | Random Sample Consensus. Subsets of input data are randomly selected and are fitted to a given parametric model. The subsets (called contain $p$ points where $p$ is dimension of parameter space. The quality of model parameters are evaluated on all the input data. Different cost functions may be used. Conventional cost is number of inliers, i.e., number of input data points which are consistent with the model. See Sect. 6.5 |
| MUSE | Minimum Unbiased Scale Estimator. MUSE randomly selects $p$-subsets and estimates fits based on $p$-subsets. It calculates an unbiased estimate of scale for each fit's $k$ smallest smallest residuals where $k$ is set to all possible values and satisfies $1 \leq k \leq N - p$. The smallest scale estimate over all possible $k$ is chosen as representative value of hypothesized fits. Finally fit from the $p$-subset with the smallest scale estimate is chosen as the optimum fit |
| ALKS | Adaptive Least $k$th Order Square. ALKS is based on least $k$th order squares (LKS) estimator. Difference between LKS and LMedS is: LKS uses $k$ data points out of $N$ data points ($p < k < N$) while LMedS uses $N/2$ data points. ALKS uses multi-step procedure in which in each step ALKS employs LKS with a different $k$ value. In order to estimate correct $k$, a random sampling technique similar to that used in RANSAC is employed. ALKS is robust against multiple structures but it cannot resist the influence of extreme outliers |
| RESC | Residual Consensus. The idea of RESC is: if a model is correctly fitted, the residuals of inliers should be small and the histogram of the residuals should concentrate within a small range in the lower part of the histogram. The RESC is very robust. Its main disadvantage is: it requires many parameters which must be tuned by user for optimal performance |

solutions are applicable then the data set with the smallest standard deviation is selected. After the estimation process the outliers are eliminated and the model is recomputed with all the inliers. Alternatively a robust cost function (e.g. a $M$ estimator or a LMedS) can be used to verify the model hypothesis concerning the data $D$. The main drawback to RANSAC is its high computational load. For data with $N$ points and $p$ unknown parameters a combination of $N/(p(N-p))$ possibilities must be considered.

Although very successful one of the problems with the RANSAC is that if the threshold for considering inliers is set too high then the robust estimate can be very poor. This undesirable situation can be remedied by minimizing a new cost function:

$$\widetilde{C} = \sum \widetilde{\rho}(e_i^2) ,$$

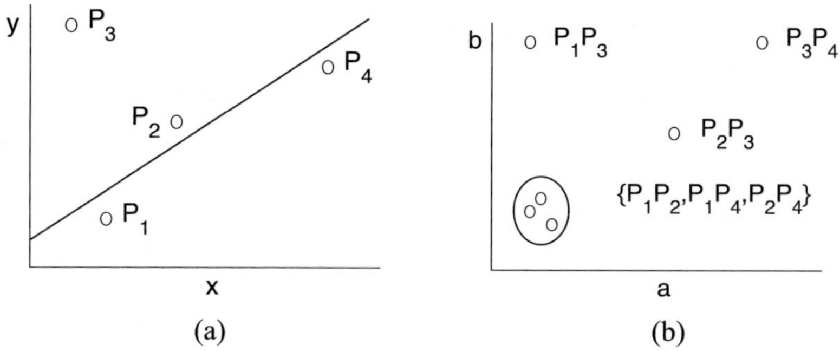


**Fig. 12.5** **a** Shows observations $P_i = (x_i, y_i)$, $i \in \{1, 2, \ldots, 4\}$, made at different times $t_i$ by sensor $S$. Observations $P_1$, $P_2$ and $P_4$ are associated with the object while $P_3$ is a false alarm. **b** Shows the corresponding six points $(a_{ij}, b_{ij})$ which represent the straight-lines $P_i P_j$ in Hough space. There is a single cluster of three points which represents the straight-lines $P_1 P_2$, $P_1 P_3$ and $P_2 P_3$ and three isolated points which represent the straight-lines $P_1 P_3$, $P_2 P_3$ and $P_4 P_3$

where

$$\widetilde{\rho}(e^2) = \begin{cases} e^2 & \text{if } e^2 < T^2 , \\ T^2 & \text{if } e^2 \geq T^2 . \end{cases}$$

We set $T = 1.96\sigma$ so that Gaussian inliers are only incorrect approximately 5 percent of the time.

The RANSAC algorithm is given in pseudo-code in Sect. 6.5.

The following example illustrates the use of the Hough transform [23] to initiate target tracks made by a moving object $O$ (Fig. 12.5).

**Example 12.14** *Target Tracking Initialization Using the Hough Transform* [5]. We consider an object $O$ which starts to move at time step 0 in a two-dimensional, $(x, y)$, space. A sensor, $S$, detects (with a probability $p_D < 1$) the position of $O$ at time steps $k$, $k \in \{0, 1, 2, \ldots\}$. In addition, the sensor may make several false detections at these times. If we assume that initially $O$ moves along a straight-line, we may use the Hough transform to identify the initial track as follows.

Consider a pair of detections $P_i = (x_i, y_i)$ and $P_j = (x_j, y_j)$ made at *different* time steps $k_i$ and $k_j$. The equation for the straight-line joining $P_i$, $P_j$ is

$$\frac{x}{a_{ij}} + \frac{y}{b_{ij}} = 1 ,$$

where

$$a_{ij} = \frac{x_i y_j - x_j y_i}{y_j - y_i} \quad \text{and} \quad b_{ij} = \frac{y_i x_j - y_j x_i}{x_j - x_i} .$$

We record each of pair of $a_{ij}$ and $b_{ij}$ values in a Hough space (i.e. the two-dimensional $(a, b)$ space). Then we identify the initial track of the object $O$ by finding the point $(a^*, b^*)$ around which there is a cluster of $(a_{ij}, b_{ij})$ values. We often only accept $(a^*, b^*)$ if the number of points in the cluster exceeds a given threshold.

The advantage of using the Hough transform is we do not specify the number of straight-line tracks. Instead, the Hough transform automatically estimates the number of tracks present.

## 12.7 Software

**MATLAB.** MATLAB COMPUTER VISION, SIGNAL PROCESSING and STATISTICS AND MACHINE LEARNING toolboxes include m-files on median, Hampel filters and RANSAC. LIBRA (Authors: Verboven and Hubert) is robust statistics toolbox, LMS (Author: Leontitsis) is a least median of squares toolbox and StMAR (Authors: Preve, Meitz, Saikkonen) is students-t mixture toolbox.

**PYTHON.** ROBPY (Authors: Leyder, Raymaekers, Rousseeuw, Servotte) is a Python library for robust statistical analysis. Additional python libraries and modules are: PyOD (outlier detection) and SMM (student t-mixture model).

## 12.8 Further Reading

Robustifying the Bayesian framework, i.e. making the Bayesian framework robust against outliers, is a controversial subject which is still being intensively investigated. Apart from the discussion in Sect. 12.3 the reader should also review Sect. 2.8. Some modern references on the subject are: [4, 9, 18, 19, 29]. The development of robust EM algorithms based on Student-$t$ functions are discussed in detail in [3]. A comprehensive review of robust subspace techniques is [20]. Special robust techniques which have been developed for computer vision which are described in [14, 21, 23, 25]. A full description of Bayesian curve fitting including robust Bayesian curve fitting is given in [7].

## Problems

**12.1** Define robust statistics. Give an example showing the importance of using robust statistics.

**12.2** Gating is often used in target tracking to eliminate outliers. What is gating? What are the relative advantages and disadvantages of gating?

**12.3** Apart from eliminating outliers, robust techniques work by reducing the influence of outliers. Explain how employing student-t function reduces the influence of outliers.

**12.4** Write down the "Good-and-Bad" likelihood function. Given $N$ independent measurements $(y_i, \sigma_i), i \in \{1, 2, .., N\}$ of a parameter $\theta$ show how we calculate the a posteriori pdf of the parameter $\theta$.

**12.5** Describe the Hough transform and its use in tracking initialization.

**12.6** Describe the method of RANSAC.

## References


1. Akinshin A (2022) Finite-sample Rousseeuw-Croux scale estimators. arXiv:2208.13459
2. D'Agostini G (1999) Sceptical combinations of experimental results: general considerations and applications to $\epsilon'/\epsilon$. European organization for nuclear research ceRN-EP/99-139
3. Archambeau C (2005) Probabilistic models in noisy environments. PhD thesis, Departement d'Electricite, Universite Catholique de Louvain, Belgium
4. Berger J (1994) An overview of robust Bayesian analysis. Test 3:5–124
5. Blackman SS, Popoli RF (1999) Design and analysis of modern tracking Systems. Published by Artech House, MA
6. Chen X, Yang J, Jun Z (2014) An improved linear discriminant analysis with L1-norm for robust feature extraction. Int Conf Patt Recogn 1585–1590
7. Denison DGT, Holmes C, Mallick BK et al (2002) Bayesian methods for nonlinear classification and regression. Published by Wiley
8. Dose V, der Linden Von W (1999) Outlier tolerant parameter estimation. In: Dose V et al (eds) Maximum entropy and bayesian methods. Published by Kluwer Academic Publishers
9. Huber PJ, Ronchetti EM (2009) Robust statistics. Published by Wiley
10. Kwak N (2008) Principal component analysis based on L1-norm maximization. IEEE Trans Patt Analy Mach Intell 30:1672–1680
11. Leys C, Ley C, Klein O, Bernard P, Licata L (2013) Detecting outliers: do not use standard deviation around the mean, use absolute deviation around the median. J Expt Social Psych 49:764–766
12. Liu Y-S, Fang Y, Ramani K (2009) Using least median of squares for structural superposition of flexible proteins. BMC Bioinform 10
13. Matinez-Otzeta JM, Rodriguez-Moreno I, Mendialdua I et al (2023) RANSAC for robotic applications: a survey. Sensors 23:327
14. Meer P (2004) Robust techniques for computer vision. In: Medioni G et al (ed) Emerging topics in computer vision, Published by Prentice-Hall

15. Menon SV, Seelamantula CS (2014) Robust Savitzky-Golay filters. Int Conf Digital Sig Proc 688-693
16. Pearson RK, Neuvo Y, Astola J et al (2016) Generalized Hampel filters. EURASIP J Adv Sig Proc 2016:Article 87
17. Rousseeuw PJ, Croux C (1993) Alternatives to the median absolute deviation. J Am Stat Assoc 88:1273–1283
18. Sivia DS, Skilling J (2006) Data analysis: a bayesian tutorial, 2nd edn. Published by Oxford University Press
19. Sivia DS (1996) Dealing with duff data. In: Sears M, Nedeljkovic NE, Sibisi S (eds) Proc Maximum Entropy Conf South Africa 131–137
20. Skocaj D (2003) Robust subspace approaches to visual learning and recognition. PhD thesis, University of Ljubljana
21. Stewart CV (1999) Robust parameter estimation in computer vision. SIAM Rev 41:513–537
22. Voloh B, Watson MR, Konig S, Womelsdorf T (2020) MAD saccade: statistically robust saccade threshold estimation via the median absolute deviation. J Eye Movement Res 12(8)
23. Wang H (2004) Robust statistics for computer vision: model fitting, image segmentation and visual motion analysis. PhD thesis, Monash University
24. Wang S, Du H, Zhang G et al (2020) Robust canonical correlation analysis based on L1-norm minimization for feature learning and image recognition. J Elec Imag 29:023001
25. Wang S, Veldhuis R., Brune C et al (2024) A survey on the robustness of computer vision models against common corruptions. arXiv:2305.06024
26. Yang X, Shen X, Long J, Chen H (2012) An improved median-based Otsu thresholding algorithm. AASRI Procedia 3:468–473
27. Zhang Y, Tian H, Deng Y et al (2019) Improving RANSAC for efficient and precise model fitting with statistical analysis. Eur J Elec Eng Comp Sci 3(5)
28. Zhaong F, Zhang J (2013) Linear discriminant analysis based on L1-norm maximization. IEEE Trans Im Proc 22:3018–3027
29. Zoubir AM, Koivunen V, Chakhchoukh Y et al (2012) Robust estimation in signal processing: a tutorial-style treatment of fundamental concepts. IEEE Sig Proc Mag 29:61–80

# Chapter 13
# Sequential Bayesian Inference and Kalman Filters

**Abstract** In this chapter we consider sequential Bayesian inference in which we consider the Bayesian estimation of a dynamic system which is changing in time. We start by considering tracking a single target observed from a single sensor. For most applications the sequential Bayesian filter is analytically intractable. In this case we consider the Kalman filter. This is a special case of linear Gaussian systems in which the equations are tractable and a closed form recursive solution available. We consider several data fusion applications of the Kalman filter including digital image stabilizer and small fast moving object (golf ball) detection and tracking. We consider several important variants of the Kalman filter including robust, augmented, extended, unscented, switching and ensemble Kalman filters. We then consider the generalization of the Kalman filter to track multiple targets. Finally we consider multi-sensor multi-temporal fusion.

## 13.1 Introduction

The subject of this chapter is *sequential Bayesian inference* in which we consider the Bayesian estimation of a dynamic system which is changing in time. Let $\boldsymbol{\theta}_k$ denote the state of the system, i.e. a vector which contains all relevant information required to describe the system, at some (discrete) time $k$. Then the goal of sequential Bayesian inference is to estimate the a posteriori pdf $p(\boldsymbol{\theta}_k|\mathbf{y}_{1:l}, I)$, by fusing together a sequence of sensor measurements $\mathbf{y}_{1:l} = \{\mathbf{y}_1, \mathbf{y}_2, \ldots, \mathbf{y}_l\}$. In this chapter we shall only consider the calculation of $p(\boldsymbol{\theta}_k|\mathbf{y}_{1:l}, I)$ for $k = l$ which is known as (sequential) Bayesian *filtering*.[1]

We start the chapter by assuming the measurements are all made by the same sensor. In this case we have *single-sensor multi-temporal data fusion*. In Sect. 13.7 we remove this restriction and consider the case when the measurements are made by multiple sensors. This is known as *multi-sensor multi-temporal data fusion*.

[1] Calculation of $p(\boldsymbol{\theta}_k|\mathbf{y}_{1:l}, I)$ for $k < l$ and $k > l$ are known, respectively, as (sequential) Bayesian *smoothing* and *forecasting*.



H. B. Mitchell, *Data Fusion*, https://doi.org/10.1007/978-3-662-71023-4_13

**Table 13.1** Applications in which sequential filter is the primary fusion algorithm

| Class | | Application |
|---|---|---|
| Early-DaO | Example 3.9 | Heart rate estimation using a Kalman Filter |
| | Example 11.6 | Non-intrusive speech quality estimation using GMM's |
| | Example 13.5 | Golf ball detection and tracking using a joint R-CNN/Kalman filter |
| | Example 13.3 | Tracking a moving target |
| | Example 13.6 | Correcting near surface temperature forecasts using a Kalman filter |
| | Example 13.12 | Multi-sensor multi-temporal data fusion: measurement fusion |
| Early-FeO | Example 13.8 | Recursive parameter estimation for a first-order autoregressive process |
| Inter-FeO | Example 13.1 | INS/Radar altimeter: a hybrid navigation algorithm |
| | Example 13.12 | Multi-sensor multi-temporal data fusion: measurement fusion |
| | Example 13.13 | Multi-sensor multi-temporal data fusion: track-to-track fusion |
| | Example 13.15 | Modified track-to-track fusion algorithm |

For many applications an estimate of $p(\boldsymbol{\theta}_k|\mathbf{y}_{1:k}, I)$ is required every time step that a measurement is received.[2] In this case it is common practice to rewrite $p(\boldsymbol{\theta}_k|\mathbf{y}_{1:k}, I)$ using a recursive formulation of Bayes' theorem:

$$p(\boldsymbol{\theta}_k|\mathbf{y}_{1:k}, I) = \frac{\overbrace{p(\mathbf{y}_k|\boldsymbol{\theta}_k, I)}^{\text{likelihood}}\overbrace{p(\boldsymbol{\theta}_k|\mathbf{y}_{1:k-1}, I)}^{a\ priori\ \text{pdf}}}{\underbrace{p(\mathbf{y}_k|\mathbf{y}_{1:k-1}, I)}_{\text{evidence}}} , \tag{13.1}$$

where

$$p(\boldsymbol{\theta}_k|\mathbf{y}_{1:k-1}, I) = \int p(\boldsymbol{\theta}_k|\boldsymbol{\theta}_{k-1}, I)p(\boldsymbol{\theta}_{k-1}|\mathbf{y}_{1:k-1}, I)d\boldsymbol{\theta}_{k-1} , \tag{13.2}$$

$$p(\mathbf{y}_k|\mathbf{y}_{1:k-1}, I) = \int p(\mathbf{y}_k|\boldsymbol{\theta}_k, I)p(\boldsymbol{\theta}_k|\mathbf{y}_{1:k-1}, I)d\boldsymbol{\theta}_k . \tag{13.3}$$

Equations (13.1)–(13.3) represent a convenient solution to the problem of (sequential) Bayesian filtering: the sensor measurements are processed sequentially and not as a batch. At each time step only the current sensor measurement, $\mathbf{y}_k$, is used and it is not necessary to maintain, or re-process, previous measurements.

The main difficulty in calculating $p(\boldsymbol{\theta}_k|\mathbf{y}_{1:k}, I)$ involves solving the integrals defined in Eq. (13.3): this forms the main topic of our discussion in this chapter.

In many applications the recursive filter is the primary fusion algorithm in a single-sensor, or a multi-sensor, multi-temporal data fusion system. In Table 13.1 we list some examples of these applications together with their classification.

[2] In this chapter we follow standard practice and assume the length of each time step is 1 s. In this case, an object moving with a velocity $v$ makes a displacement of size $v$ in one time step.

## 13.2 Recursive Filter

The recursive filter is defined in Eqs. (13.1)–(13.3). These equations may be derived as follows.

**Bayes' Theorem.** Equation (13.1) represents a recursive formulation of Bayes' theorem. It is obtained by applying Bayes' theorem to the last measurement in $p(\boldsymbol{\theta}_k|\mathbf{y}_{1:k}, I)$:

$$\begin{aligned} p(\boldsymbol{\theta}_k|\mathbf{y}_{1:k}, I) &= p(\boldsymbol{\theta}_k|\mathbf{y}_k, \mathbf{y}_{1:k-1}, I) , \\ &= \frac{p(\boldsymbol{\theta}_k, \mathbf{y}_k, \mathbf{y}_{1:k-1}, I)}{p(\boldsymbol{\theta}_k, \mathbf{y}_{1:k-1}, I)} , \\ &= \frac{p(\mathbf{y}_k|\mathbf{y}_{1:k-1}, \boldsymbol{\theta}_k, I)p(\mathbf{y}_{1:k-1}|\boldsymbol{\theta}_k, I)p(\boldsymbol{\theta}_k|I)}{p(\mathbf{y}_k|\mathbf{y}_{1:k-1}, I)p(\mathbf{y}_{1:k-1}|I)} , \\ &= \frac{p(\mathbf{y}_k|\mathbf{y}_{1:k-1}, \boldsymbol{\theta}_k, I)p(\boldsymbol{\theta}_k|\mathbf{y}_{1:k-1}, I)p(\mathbf{y}_{1:k-1}|I)}{p(\mathbf{y}_k|\mathbf{y}_{1:k-1}, I)p(\mathbf{y}_{1:k-1}, I)} , \\ &= \frac{p(\mathbf{y}_k|\boldsymbol{\theta}_k, I)p(\boldsymbol{\theta}_k|\mathbf{y}_{1:k-1}, I)}{p(\mathbf{y}_k|\mathbf{y}_{1:k-1}, I)} , \end{aligned} \tag{13.4}$$

where the last equation follows from the assumption that the $\mathbf{y}_k$ obey a *Markov measurement model* in which $\mathbf{y}_k$ is independent of the previous states $\boldsymbol{\theta}_{1:k-1}$ and the previous measurements $\mathbf{y}_{1:k-1}$, i.e.

$$p(\mathbf{y}_k|\boldsymbol{\theta}_k, \mathbf{y}_{1:k-1}, I) = p(\mathbf{y}_k|\boldsymbol{\theta}_k, I) . \tag{13.5}$$

**A Prior pdf.** Equation (13.2) represents an integral formulation of the a priori pdf. It is obtained by marginalizing $\boldsymbol{\theta}_{k-1}$ out of the joint pdf $p(\boldsymbol{\theta}_k, \boldsymbol{\theta}_{k-1}|\mathbf{y}_{1:k-1}, I)$:

$$\begin{aligned} p(\boldsymbol{\theta}_k|\mathbf{y}_{1:k-1}, I) &= \int p(\boldsymbol{\theta}_k, \boldsymbol{\theta}_{k-1}|\mathbf{y}_{1:k-1}, I)d\boldsymbol{\theta}_{k-1} , \\ &= \int p(\boldsymbol{\theta}_k|\boldsymbol{\theta}_{k-1}, \mathbf{y}_{1:k-1}, I)p(\boldsymbol{\theta}_{k-1}|\mathbf{y}_{1:k-1}, I)d\boldsymbol{\theta}_{k-1} , \\ &= \int p(\boldsymbol{\theta}_k|\boldsymbol{\theta}_{k-1}, I)p(\boldsymbol{\theta}_{k-1}|\mathbf{y}_{1:k-1}, I)d\boldsymbol{\theta}_{k-1} , \end{aligned} \tag{13.6}$$

where the last equation follows from the assumption that the $\boldsymbol{\theta}_k$ obey a *Markov process model* in which the $\boldsymbol{\theta}_k$ depends only on $\boldsymbol{\theta}_{k-1}$:

$$p(\boldsymbol{\theta}_k|\boldsymbol{\theta}_{k-1}, \mathbf{y}_{1:k-1}, I) = p(\boldsymbol{\theta}_k|\boldsymbol{\theta}_{k-1}, I) . \tag{13.7}$$

**Bayesian Evidence.** Equation (13.3) represents an integral formulation of the Bayesian *evidence*. It follows directly from the definition of evidence which acts as a normalization factor in Bayes' rule.

In order to carry out the integrals in Eqs. (13.2) and (13.3) we rewrite the Markov process and measurement models as two stochastic equations:

$$\boldsymbol{\theta}_k = \mathbf{F}_k(\boldsymbol{\theta}_{k-1}, \mathbf{v}_{k-1}) , \tag{13.8}$$

$$\mathbf{y}_k = \mathbf{H}_k(\boldsymbol{\theta}_k, \mathbf{w}_k) , \tag{13.9}$$

where $\mathbf{v}_{k-1}$ is a noise term which represents the stochastic disturbances in the system and $\mathbf{w}_k$ is a noise term which represents the errors in the measurement process. Substituting Eq. (13.8) in Eq. (13.2) gives

$$\begin{aligned} p(\boldsymbol{\theta}_k|\boldsymbol{\theta}_{k-1}, I) &= \int p(\boldsymbol{\theta}_k|\boldsymbol{\theta}_{k-1}, \mathbf{v}_{k-1}, I) p(\mathbf{v}_{k-1}|\boldsymbol{\theta}_{k-1}, I) d\mathbf{v}_{k-1} , \\ &= \int \delta\big(\boldsymbol{\theta}_k - \mathbf{F}_k(\boldsymbol{\theta}_{k-1}, \mathbf{v}_{k-1})\big) p(\mathbf{v}_{k-1}|I) d\mathbf{v}_{k-1} , \end{aligned} \tag{13.10}$$

where the last equation follows from the assumption that the noise is independent of the state. Likewise, substituting Eq. (13.9) in Eq. (13.3) gives

$$\begin{aligned} p(\mathbf{y}_{1:k}|\boldsymbol{\theta}_k, I) &= \int p(\mathbf{y}_{1:k}|\boldsymbol{\theta}_k, \mathbf{w}_k, I) p(\mathbf{w}_k|\boldsymbol{\theta}_k, I) d\mathbf{w}_k , \\ &= \int \delta\big(\mathbf{y}_k - h_k(\boldsymbol{\theta}_k, \mathbf{w}_k)\big) p(\mathbf{w}_k|I) d\mathbf{w}_k . \end{aligned} \tag{13.11}$$

In Fig. 13.1 we show the main processing steps in the Bayesian recursive filter. The filter is initialized by specifying an *a priori* pdf, $p(\boldsymbol{\theta}_0|\mathbf{y}_0, I)$, where $\mathbf{y}_0$ denotes that no measurements have been made. The filter then operates recursively, performing a single cycle each time a new measurement becomes available. Suppose the *a*

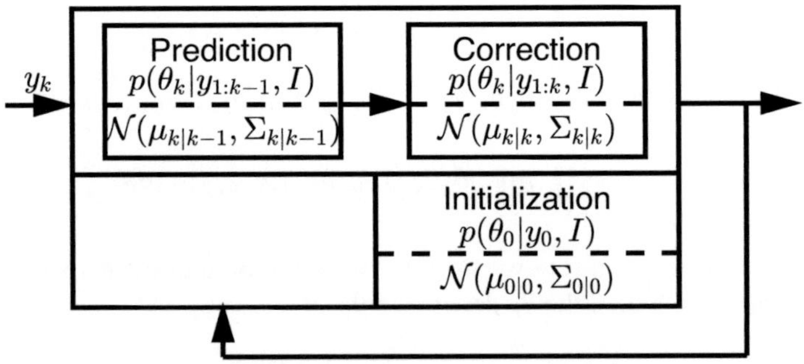


**Fig. 13.1** Shows the main processing steps (initialization, prediction and correction) in a recursive filter (RF) alongside the corresponding steps in a Kalman filter (KF). The equations for the KF are written below the dashed lines. **Initialization**: Performed one time. **Prediction**: Calculate predicted $p(\boldsymbol{\theta}|\mathbf{y}_{1:k-1}, I)$ (RF), $\mathcal{N}(\mu_{k|k-1}, \Sigma_{k|k-1})$ (KF) using process model $\boldsymbol{\theta}_k = \mathbf{F}_{k-1}(\boldsymbol{\theta}_{k-1}, \mathbf{v}_{k-1})$ and *a posteriori* pdf (from previous time step). **Correction**: Calculate *a posteriori* $p(\boldsymbol{\theta}_k|\mathbf{y}_{1:k}, I)$ (RF), $\mathcal{N}(\mu_{k|k}, \Sigma_{k|k})$ (KF) by correcting predicted pdf using current measurement $\mathbf{y}_k$

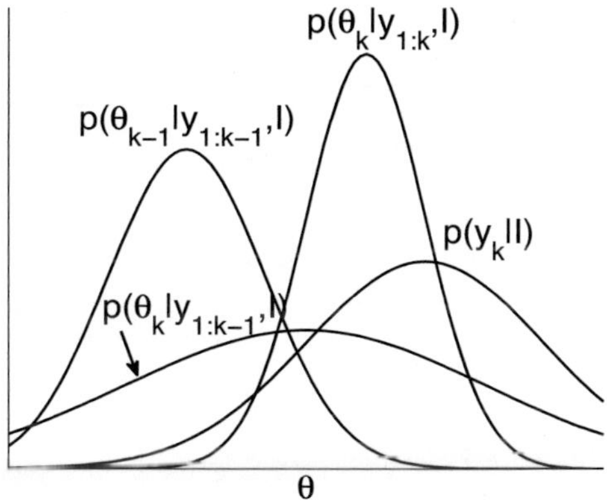


**Fig. 13.2** Shows the different pdf's in a generic recursive filter. In prediction step we use process model (13.7) to forward propagate $p(\boldsymbol{\theta}_{k-1}|\mathbf{y}_{1:k-1}, I)$ to next time step $k$. The result is a predicted pdf which acts as *a priori* pdf $p(\boldsymbol{\theta}_k|\mathbf{y}_{1:k-1}, I)$ for time step $k$. Because of noise $\mathbf{v}_{k-1}$, the *a priori* pdf is a deformed, spread-out version of $p(\boldsymbol{\theta}_{1-k}|\mathbf{y}_{1:k-1}, I)$. In correction step, we use sensor measurement $p(\mathbf{y}_k|I)$ to modify *a priori* pdf. The result is a sharp *a posteriori* pdf $p(\boldsymbol{\theta}_k|\mathbf{y}_{1:k}, I)$ for time step $k$

*posteriori* pdf, $p(\boldsymbol{\theta}_{k-1}|\mathbf{y}_{1:k-1}, I)$, has been estimated at time step $(k-1)$ and that a new measurement has become available at time step $k$. Then we calculate a *predicted* pdf, $p(\boldsymbol{\theta}_k|\mathbf{y}_{1:k-1}, I)$, by propagating $p(\boldsymbol{\theta}_{k-1}|\mathbf{y}_{1:k-1}, I)$ from the time step $(k-1)$ to $k$. The predicted pdf acts as the *a priori* pdf for the time step $k$ and is calculated using a *process model* which describes the evolution of the system. In general the predicted pdf will contain errors due to inaccuracies in the process model and noise in the input measurements. To eliminate these errors we correct $p(\boldsymbol{\theta}_k|\mathbf{y}_{1:k-1}, I)$ using the latest sensor measurement $\mathbf{y}_k$. This is achieved by using Bayes' theorem which is a mechanism for updating our knowledge in the light of new information (see Chap. 10).

Figure 13.2 shows the shapes of the different pdf's in a generic recursive filter.

The following example illustrates Eqs. (13.1)–(13.3) in terrain-aided navigation.

**Example 13.1** *INS/Radar Altimeter: A Hybrid Navigation Principle* [12]. In terrain-aided navigation we combine radar altimeter measurements with inertial measurements (INS) to achieve an optimal (minimum mean square error) estimate of the position of the aircraft. Terrain-aided navigation works as follows: At each time step $k$ we use INS measurements (accelerometer and gyrometer) to compute the INS estimate, $\boldsymbol{\theta}_k$, of the aircraft. We then use sequential Bayesian inference to fuse $\boldsymbol{\theta}_k$ with the radar altimeter measurements $y_k$. The output is the INS drift $\delta\boldsymbol{\theta}_k$ (i.e. error in $\boldsymbol{\theta}_k$). The drift $\delta\boldsymbol{\theta}_k$ is then used to correct $\boldsymbol{\theta}_k$ (see Fig. 13.3). Mathematically, algorithm works as follows: We write INS estimate at time step $k$ as a linear Gaussian Eq. (see Sect. 11.6)

$$\boldsymbol{\theta}_k = \mathbf{F}_k\boldsymbol{\theta}_{k-1} + \mathbf{v}_{k-1} ,$$

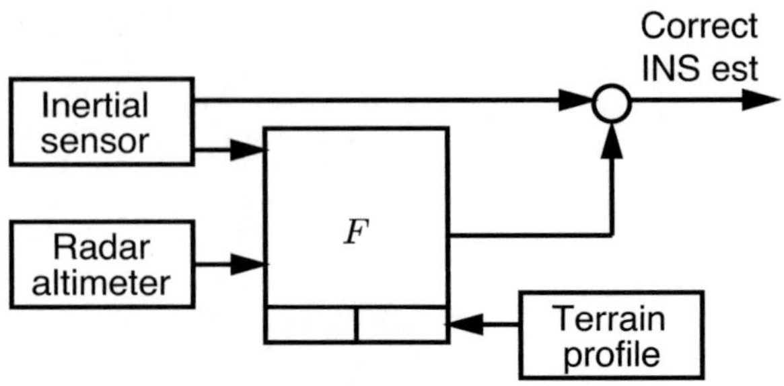


**Fig. 13.3** Shows the sensor-fusion filter for terrain-aided navigation. The sensor-fusion filter outputs the INS drift estimate which is then removed from the initial INS estimate to give a corrected INS estimate

where $\boldsymbol{\theta}_k$ denotes INS estimates (latitude, longitude, altitude, roll, pitch, yaw and speed) and $\mathbf{v}_{k-1} \sim \mathcal{N}(0, \mathbf{Q}_{k-1})$ is the noise vector in $\boldsymbol{\theta}_{k-1}$.

The scalar radar altimeter measurement equation is given by

$$y_k = z_k - h(l_k, L_k) + w_k \ ,$$

where $l_k$, $L_k$ and $z_k$ are, respectively, longitude, latitude and altitude of the aircraft. The function $h$ represents the terrain profile which is stored on-board the aircraft. The noise $w_k \sim \mathcal{N}(0, R_k)$ represents measurement error.

We compute *a posteriori* pdf $p(\boldsymbol{\theta}_k|y_{1:k}, I)$ using Eqs. (13.1, 13.3). The INS drift is given by

$$\delta\boldsymbol{\theta}_k = \boldsymbol{\mu}_k - \boldsymbol{\theta}_k \ ,$$

where $\boldsymbol{\mu}_k = \int \boldsymbol{\theta}_k p(\boldsymbol{\theta}_k|y_{1:k}, I) d\boldsymbol{\theta}_k$ is the mean INS value at time step $k$.

## 13.3 Kalman Filter (KF)

For most applications, Eqs. (13.2) and (13.3) are analytically intractable and approximate solutions must be used. However, for the special case of linear Gaussian systems (Sect. 11.6), the equations are tractable and a closed-form recursive solution for the sequential Bayesian filter is available. This is the *Kalman* filter (KF) [5] and because of its computational efficiency, it quickly established itself as the favourite algorithm for sequential Bayesian inference.

Mathematically, KF assumes linear Gaussian process and measurement models:

$$\boldsymbol{\theta}_k = \mathbf{F}_k \boldsymbol{\theta}_{k-1} + \mathbf{v}_{k-1} \ , \tag{13.12}$$

$$\mathbf{y}_k = \mathbf{H}_k \boldsymbol{\theta}_k + \mathbf{w}_k \ , \tag{13.13}$$

where $\mathbf{v}_{k-1} \sim \mathcal{N}(0, \mathbf{Q}_{k-1})$ and $\mathbf{w}_k \sim \mathcal{N}(0, \mathbf{R}_k)$ are independent Gaussian noise sources. The following example illustrates the calculation of $\mathbf{F}_k$ and $\mathbf{H}_k$ for a digital image stabilizer. For more examples, see Sect. 11.6.

**Example 13.2** *Digital Image Stabilizer Using KF* [33]. Hand-held cameras usually suffer from image/video instability due to unintentional shaking. Digital image stabilization is a method which aims to remove the camera shakes while keeping the intentional motion of the camera intact. We approximate the camera trajectory as

$$\boldsymbol{\theta}_k = \mathbf{F}\boldsymbol{\theta}_{k-1} + \mathbf{v}_{k-1} \quad \text{and} \quad \mathbf{y}_k = \mathbf{H}\boldsymbol{\theta}_k + \mathbf{w}_k \ ,$$

where $\boldsymbol{\theta}_k$ denotes the "state" of the camera (camera position, velocity and acceleration) at time step $k$ and $\mathbf{y}_k$ denotes the observed camera position at time step $k$. Then, assuming a panning/tilting model for the intentional motion of the camera, we have

$$\begin{pmatrix} \phi_k \\ x_k \\ u_k \\ y_k \\ v_k \end{pmatrix} = \begin{pmatrix} 1\,0\,0\,0\,0 \\ 0\,1\,1\,0\,0 \\ 0\,0\,1\,0\,0 \\ 0\,0\,0\,1\,1 \\ 0\,0\,0\,0\,1 \end{pmatrix} \begin{pmatrix} \phi_{k-1} \\ x_{k-1} \\ u_{k-1} \\ y_{k-1} \\ v_{k-1} \end{pmatrix} + \begin{pmatrix} m_k \\ 0 \\ n_k \\ 0 \\ n_k \end{pmatrix} ,$$

where $\phi_k$, $x_k$, $y_k$, $u_k$ and $v_k$ describe the intentional rotation, $x$ and $y$ translation and $x$ and $y$ translational velocities of the camera at time step $k$. The camera rotation is caused by the tilt of the camera and described by

$$\phi_k = \phi_{k-1} + m_{k-1} \ ,$$

where $m_{k-1} \sim \mathcal{N}(0, \Sigma_{\text{TILT}})$ is white noise with variance $\Sigma_{\text{TILT}}$.

The $x$ and $y$ translations are caused by the panning of the camera which is assumed to have a constant velocity $(u_k, v_k)$ with small random perturbations over time:

$$x_k = x_{k-1} + u_k \ , \quad y_k = y_{k-1} + v_k \ ,$$
$$u_k = u_{k-1} + n_k \ , \quad v_k = v_{k-1} + n_k \ ,$$

and $n_k \sim \mathcal{N}(0, \Sigma_{\text{PAN}})$ is white noise with variance $\Sigma_{\text{PAN}}$.

In general, the translational velocities due to the camera tilt, $u_k$ and $v_k$, are not observed. In this case, the matrix $\mathbf{F}_k$ is a $3 \times 5$ matrix:

$$\begin{pmatrix} \phi_k^0 \\ x_k^0 \\ y_k^0 \end{pmatrix} = \begin{pmatrix} 1\,0\,0\,0\,0 \\ 0\,1\,0\,0\,0 \\ 0\,0\,0\,1\,1 \end{pmatrix} \begin{pmatrix} \phi_k \\ x_k \\ u_k \\ y_k \\ v_k \end{pmatrix} + \begin{pmatrix} m_k^0 \\ n_k^0 \\ n_k^0 \end{pmatrix} ,$$

where superscript "0" denotes variables for observed or global motion; $m_k^0 = N(0, \Sigma_{\text{TILT}}^0)$ and $n_k^0 = N(0, \Sigma_{\text{PAN}}^0)$ are white noise with variance $\Sigma_{\text{TILT}}^0$ and $\Sigma_k^0$ standing for rotation and translation measurement noise.

If we assume the initial pdf is Gaussian, $\pi(\boldsymbol{\theta}_0|\mathbf{y}_0, I) \sim \mathcal{N}(\boldsymbol{\mu}_{0|0}, \boldsymbol{\Sigma}_{0|0})$, then both the predicted pdf (Eq. 13.2) and the *a posteriori* pdf (Eq. 13.1) are Gaussian:

$$\begin{aligned} p(\boldsymbol{\theta}_k|\mathbf{y}_{1:k-1}, I) &\sim \mathcal{N}(\boldsymbol{\mu}_{k|k-1}, \boldsymbol{\Sigma}_{k|k-1}) \ , \\ p(\boldsymbol{\theta}_k|\mathbf{y}_{1:k}, I) &\sim \mathcal{N}(\boldsymbol{\mu}_{k|k}, \boldsymbol{\Sigma}_{k|k}) \ , \end{aligned} \tag{13.14}$$

where

$$\begin{aligned} \boldsymbol{\mu}_{k|k-1} &= \mathbf{F}_k \boldsymbol{\mu}_{k-1|k-1} \ , \\ \boldsymbol{\Sigma}_{k|k-1} &= \mathbf{Q}_{k-1} + \mathbf{F}_k \boldsymbol{\Sigma}_{k-1|k-1} \mathbf{F}_k^T \ , \\ \boldsymbol{\mu}_{k|k} &= \boldsymbol{\mu}_{k|k-1} + \mathbf{K}_k (\mathbf{y}_k - \mathbf{H}_k \boldsymbol{\mu}_{k|k-1}) \ , \\ \boldsymbol{\Sigma}_{k|k} &= \boldsymbol{\Sigma}_{k|k-1} - \mathbf{K}_k \mathbf{H}_k \boldsymbol{\Sigma}_{k|k-1} \ , \end{aligned} \tag{13.15}$$

and $\mathbf{K}_k$ is the *Kalman gain matrix*,

$$\mathbf{K}_k = \boldsymbol{\Sigma}_{k|k-1} \mathbf{H}_k^T (\mathbf{H}_k \boldsymbol{\Sigma}_{k|k-1} \mathbf{H}_k^T + \mathbf{R}_k)^{-1} \ . \tag{13.16}$$

For reasons of numerical stability we often write

$$\boldsymbol{\Sigma}_{k|k} = (\mathbf{I} - \mathbf{K}_k \mathbf{H}_k) \boldsymbol{\Sigma}_{k|k-1} (\mathbf{I} - \mathbf{K}_k \mathbf{H}_k)^T + \mathbf{K}_k \mathbf{R}_k \mathbf{K}_k^T \ , \tag{13.17}$$

where $\mathbf{I}$ is the unit matrix.

The following example describes the tracking of a moving target which is subject to stochastic accelerations.

**Example 13.3** *Tracking a Moving Target* [3]. We consider a target moving in *one-dimension* at a nominal constant velocity while being subjected to *stochastic* accelerations. If $\theta_k$ denotes the true position of the target at the $k$th time step, then the corresponding matrix equation of motion of the target is:

$$\begin{pmatrix} \theta_k \\ \dot{\theta}_k \end{pmatrix} = \begin{pmatrix} 1 & T \\ 0 & 1 \end{pmatrix} \begin{pmatrix} \theta_{k-1} \\ \dot{\theta}_{k-1} \end{pmatrix} + \begin{pmatrix} \frac{T^2}{2} \\ T \end{pmatrix} \ddot{\theta}_{k-1} \ ,$$

where $\ddot{\theta}_k \sim \mathcal{N}(0, \sigma^2)$ is the (stochastic) acceleration which acts upon the target at time step $k$ and $T$ is the interval between successive time steps. The corresponding process model is

$$\begin{pmatrix} \theta_k \\ \dot{\theta}_k \end{pmatrix} = \begin{pmatrix} 1 & T \\ 0 & 1 \end{pmatrix} \begin{pmatrix} \theta_k \\ \dot{\theta}_k \end{pmatrix} + \mathbf{v}_{k-1} \ .$$

Equating the matrix equation of motion model with the process model leads to the identification of $\begin{pmatrix} \frac{T^2}{2} \\ T \end{pmatrix} \ddot{\theta}_k$ with the process noise $\mathbf{v}_{k-1} \sim \mathcal{N}(\mathbf{0}, \mathbf{Q})$. In this case,

$$\mathbf{Q} = \sigma^2 \begin{pmatrix} \frac{T^4}{4} & \frac{T^3}{2} \\ \frac{T^3}{2} & T^2 \end{pmatrix} \ .$$

### Kalman Filter and Deep Learning Object Detection

The KF is used in a very wide variety of data fusion applications see e.g. Example 3.9. The next two examples illustrate the fusion of CNN object detection with KF tracking. In the first example we track the bounding box of a slow moving object. The bounding box itself is detected using YOLO ("You-only-look-once") object detection network.

**Example 13.4** *Detection and tracking of chickens using YOLO and Kalman filter* [31] Monitoring the movement of birds is an important indicator of their well-being. In [31] chickens were monitored in a real-world farm environment using low-light low-resolution gray-scale images. In each frame chickens were detected and the corresponding bounding box calculated for each detected chicken.

The sampling frequency of the camera was 30 frames/sec. This means the movement of the chickens in two adjacent frames is small. In this case we use a four-dimensional state vector $\mathbf{x}$ for each chicken, where

$$\mathbf{x}_t = (p_x, p_y, \dot{p}_x, \dot{p}_y)^T \ ,$$

where $p_x$, $\dot{p}_x$ and $p_y$, $\dot{p}_y$ are, respectively, the center position and velocity of the bounding box along the $x$ and $y$ axes at time frame $t$.

We use linear Gaussian process and process and measurement models:

$$\mathbf{x}_t = \begin{pmatrix} 1 & 0 & T & 0 \\ 0 & 1 & 0 & T \\ 0 & 0 & 1 & 0 \\ 0 & 0 & 0 & 1 \end{pmatrix} \mathbf{x}_{i-1} + \mathbf{v}_{t-1} \quad \text{and} \quad \mathbf{y}_t = \begin{pmatrix} 1 & 0 \\ 0 & 1 \\ 0 & 0 \\ 0 & 0 \end{pmatrix} \mathbf{x}_t + \mathbf{w}_t \ ,$$

where $\mathbf{v}_{t-1} \sim \mathcal{N}(0, \mathbf{Q}_{t-1})$ and $\mathbf{w}_t \sim \mathcal{N}(0, \mathbf{R}_t)$.

For optimal KF performance, it is important to select appropriate $Q$ and $R$ matrices. The optimal $Q$ and $R$ matrices were found off-line in a series of experimental tests.

In the second example we track the bounding box of a very fast moving object. The bounding box itself is detected using faster R-CNN object detection network.

**Example 13.5** *Golf ball detection and tracking using joint R-CNN/Kalman Filter* [35]. Golf is a popular sport that has attracted a huge number of participants and audiences. In teaching golf, teachers often use video analysis. However, detecting the fast-moving tiny golf ball is difficult, since the golf ball can fly extremely fast and quickly disappear from the camera field of view.

In [35] the authors describe an efficient and effective scheme for real-time detection and tracking of a small golf ball. For each video frame we employ a KF to predict the estimated location of the golf ball in the next frame. The area centered at the predicted location is cropped and sent to the faster R-CNN detector (cf. Example 5.2) for golf ball detection. The KF then uses the detected location of the golf ball to update the state of the system.

In this application the golf balls are detected in small cropped image patches and not the entire input image. This increases the detection accuracy without sacrificing computational speed.

### Parameter Estimation

The KF as defined in Eqs. (13.14)–(13.16). contains the parameters $\boldsymbol{\mu}_{0|0}$, $\boldsymbol{\Sigma}_{0|0}$, $\mathbf{Q}_k$ and $\mathbf{R}_k$ which must be defined before running the filter. In general the values chosen for $\boldsymbol{\mu}_{0|0}$ and $\boldsymbol{\Sigma}_{0|0}$ do not seriously affect the medium and long-term results obtained with the filter, i.e. the values $\boldsymbol{\mu}_{k|k}$ and $\boldsymbol{\Sigma}_{k|k}$ for $k >> 0$. For this reason we often use simple *a priori* values for $\boldsymbol{\mu}_{0|0}$ and $\boldsymbol{\Sigma}_{0|0}$ or use a track initiation filter (see Example 12.14).

Regarding $\mathbf{Q}_k$ and $\mathbf{R}_k$, the way that $\mathbf{Q}_k$ and $\mathbf{R}_k$ are specified crucially affects $\boldsymbol{\mu}_{k|k}$ and $\boldsymbol{\Sigma}_{k|k}$. In mathematical terms, the values of $\mathbf{Q}_k$ and $\mathbf{R}_k$ affect the Kalman gain $\mathbf{K}_k$ which in turn affects the capability of the filter to adjust itself to changing conditions. A detailed discussion regarding the identification of the noise covariances $\mathbf{Q}_k$ and $\mathbf{R}_k$ is given in [34].

The following example illustrates a simple, but effective, procedure for *adaptively* calculating $\mathbf{Q}_k$ and $\mathbf{R}_k$.

**Example 13.6** *Correcting near surface temperature forecasts using a KF* [20]. Numerical weather prediction models often exhibit systematic errors in the forecasts of near surface weather parameters. One such parameter is the near surface temperature forecast $T$ which is a commonly biased variable, where the magnitude of the bias depends on many factors, including geographical location and the season. To correct for the systematic error in $T$ let the measurement $y_k$ be the difference between an observed temperature and the weather prediction model forecast and let the state vector $\theta_k$ be the systematic part of this error. We assume the changes in $\theta_k$ are random. The corresponding process and measurement models are then

$$\theta_k = \theta_{k-1} + v_{k-1} \quad \text{and} \quad y_k = \theta_k + w_k \ ,$$

where $v_k \sim \mathcal{N}(0, Q_k)$ and $w_k \sim \mathcal{N}(0, R_k)$. In [20] the authors estimate $Q_k$ and $R_k$ using the sample variances of the $N$ latest $v_{k-n} = \theta_{k-n} - \theta_{k-n-1}$ and $w_{k-n} = y_{k-n} - \theta_{k-n}$ values:

$$Q_k = \frac{1}{N-1} \sum_n \big((\theta_{k-n} - \theta_{k-n-1}) - \overline{Q}_k\big)^2 \ ,$$

$$R_k = \frac{1}{N-1} \sum_n \big((y_{k-n} - \theta_{k-n}) - \overline{R}_k\big)^2 \ ,$$

where $\overline{Q}_k = \frac{1}{N} \sum_n (\theta_{k-n} - \theta_{k-n-1})$ and $\overline{R}_k = \frac{1}{N} \sum_n (y_{k-n} - \theta_{k-n})$ .

### Model Inaccuracies

In the KF we model the system using a linear Gaussian process model (Eq. 13.12) and a linear Gaussian measurement model (Eq. 13.13). In practice most systems can never be perfectly modeled and this will set a lower limit on the final performance that can be obtained. However, even when the conditions are far from Gaussian, the KF can often give sensible results by giving more weight to the input measurements, i.e. by increasing the processes noise covariance $\mathbf{Q}$.

The following example illustrates an important case in which there is a clear mismatch between the assumed process model and the true system dynamics.

**Example 13.7** *Maneuvering Target* [3]. We track a target moving in one-dimension with a nominally constant velocity which is subject to stochastic accelerations. The corresponding process model (see Example 13.3) is

$$\begin{pmatrix} \theta_k \\ \dot{\theta}_k \end{pmatrix} = \begin{pmatrix} 1 & T \\ 0 & 1 \end{pmatrix} \begin{pmatrix} \theta_{k-1} \\ \dot{\theta}_{k-1} \end{pmatrix} + \mathbf{v}_{k-1} ,$$

where $\mathbf{v}_{k-1} \sim \mathcal{N}(\mathbf{0}, \mathbf{Q})$ and $\mathbf{Q} = \sigma^2 \begin{pmatrix} \frac{T^4}{4} & \frac{T^3}{2} \\ \frac{T^3}{2} & T^2 \end{pmatrix}$. If the process model is correct, the KF will provide optimum estimates of the target's position and velocity. However, if the target initiates and sustains a *driver-induced maneuver*, then the above model is incorrect. This is because a driver-induced maneuver is not well-modeled as a stochastic random variable.

One solution to this problem is to use a maneuver detector [3]. As long as the maneuver detector has not detected a target maneuver we continue to use the above process model. However, once a maneuver has been detected, we initialize a *new* recursive filter whose process model describes a target maneuver.

#### KalmanNet

Recently Revach et al. [28, 32] have described a novel approach to handle model and noise statistic uncertainties in the KF. The method (known as KalmanNet) combines deep learning with the classic KF. The KalmanNet replaces specific noise-model-dependent computations in the KF with a trainable neural network. See Sect. 13.5.

## 13.4 Kalman Filter Variants

The KF assumes a single linear Gaussian process model Eq. (13.12) and measurement model (Eq. 13.13). In practice these models may not adequately represent the truth in which case more sophisticated filtering techniques are necessary. Some standard non-linear filter techniques which involve modifications to the KF are listed in Table 13.2.

#### Robust Kalman Filter *

The KF assumes Gaussian process and measurement noise. The filter is therefore sensitive to outliers (see Chap. 12) which may be present in the measurement $\mathbf{y}_k$. If the $\mathbf{y}_k$ are thought to be contaminated by outliers we may limit their effect by gating

**Table 13.2** Kalman filter and its modifications

| Filter | Description |
|---|---|
| Kalman filter | Assumes linear Gaussian process and measurement models |
| Robust KF | Assumes linear Gaussian process and measurement models contaminated with outliers. Use robust techniques (Chap. 12) to reduce influence of outliers [9] |
| Extended KF | First-order local linearization of non-linear Gaussian process and measurement models |
| Unscented KF | Uses a deterministic set of weighted points $(\omega_k^{(m)}, \Omega_k^{(m)}, \boldsymbol{\theta}_k^{(m)})$, $m \in \{1, 2, \ldots, M\}$, to represent *a priori* pdf $p(\boldsymbol{\theta}_{k-1} \mid \mathbf{y}_{1:k-1}, I)$ |
| Ensemble KF | Used when dimensions of $\boldsymbol{\theta}_k$ is too large to permit conventional calculation of covariance matrix. Instead estimate covariance matrices using ensemble of trajectories |

(see Example 12.2) or by using a robust KF which is designed to be insensitive to outliers. A simple robust KF may be obtained by simply replacing the Kalman gain matrix $\mathbf{K}_k$ with the following expression:

$$\mathbf{K}_k = \boldsymbol{\Sigma}_{k|k-1}\mathbf{H}_k^T(\mathbf{H}_k\boldsymbol{\Sigma}_{k|k-1}\mathbf{H}_k^T + \mathbf{R}_k^{1/2}\mathbf{W}_k\mathbf{R}_k^{1/2})^{-1} , \tag{13.18}$$

where $\mathbf{W}_k = \text{diag}(W_k^{(1)}, W_k^{(2)}, \ldots, W_k^{(M)})$ denotes a set of weights given by

$$W_k^{(m)} = \frac{\psi_H\big([\mathbf{R}_k^{-1/2}(\mathbf{y}_k - \mathbf{H}_k\boldsymbol{\mu}_{k|k-1})]^{(m)}\big)}{[\mathbf{R}_k^{-1/2}(\mathbf{y}_k - \mathbf{H}_k\boldsymbol{\mu}_{k|k-1})]^{(m)}} , \tag{13.19}$$

where $\mathbf{X}^{(m)}$ is the $m$th component of $\mathbf{X}$ and $\psi_H(z)$ is the Huber influence function (see Example 13.8) [14].

**Example 13.8** *Recursive Parameter Estimation for a First-Order Autoregressive (AR) Process* [9]. Consider a first-order AR process:

$$y_k = \theta y_{k-1} + w_{k-1} .$$

We wish to recursively estimate $\theta$ given the sequence of measurements $y_{1:k}$ which may be contaminated by outliers. In this case, the corresponding process and measurement equations are:

$$\theta_k = \theta_{k-1} \quad \text{and} \quad y_k = y_{k-1}\theta_k + w_k ,$$

where $F = 1$, $H = y_{k-1}$, $v_k = 0$ and we use the "Good-and-Bad" model (Sect. 12.3.2) to represent the measurement noise $w_k$:

$$w_k \sim (1 - \alpha)\mathcal{N}(0, \sigma^2) + \alpha\mathcal{N}(0, \beta\sigma^2) ,$$

where $\alpha$ is the estimated fraction of outliers and $\beta >> 1$. The *a posteriori* pdf is $p(\theta_k|y_{1:k}, I) \sim \mathcal{N}(\mu_{k|k}, \Sigma_{k|k})$, where

$$\mu_{k|k} = \mu_{k-1|k-1} + \frac{1}{\sigma}\Sigma_{k|k-1} y_{k-1} \psi_H \left( \frac{\sigma(y_k - y_{k-1}\mu_{k-1|k-1})}{\Sigma_{k|k-1} y_{k-1}^2 + \sigma^2} \right) ,$$
$$\Sigma_{k|k} = \frac{\Sigma_{k|k-1}\sigma^2}{\Sigma_{k|k-1} y_{k-1}^2 + \sigma^2} ,$$
$$\psi_H(z) = \begin{cases} z & \text{if } |z| \le c , \\ cz/|z| & \text{otherwise} . \end{cases}$$

The recommended choice of $c$ is the $\alpha$-quantile of $\mathcal{N}(0, 1)$ e. g. $c = 1.645$ for an assumed 5% contamination.

**Extended Kalman Filter (EKF)**

In the extended KF (EKF) we apply the KF to non-linear process and measurement models by linearizing the models. We assume a non-linear Gaussian process model and a non-linear measurement model:

$$\begin{aligned} \boldsymbol{\theta}_k &= \mathbf{F}_k(\boldsymbol{\theta}_{k-1}) + \mathbf{v}_{k-1} , \\ \mathbf{y}_k &= \mathbf{H}_k(\boldsymbol{\theta}_k) + \mathbf{w}_k , \end{aligned} \tag{13.20}$$

where $\mathbf{v}_{k-1} \sim \mathcal{N}(0, \mathbf{Q}_{k-1})$ and $\mathbf{w}_k \sim \mathcal{N}(0, \mathbf{R}_k)$ are independent Gaussian noise sources. We linearize Eq. (13.20) by performing Taylor series expansions:

$$\begin{aligned} \boldsymbol{\theta}_k &\approx \mathbf{F}_k(\boldsymbol{\mu}_{k-1|k-1}) + \widetilde{\mathbf{F}}_k(\boldsymbol{\theta}_{k-1} - \boldsymbol{\mu}_{k-1|k-1}) + \mathbf{v}_{k-1} , \\ \mathbf{y}_k &\approx \mathbf{H}_k(\boldsymbol{\mu}_{k|k-1}) + \widetilde{\mathbf{H}}_k(\boldsymbol{\theta}_k - \boldsymbol{\mu}_{k|k-1}) + \mathbf{w}_k , \end{aligned} \tag{13.21}$$

where $\widetilde{\mathbf{F}}_k = \frac{\partial \mathbf{F}_k(\boldsymbol{\theta})}{\partial \boldsymbol{\theta}}\Big|_{\boldsymbol{\theta}=\boldsymbol{\mu}_{k-1|k-1}}$ and $\widetilde{\mathbf{H}}_k = \frac{\partial \mathbf{H}_k(\boldsymbol{\theta})}{\partial \boldsymbol{\theta}}\Big|_{\boldsymbol{\theta}=\boldsymbol{\mu}_{k|k-1}}$. By substituting $\boldsymbol{\alpha}_k = \mathbf{F}_k(\boldsymbol{\mu}_{k-1|k-1}) - \widetilde{\mathbf{F}}_k \boldsymbol{\mu}_{k-1|k-1}$ and $\boldsymbol{\beta}_k = \mathbf{H}_k(\boldsymbol{\mu}_{k|k-1}) - \widetilde{\mathbf{H}}_k \boldsymbol{\mu}_{k|k-1}$ into Eq. (13.21) we obtain the following linear Gaussian models:

$$\begin{aligned} \boldsymbol{\theta}_k &= \boldsymbol{\alpha}_k + \widetilde{\mathbf{F}}_k \boldsymbol{\theta}_{k-1} + \mathbf{v}_{k-1} , \\ \mathbf{y}_k &= \boldsymbol{\beta}_k + \widetilde{\mathbf{H}}_k \boldsymbol{\theta}_k + \mathbf{w}_k , \end{aligned} \tag{13.22}$$

which may be filtered using the standard KF. The corresponding EKF equations are

$$\begin{aligned} p(\boldsymbol{\theta}_k|\mathbf{y}_{1:k-1}, I) &\approx \mathcal{N}(\boldsymbol{\mu}_{k|k-1}, \boldsymbol{\Sigma}_{k|k-1}) , \\ p(\boldsymbol{\theta}_k|\mathbf{y}_{1:k}, I) &\approx \mathcal{N}(\boldsymbol{\mu}_{k|k}, \boldsymbol{\Sigma}_{k|k}) , \end{aligned} \tag{13.23}$$

where

$$\begin{aligned}
\boldsymbol{\mu}_{k|k-1} &= \mathbf{F}_k(\boldsymbol{\mu}_{k-1|k-1}) , \\
\boldsymbol{\Sigma}_{k|k-1} &= \mathbf{Q}_{k-1} + \widetilde{\mathbf{F}}_k \boldsymbol{\Sigma}_{k-1|k-1} \widetilde{\mathbf{F}}_k^T , \\
\boldsymbol{\mu}_{k|k} &= \boldsymbol{\mu}_{k|k-1} + \mathbf{K}_k(\mathbf{y}_k - \mathbf{H}_k(\boldsymbol{\mu}_{k|k-1})) , . \\
\boldsymbol{\Sigma}_{k|k} &= \boldsymbol{\Sigma}_{k|k-1} - \mathbf{K}_k \widetilde{\mathbf{H}}_k \boldsymbol{\Sigma}_{k|k-1} , \\
\mathbf{K}_k &= \boldsymbol{\Sigma}_{k|k-1} \widetilde{\mathbf{H}}_k^T (\widetilde{\mathbf{H}}_k \boldsymbol{\Sigma}_{k|k-1} \widetilde{\mathbf{H}}_k^T + \mathbf{R}_k)^{-1}
\end{aligned} \tag{13.24}$$

The EKF is widely used since it enables us to use the KF framework to perform sequential Bayesian inference on non-linear Gaussian systems. Its main drawbacks are that it requires the *analytical* computation of $\widetilde{\mathbf{F}}_k$ and $\widetilde{\mathbf{H}}_k$ and it is only accurate to first-order.

### Unscented Kalman Filter (UKF)*

In the unscented KF (UKF) we retain the Gaussian approximations in Eq. (13.23) but do not perform a Taylor series expansion on the process and measurement models $\mathbf{F}_{k-1}(\boldsymbol{\theta}_{k-1})$ and $\mathbf{H}_k(\boldsymbol{\theta}_k)$. Instead $\boldsymbol{\theta}_{k-1}$ and $\boldsymbol{\theta}_k$ are specified using a minimal set of carefully chosen sample points. These points (known as "sigma-points") completely capture the true mean and covariance of $\boldsymbol{\theta}_{k-1}$ and $\boldsymbol{\theta}_k$ and when propagated, respectively, through the non-linear process and measurement models, they capture the *a posteriori* means and covariances to third-order[3]. The UKF is a straightforward extension of the unscented transformation (UT) to the recursive estimation in (Eq. 13.20), where the state random variable is redefined as the concatenation of the original state $\boldsymbol{\theta}$ and the noise variables $\mathbf{v}$ and $\mathbf{w}$: $\boldsymbol{\phi}_k = (\boldsymbol{\theta}_k^T, \mathbf{v}_k^T, \mathbf{w}_k^T)^T$. The UT sigma point selection is applied to the new augmented state $\boldsymbol{\phi}_{k-1}$ to calculate the corresponding sigma points. The final UKF equations are:

$$\begin{aligned}
\boldsymbol{\mu}_{k|k-1} &= \sum_{m=1}^{2M} \omega_{k-1}^{(m)} \hat{\boldsymbol{\theta}}_k^{(m)} , \\
\boldsymbol{\Sigma}_{k|k-1} &= \sum_{m=0}^{2M} \Omega_{k-1}^{(m)} (\hat{\boldsymbol{\theta}}_k^{(m)} - \boldsymbol{\mu}_{k|k-1})(\hat{\boldsymbol{\theta}}_k^{(m)} - \boldsymbol{\mu}_{k|k-1})^T + \mathbf{Q}_{k-1} , \\
\boldsymbol{\mu}_{k|k} &= \boldsymbol{\mu}_{k|k-1} + \mathbf{K}_k(\mathbf{y}_k - \overline{\mathbf{y}}_k) , \\
\boldsymbol{\Sigma}_{k|k} &= \boldsymbol{\Sigma}_{k|k-1} + \mathbf{K}_k \mathbf{S}_k \mathbf{K}_k^T ,
\end{aligned} \tag{13.25}$$

[3] This is true for any non-linear transformation and is the basis of the *unscented transformation* (UT) [15]. The UT is a method for calculating the statistics of a random variable which undergoes a non-linear transformation. Consider propagating a $M$-dimensional random variable $\mathbf{x} = (x^{(1)}, x^{(2)}, \ldots, x^{(M)})^T$ (mean vector $\overline{\mathbf{x}}$ and covariance matrix $\mathbf{R}$) though a non-linear function $\mathbf{y} = g(\mathbf{x})$. Then the mean and covariance of $\mathbf{y}$ are: $\overline{\mathbf{y}} \approx \sum_{m=0}^{2M} \omega^{(m)} y^{(m)}$ and $\mathbf{S} \approx \sum_{m=0}^{2M} \Omega^{(m)} (\mathbf{y}^{(m)} - \bar{\mathbf{y}})(\mathbf{y}^{(m)} - \bar{\mathbf{y}})^T$, where $\mathbf{x}^{(m)}$, $m \in \{0, 1, \ldots, 2M\}$, are a set of deterministically selected sigma-points with weights $\omega^{(m)}$ and $\Omega^{(m)}$ and $\mathbf{y}^{(m)} = g(\mathbf{x}^{(m)})$.

**Table 13.3** Sigma-Points and their weights

| Variable | Formula |
|---|---|
| Sigma-Points $\boldsymbol{\theta}^{(m)}$ | $\boldsymbol{\theta}^{(0)} = \boldsymbol{\mu}$; $\boldsymbol{\theta}^{(m)} = \boldsymbol{\mu} + [\sqrt{(n+\lambda)\boldsymbol{\Sigma}}]_m$; $\boldsymbol{\theta}^{(m+M)} = \boldsymbol{\mu} - [\sqrt{(n+\lambda)\boldsymbol{\Sigma}}]_m$, $m \in \{1, 2, \ldots, M\}$; where $\lambda = \alpha^2(M+K)$ and $[\sqrt{(n+\lambda)\boldsymbol{\Sigma}}]_m$ is the $m$th column of the matrix square root of $(n+\lambda)\boldsymbol{\Sigma}$ |
| Weights $\omega^{(m)}$ | $\omega^{(0)} = \lambda/(M+\lambda)$, $\omega^{(m)} = \lambda/(2(M+\lambda))$, $m \in \{1, 2, \ldots, 2M\}$ |
| Weights $\Omega^{(m)}$ | $\Omega^{(0)} = \lambda/(M+\lambda) + (1-\alpha^2+\beta)$, $\Omega^{(m)} = \lambda/(2(M+\lambda))$, $m \in \{1, 2, \ldots, 2M\}$ |

where

$$
\begin{aligned}
\hat{\boldsymbol{\theta}}_{k-1}^{(m)} &= f_k(\boldsymbol{\theta}_{k-1}^{(m)}) \ , \\
\hat{\mathbf{y}}_k^{(m)} &= h_k(\overline{\boldsymbol{\theta}}_{k-1}^{(m)}) \ , \\
\overline{\mathbf{y}}_k &= \sum_{m=0}^{2M} \omega_{k-1}^{(m)} \hat{\mathbf{y}}_k^{(m)} \ , \\
\mathbf{S}_k &= \sum_{m=0}^{2M} \Omega_{k-1}^{(m)} (\hat{\mathbf{y}}_k^{(m)} - \overline{\mathbf{y}}_k)(\hat{\mathbf{y}}_k^{(m)} - \overline{\mathbf{y}}_k)^T + \mathbf{R}_k \ , \\
\mathbf{K}_k &= \sum_{m=0}^{2M} \Omega_{k-1}^{(m)} (\hat{\boldsymbol{\theta}}_k^{(m)} - \boldsymbol{\mu}_{k|k-1})(\hat{\mathbf{y}}_k^{(m)} - \overline{\mathbf{y}}_k)^T \mathbf{S}_k^{-1} \ .
\end{aligned}
\tag{13.26}
$$

The *sigma*-points $\boldsymbol{\theta}^{(m)}$, $m \in \{0, 1, \ldots, 2M\}$, and their weights, $\omega^{(m)}$ and $\Omega^{(m)}$, are selected in deterministic fashion (see Table 13.3).

### Particle Filter*

The above techniques relax some of the Gaussian linear assumptions which underlie the KF. However, in many practical applications these techniques are not sufficiently accurate to model all aspects of the *a posteriori* pdf $p(\boldsymbol{\theta}_k|\mathbf{y}_{1:k}, I)$. If this happens then we may use a different type of filter based on a Monte Carlo approximation. This filter is known as a *particle filter* filter [1].

The key idea underlying the particle filter is to represent the pdf $p(\boldsymbol{\theta}_k|\mathbf{y}_k, I)$ by a set of $N$ weighted random samples, or particles, which constitutes a discrete approximation of the pdf. In this regard the particle filter is similar to the sigma filter. The key difference being that in the former we may use several thousand particles while in the sigma filter we use a very small number of deterministically selected particles.

## 13.5 Kalman Filter and Deep Learning*

An important innovation in the Kalman filter has been the development of the *KalmanNet* in which the KF is combined with the methods of deep learning. The idea behind KalmanNet is to replace specific computations in the KF which are noise or model-dependent with a trainable neural network.

### KalmanNet

We illustrate the idea underlying KalmanNet by considering a KF with linear Gaussian process and measurement models (cf. Eqs. 13.12–13.13):

$$\mathbf{x}_k = \mathbf{F}_k \mathbf{x}_{k-1} + \mathbf{v}_{k-1} \quad \text{and} \quad \mathbf{y}_k = \mathbf{H}_k \mathbf{x}_k + \mathbf{w}_k \ ,$$

where

$$\mathbf{v}_k \sim \mathcal{N}(\mathbf{0}, \mathbf{Q}) \quad \text{and} \quad \mathbf{w}_k \sim \mathcal{N}(\mathbf{0}, \mathbf{R}) \ .$$

However we now assume *no* knowledge of the noise statistics $\mathbf{Q}$ and $\mathbf{R}$ and only *approximate* knowledge of the process and measurement model matrices $\mathbf{F}$ and $\mathbf{H}$.

Inspection of the equations underlying the KF (Eqs. 13.14–13.16) shows that knowledge of the noise statistics only enters the KF through the Kalman gain $\mathbf{K}_k$ as defined in Eq. (13.16). Without any knowledge of the noise statistics we clearly cannot calculate $\mathbf{K}_k$. However, by using a training set of labeled trajectories, we may *learn* an appropriate value of $\mathbf{K}_k$. This is the idea underlying KalmanNet.

Let $\mathbf{K}_k(\Theta)$ denote the *learned* Kalman gain matrix with train parameters $\Theta$. The corresponding dataset $\mathcal{D}$ is composed of $L$ different trajectories $(\mathbf{X}^{(l)}, \mathbf{Y}^{(l)})$, $l \in \{1, 2, \ldots, L\}$, where

$$\mathbf{X}^{(l)} = (\mathbf{x}_1^{(l)}, \mathbf{x}_2^{(l)}, \ldots, \mathbf{x}_M^{(l)}) \quad \text{and} \quad \mathbf{Y}^{(l)} = (\mathbf{y}_1^{(l)}, \mathbf{y}_2^{(l)}, \ldots, \mathbf{y}_M^{(l)})$$

are sequences of (true) state and measurement vectors, for trajectory $l$.

We define an empirical loss function for each trajectory $l$ as:

$$\mathcal{L}^{(l)} = \frac{1}{M} \sum_{m=1}^{M} \|\hat{\mathbf{x}}_m^{(l)} - \mathbf{x}_m^{(l)}\|^2 + \gamma \|\boldsymbol{\Theta}\|^2 \ ,$$

where $\hat{\mathbf{x}}_m^{(l)}$, $m \in \{1, 2, \ldots, M\}$, is the output of the KalmanNet and $\gamma$ is a regularization coefficient. To optimize the parameters $\boldsymbol{\Theta}$, we minimize $\mathcal{L}^{(l)}$ averaged over a random number of trajectories.

Experimentally KalmanNet was found to be robust against inaccurate process and measurement models and inaccurate noise statistics.

### Split-KalmanNet*

The Kalman gain matrix $\mathbf{K}_k$ in Eq. (13.16) depends upon both the noise matrices $\mathbf{Q}_k$ and $\mathbf{R}_k$. As a consequence, when the uncertainties in $\mathbf{Q}_k$ and $\mathbf{R}_k$ are heterogeneous, the original KalmanNet may be less robust because $\mathbf{K}_k(\Theta)$ cannot separate the statistical mismatch effects in $\mathbf{Q}_k$ and $\mathbf{R}_k$.

In Split-KalmanNet [8] the authors resolve the statistical mismatch effects by writing $\mathbf{K}_k(\Theta)$ as the following product:

$$\mathbf{K}_k(\Theta) = \mathbf{K}_k^{(1)}(\Theta^{(1)})\mathbf{H}_k^T\mathbf{K}_k^{(2)}(\Theta^{(2)}) ,$$

where the matrices $\mathbf{K}_k^{(1)}(\Theta^{(1)})$ and $\mathbf{K}_k^{(2)}(\Theta^{(2)})$ are separately trained using *two* neural networks.

Experimentally split-KalmanNet was found to robust and significantly improve tracking error compared to traditional Kalman filter methods.

### Multi-model KalmanNet*

In the multi-model KalmanNet, Han et al. [21] extended the original KalmanNet by employing an ensemble of multiple KalmanNets; each KalmanNet having a different motion model. The estimates of these KalmanNets are averaged using weights that are predicted by a multi-layer perceptron that learns to evaluate the similarity between the predicted and observed target motion.

## 13.6 Multi-Target Tracking

An important application of the recursive filter is the tracking of multiple targets. If the targets are widely separated one from the other and there are no false, or spurious, measurements then we may track the targets using a separate KF for each target (cf. Examples 13.3 and 13.5). However, in most real-world systems, the targets are not sufficiently separated and/or there are false and spurious measurements. In this case, there is a double ambiguity regarding the measurements:

- Origin: It is not known which measurement corresponds to which target.
- Nature: It is not known whether a measurement is "true" and originates from a true, but unknown, target or whether a measurement is "false".

This problem is known as the *data association* problem (cf. Sects. 8.1 and 8.2). It is clear that the data association problem must be solved before we can use the KF. In many data fusion applications the solution of the data association problem is the primary fusion algorithm. See Table 13.4.

**Table 13.4** Applications in which solving the data association problem is the primary fusion algorithm

| Class | | Application |
|---|---|---|
| Early-FeO | Example 13.10 | The strongest neighbour (SN) filter |
| | Example 13.11 | Covariance Union Filter |
| Inter-FeO | Example 4.12 | Shape Context and Inner Distance |
| | Example 6.10 | Global image registration using RANSAC |
| | Example 6.14 | Local point matching (LPM) |
| | Example 8.4 | Closed contour matching using the shape context |
| | Example 13.9 | Tracking Pedestrians Across Non-Overlapping Cameras |

The following example illustrates the use of data association in an application involving tracking people in a surveillance area with multiple non-overlapping cameras.

**Example 13.9** *Tracking Pedestrians Across Non-Overlapping Cameras* [17]. In visual surveillance it is not possible to monitor a wide area using a single camera because the field of view (FOV) of one camera is finite and often the structure of the scene limits the visible area in one camera view. For these reasons it is common practice to employ a network of cameras for a surveillance systems. In this case, a major task is to track objects moving across multiple FOV's. If the FOV's do not overlap we use data association to join the tracks from the same objects. The idea is: we estimate the likelihood that $i$th track $T_i$ observed in camera $C_i$ is associated with the $j$th track $T_j$ observed in camera $C_j$ by measuring the similarity $S(T_i, T_j)$ between $T_i$ and $T_j$.

In general, the solution of the data association problem can be *hard* or *soft*. In a hard solution we assign a given measurement to the system. In this case, the state of the system, $\boldsymbol{\theta}_k$, is updated assuming that the assigned measurement is the true measurement. On the other hand, in a soft solution we assign more than one measurement to the system, each measurement having its own assignment probability. In this case, we use each measurement which has an assignment probability greater than zero to update $\boldsymbol{\theta}_k$. The collection of updated states is then combined using the assignment probabilities.

In general, different applications will require different data association algorithms. However, the algorithms given in Table 13.5 have been found useful in a wide-range of applications although they were originally developed for single-target tracking [3].

The algorithms listed in Table 13.5 are then combined with the KF to create new recursive filters. We now consider three such filters.

**Table 13.5** Data association algorithms

| Name | Description |
|---|---|
| NN | Nearest Neighbor. Assumes measurement that is closest to predicted state of system is the true measurement. If the true measurement is *not* detected at time step $k$, then the NN is a false measurement which may cause erratic behaviour in KF. To avoid this we may restrict ourselves to *validated* measurements. In general, performance of NN and of validated NN algorithms is low |
| PDA | Probabilistic Data Association. State of system is separately updated using each validated measurement in turn. The collection of updated states are then combined together to form the final state of system. Each validated measurement $\mathbf{y}_k^{(m)}$, $m \in \{1, 2, \ldots, M_k\}$, is given a weight $\beta^{(m)}$ which is the probability that $\mathbf{y}_k^{(m)}$ is the true measurement. Also computed is a weight $\beta^{(0)}$ which is the probability that none of the validated measurements are the true measurement. These events encompass all possible interpretations of the data, so $\sum_{m=0}^{M_k} \beta^{(m)} = 1$ |
| MHT | Multiple Hypothesis Testing. Enumerates all possible hard assignments taken over all time steps. The number of possible assignments increases exponentially with time. As a consequence at each time step the MHT algorithm discards assignments which have a very low probability of being true |

Validated measurements are measurements which fall inside a given region ("validation gate"). Generally the validation gate is centered on the predicted state of the system. See Example 12.2

## Nearest Neighbour (NN) Filter

Let $\widetilde{\mathbf{y}}_k = \{\mathbf{y}_k^{(1)}, \mathbf{y}_k^{(2)}, \ldots, \mathbf{y}_k^{(M_k)}\}$, denote the set of $M_k$ measurements received at time step $k$ In the NN filter we assume the true measurement is

$$\mathbf{y}_k^{(*)} = \arg\min_m (D^{(m)}) \tag{13.27}$$

where

$$D^{(m)} = (\mathbf{y}_k^{(m)} - \hat{\mathbf{y}}_{k|k-1})^T \hat{\mathbf{S}}_{k|k-1}^{-1} (\mathbf{y}_k^{(m)} - \hat{\mathbf{y}}_{k|k-1}) , \tag{13.28}$$

is the normalized distance between $\mathbf{y}_k^{(m)}$ and the predicted measurement $\hat{\mathbf{y}}_{k|k-1}$, and $\hat{\mathbf{S}}_{k|k-1}$ is the predicted covariance matrix $\mathbf{H}_k^T \boldsymbol{\Sigma}_{k|k-1} \mathbf{H}_k^T + \mathbf{R}_k$.

In validated NN we only use *validated* measurements (i.e. measurements $\mathbf{y}_k^{(*)}$ for which $D^{(*)} \leq \gamma$, where $\gamma$ is a given threshold. The corresponding NN filter combines the KF with a validated NN solution of the data association problem:

$$\begin{aligned}
\boldsymbol{\mu}_{k|k-1} &= \mathbf{F}_k \boldsymbol{\mu}_{k-1|k-1} , \\
\boldsymbol{\Sigma}_{k|k-1} &= \mathbf{Q}_{k-1} + \mathbf{F}_k \boldsymbol{\Sigma}_{k-1|k-1} \mathbf{F}_k^T , \\
\boldsymbol{\mu}_{k|k} &= \begin{cases} \boldsymbol{\mu}_{k|k-1} + \mathbf{K}_k(\mathbf{y}_k^{(*)} - \mathbf{H}_k \boldsymbol{\mu}_{k|k-1}) & \text{if } m_k \geq 1 , \\ \boldsymbol{\mu}_{k|k-1} & \text{otherwise} , \end{cases} \\
\boldsymbol{\Sigma}_{k|k} &= \begin{cases} \boldsymbol{\Sigma}_{k|k-1} - \mathbf{K}_k \mathbf{H}_k \boldsymbol{\Sigma}_{k|k-1} & \text{if } m_k \geq 1 , \\ \boldsymbol{\Sigma}_{k|k-1} & \text{otherwise} , \end{cases} \\
\mathbf{K}_k &= \boldsymbol{\Sigma}_{k|k-1} \mathbf{H}_k^T (\mathbf{H}_k \boldsymbol{\Sigma}_{k|k-1} \mathbf{H}_k^T + \mathbf{R}_k)^{-1} ,
\end{aligned} \tag{13.29}$$

where $m_k$ is the number of validated measurements at time step $k$. A variant of the NN filter is the strongest neighbour (SN) filter.

**Example 13.10** *The strongest neighbour (SN) filter.* The SN filter is identical to the NN filter except we identify the true measurement as $\mathbf{y}^{(*)}$, where $\mathbf{y}^{(*)}$ is the measurement with the largest *validated* amplitude. A further refinement of the SN filter is the probabilistic SN filter [22]. In the probabilistic SN filter we weight the strongest neighbour measurement (which falls in the validation gate) with the probability that it is the true measurement. This is in opposition to the SN filter which makes the unrealistic assumption that the SN measurement is *always* the true measurement. The result is a filter which has substantially better performance than the NN, SN and PDA filters.

In the probabilistic SN filter we take into account the possibility that the measurement with the largest validated amplitude may not be the true measurement. The corresponding *probabilistic strongest neighbour filter* equations are identical to Eq. (13.29) except for the *a posteriori* covariance matrix which is replaced by

$$\boldsymbol{\Sigma}_{k|k} = \begin{cases} \boldsymbol{\Sigma}_{k|k-1} + \mathbf{K}_k(\mathbf{y}_k - \mathbf{H}_k\boldsymbol{\mu}_{k|k-1}) & \text{if } m_k \geq 1 \,, \\ \quad \times\, (\mathbf{y} - \mathbf{H}_k\boldsymbol{\mu}_{k|k-1})^T \mathbf{K}_k^T P_T P_F & \\ \quad + \left(\frac{P_D P_G P_A(1-c)}{1-P_D P_G P_A} P_T - P_F\right) \mathbf{K}\boldsymbol{\Sigma}_{k|k-1}\mathbf{K}_k^T & \\ \boldsymbol{\Sigma}_{k|k-1} + \frac{P_D P_G(1-c)}{1-P_D P_G} \mathbf{K}\boldsymbol{\Sigma}_{k|k-1}\mathbf{K}_k^T & \text{otherwise} \,, \end{cases}$$

where $P_D$ is the probability of the true target being detected and $c$ is a constant determined by the dimensionality of the measurements. In 2D $c = 1 - \frac{1}{2} P_G^{-1} \gamma \exp(-\gamma/2)$.

### Covariance Union (CU) Filter*

Another important modification of the NN filter is the covariance union (CU) filter.

**Example 13.11** *Covariance Union (CU) Filter* [16]. The CU filter is similar to the NN and SN filters with one crucial difference: The measurement which is selected by the covariance filter is a "virtual" measurement characterized by a "virtual" value $\mathbf{y}_k^{(*)}$ and a "virtual" covariance matrix $\boldsymbol{\Sigma}_k^{(*)}$, where $\mathbf{y}_k^{(*)}$ and $\boldsymbol{\Sigma}_k^{(*)}$ are chosen to be *consistent* with all validated measurements. For two observations $O^{(1)} = (\mathbf{y}_k^{(1)}, \mathbf{R}_k^{(1)})$ and $O^{(2)} = (\mathbf{y}_k^{(2)}, \mathbf{R}_k^{(2)})$ (Fig. 13.4) we have $\widetilde{O} = \left(\mathbf{y}_k^{(*)}, \mathbf{R}_k^{(*)}\right)$ as follows [16]:

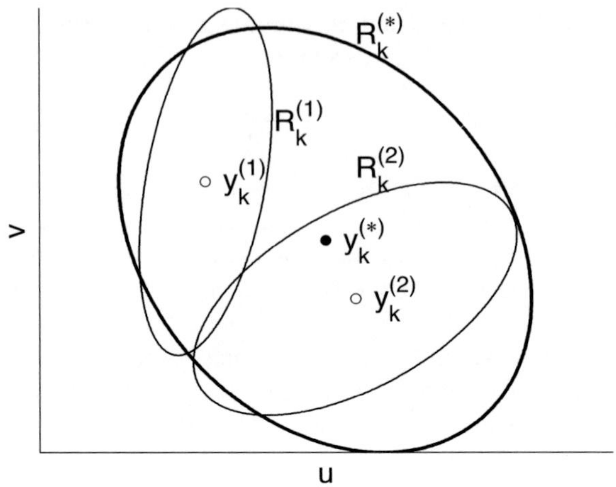


**Fig. 13.4** Illustrates CU of observations $O_1 = (\mathbf{y}_k^{(1)}, \mathbf{R}_k^{(1)})$ and $O_2 = (\mathbf{y}_k^{(2)}, \mathbf{R}_k^{(2)})$ in 2D $(u, v)$. The observation means, $\mathbf{y}_k^{(1)}$ and $\mathbf{y}_k^{(2)}$, are shown by small open circles and the observation covariances, $\mathbf{R}_k^{(1)}$ and $\mathbf{R}_k^{(2)}$, are shown by two ellipses drawn with thin solid lines. The mean, $\mathbf{y}_k^{(*)}$, and covariance, $\mathbf{R}_k^{(*)}$, of the CU are shown, respectively, by a small filled circle and an ellipse drawn with a thick solid line

$$\mathbf{y}_k^{(*)} = \frac{1}{2}(\mathbf{y}_k^{(1)} + \mathbf{y}_k^{(2)}) ,$$
$$\mathbf{R}_k^{(*)} = \mathbf{S}_k^T \mathbf{V}_k \max(\mathbf{D}_k, \mathbf{I}) \mathbf{V}_k^T \mathbf{S}_k ,$$

where $\mathbf{S}_k = |\mathbf{R}_k^{(2)}|^{1/2}$, and $\max(\mathbf{A}, \mathbf{B})$ denotes the element-by-element maximum of the matrices $\mathbf{A}$ and $\mathbf{B}$, and $\mathbf{V}_k$ and $\mathbf{D}_k$ are, respectively, the matrices of the eigenvectors and eigenvalues of $(\mathbf{S}_k^{-1})^T \mathbf{R}_k^{(1)} \mathbf{S}_k^{-1}$.

## Probabilistic Data Association Filter (PDAF)*

The NN, SN and CU filters work by choosing a *single* validated measurement $\mathbf{y}_k^{(*)}$[4] which is then used to update the state vector $\boldsymbol{\theta}_k$. The probabilistic data association filter (PDAF) [3] works differently: Let $\beta^{(m)}$ be the probability that $\mathbf{y}_k^{(m)}$ is the true measurement at the $k$th time step. Then the PDAF uses *all* of the validated measurements, $\mathbf{y}_k^{(m)}$, $m \in \{1, 2, \ldots, M_k\}$, and their probabilities $\beta^{(m)}$ to update $\boldsymbol{\theta}_k$. The corresponding PDAF equations are:

[4] In the NN and SN filters we use a "real" measurement as $\mathbf{y}_k^{(*)}$ while in the CU filter we use a "virtual" measurement for $\mathbf{y}_k^{(*)}$.

$$
\begin{aligned}
\boldsymbol{\mu}_{k|k-1} &= \mathbf{F}_k \boldsymbol{\mu}_{k-1|k-1} , \\
\boldsymbol{\Sigma}_{k|k-1} &= \mathbf{Q}_{k-1} + \mathbf{F}_k \boldsymbol{\Sigma}_{k-1|k-1} \mathbf{F}_k^T , \\
\boldsymbol{\mu}_{k|k} &= \boldsymbol{\mu}_{k|k-1} + \mathbf{K}_k \sum_{m=1}^{m_k} \beta^{(m)} (\mathbf{y}_k^{(m)} - \mathbf{H}_k \boldsymbol{\mu}_{k|k-1}) , \\
\boldsymbol{\Sigma}_{k|k} &= \beta^{(0)} \boldsymbol{\Sigma}_{k|k-1} + (1 - \beta^{(0)})(\boldsymbol{\Sigma}_{k|k-1} - \mathbf{K}_k \hat{\mathbf{S}}_{k|k-1} \mathbf{K}_k) + \Delta \boldsymbol{\Sigma}_{k|k-1} ,
\end{aligned}
\tag{13.30}
$$

where $\beta^{(0)} = 1 - \sum_{m=1}^{m_k} \beta^{(m)}$ is the probability that none of the $\mathbf{y}_k^{(m)}$, $m \in \{1, 2, \ldots, m_k\}$, is the true measurement and the correction term, $\Delta \boldsymbol{\Sigma}_{k|k-1}$, is given by

$$
\begin{aligned}
\Delta \boldsymbol{\Sigma}_{k|k-1} = {} & \mathbf{K}_k \Big( \sum_m \beta^{(m)} \mathbf{v}_k^{(m)} (\mathbf{k}^{(m)})^T - \mathbf{v}_k \mathbf{v}_k^T \Big) \mathbf{K}_k^T - \sum_m \beta^{(m)} \mathbf{v}_k^{(m)} (\mathbf{v}_k^{(m)})^T \\
& - \sum_m \beta^{(m)} \mathbf{v}_k^{(m)} \Big( \sum_m \beta^{(m)} (\mathbf{v}_k^{(m)})^T \mathbf{v}_k \mathbf{v}_k^T \Big) \mathbf{K}_k^T ,
\end{aligned}
\tag{13.31}
$$

where

$$
\mathbf{v}_k = \sum_{m=1}^{m_k} \beta^{(m)} \mathbf{v}_k^{(m)} \quad \text{and} \quad \mathbf{v}_k^{(m)} = \mathbf{y}_k^{(m)} - \mathbf{H}_k \boldsymbol{\mu}_{k|k-1} .
$$

## 13.7 Multi-sensor Multi-temporal (MSMT) Data Fusion

Until now we have considered multi-temporal data fusion in which use a recursive filter to fuse together a sequence of measurements $\mathbf{y}_{1:k}$ made using a single sensor $S$. We now extend this analysis and consider multi-sensor multi-temporal (MSMT) data fusion [11, 18, 29] in which we fuse together several sequences of measurements $\mathbf{y}_{1:k}^{(m)}$, $m \in \{1, 2, \ldots, M\}$, each sequence $\mathbf{y}_{1:k}^{(m)}$ being made with a different sensor $S^{(m)}$. We start with the simplest method of MSMT data fusion which is known as *measurement fusion*.

### *13.7.1 Measurement Fusion*

In measurement fusion we place all the measurements $\mathbf{y}_k^{(m)}$, $m \in \{1, 2, \ldots, M\}$, obtained at time step $k$ into a single augmented measurement vector $\widetilde{\mathbf{y}}_k = ((\mathbf{y}_k^{(1)})^T, (\mathbf{y}_k^{(2)})^T, \ldots, (\mathbf{y}_k^{(M)})^T)^T$. We then fuse the $\widetilde{\mathbf{y}}_{1:k} = (\widetilde{\mathbf{y}}_1, \widetilde{\mathbf{y}}_2, \ldots, \widetilde{\mathbf{y}}_k)^T$ using a single recursive filter. In the following example we illustrate the technique using a KF.

**Example 13.12** *MSMT Data Fusion: Measurement Fusion* [13]. We consider the tracking of a single target. Let $\boldsymbol{\theta}_k$ denote the filtered state of the target at time step $k$ and $\mathbf{y}_k^{(m)}$ denotes the corresponding measured position of the target as recorded by two sensors $S^{(1)}$ and $S^{(2)}$. The corresponding process and measurement equations are:

$$\begin{aligned}\boldsymbol{\theta}_k &= \mathbf{F}_k\boldsymbol{\theta}_{k-1} + \mathbf{v}_{k-1} ,\\ \mathbf{y}_k^{(1)} &= \mathbf{H}_k^{(1)}\boldsymbol{\theta}_k + \mathbf{w}_k^{(1)} ,\\ \mathbf{y}_k^{(2)} &= \mathbf{H}_k^{(2)}\boldsymbol{\theta}_k + \mathbf{w}_k^{(2)} ,\end{aligned}$$

where $\mathbf{v}_{k-1} \sim \mathcal{N}(0, \mathbf{Q}_k)$, $\mathbf{w}_k^{(1)} \sim \mathcal{N}(0, \mathbf{R}_k^{(1)})$ and $\mathbf{w}_k^{(2)} \sim \mathcal{N}(0, \mathbf{R}_k^{(2)})$.

We estimate $p(\boldsymbol{\theta}_k|\mathbf{y}_{1:k}^{(1:2)}, I) \approx \mathcal{N}(\boldsymbol{\mu}_{k|k}, \boldsymbol{\Sigma}_{k|k})$ using the KF equations:

$$\begin{aligned}\boldsymbol{\mu}_{k|k-1} &= \mathbf{F}_k\boldsymbol{\mu}_{k-1|k-1} ,\\ \boldsymbol{\Sigma}_{k|k-1} &= \mathbf{Q}_{k-1} + \mathbf{F}_k\boldsymbol{\Sigma}_{k-1|k-1}\mathbf{F}_k^T ,\\ \boldsymbol{\mu}_{k|k} &= \boldsymbol{\mu}_{k|k-1} + \widetilde{\mathbf{K}}_k(\mathbf{y}_k - \widetilde{\mathbf{H}}_k\boldsymbol{\mu}_{k|k-1}) ,\\ \boldsymbol{\Sigma}_{k|k} &= \boldsymbol{\Sigma}_{k|k-1} - \widetilde{\mathbf{K}}_k\widetilde{\mathbf{H}}_k\boldsymbol{\Sigma}_{k|k-1} ,\\ \widetilde{\mathbf{K}}_k &= \boldsymbol{\Sigma}_{k|k-1}\widetilde{\mathbf{H}}_k^T(\widetilde{\mathbf{H}}_k\boldsymbol{\Sigma}_{k|k-1}\widetilde{\mathbf{H}}_k^T + \widetilde{\mathbf{R}}_k)^{-1} ,\end{aligned}$$

where $\widetilde{\mathbf{H}}_k = \left(\mathbf{H}_k^{(1)}\mathbf{H}_k^{(2)}\right)$, $\widetilde{\mathbf{w}}_k = \begin{pmatrix}\mathbf{w}_k^{(1)}\\ \mathbf{w}_k^{(2)}\end{pmatrix} \sim \mathcal{N}(\mathbf{0}, \widetilde{R}_k)$ , $\widetilde{\mathbf{R}}_k = \begin{pmatrix}\mathbf{R}_k^{(1)} & \mathbf{0}\\ \mathbf{0} & \mathbf{R}_k^{(2)}\end{pmatrix}$.

Figure 13.5 is a graphical illustration of the process of measurement fusion for the two sensors $S^{(1)}$ and $S^{(2)}$.

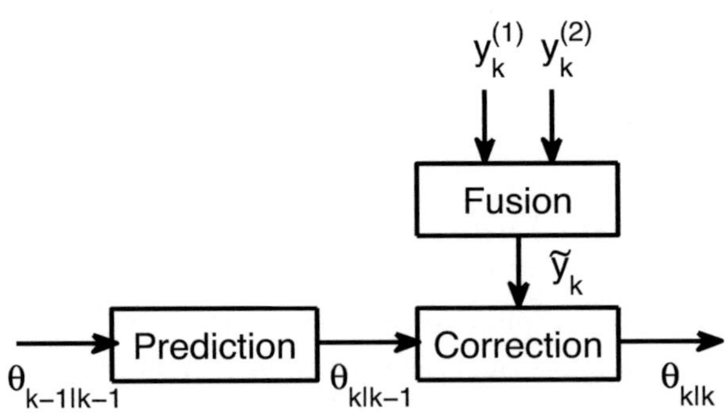


**Fig. 13.5** Illustrates the measurement fusion process for two measurement sequencies $\mathbf{y}_{1:k}^{(1)}$ and $\mathbf{y}_{1:k}^{(2)}$. The individual measurements are placed in an augmented measurement sequence $\widetilde{\mathbf{y}}_{1:k} = (\widetilde{\mathbf{y}}_1, \widetilde{\mathbf{y}}_2, \ldots, \widetilde{\mathbf{y}}_k)$, where $\widetilde{\mathbf{y}}_l = \left((\mathbf{y}_l^{(1)})^T, (\mathbf{y}_l^{(2)})^T\right)^T$. The augmented measurement vector is then fused using a single KF

Although theoretically optimum, measurement fusion requires a *centralized* architecture in which all the measurements $\mathbf{y}_k^{(m)}$, $m \in \{1, 2, \ldots, M\}$, are transmitted to a single fusion node. The difficulties and drawbacks associated with this architecture were described in Sect. 3.4.1. The most serious drawback is its sensitivity to errors involved in forming a common representational format, i.e. to the errors in spatial and temporal alignment of the sensors $S^{(m)}$. An alternative method of MSMT data fusion is to separately filter the measurements received from each sensor and then fuse the results together. This is known as *track-to-track* fusion and is considered next. Its main advantage is that it is relatively insensitive to the errors in spatial and temporal alignment of the sensors.

### *13.7.2 Track-To-Track Fusion*

In track-to-track fusion we first perform multi-temporal fusion using $M$ recursive filters, in which the $m$th filter fuses the measurements $\mathbf{y}_{1:k}^{(m)} = \{y_1^{(m)}, y_2^{(m)}, \ldots, y_k^{(m)}\}$ made by the $m$th sensor $S^{(m)}$. Let $p(\boldsymbol{\theta}_k^{(m)}|\mathbf{y}_{1:k}^{(m)}, I)$ be the corresponding *a posteriori* pdf. Then, at each time step $k$, we fuse together the $M$ *a posteriori* pdf's $p(\boldsymbol{\theta}_k^{(m)}|\mathbf{y}_{1:k}^{(m)}, I)$, $m \in \{1, 2, \ldots, M\}$, using a linear least squares (minimum mean square error) estimate (see Sect. 11.5). Track-to-track fusion uses a hierarchical architecture (Sect. 3.4.3) containing $M$ local fusion nodes and one central fusion node. In general track-to-track fusion has a lower performance [7, 13] than the corresponding measurement fusion algorithm (Sect. 13.7.1). The reason for the reduction in performance is that the track-to-track fusion is performed using a *linear* estimate.

The following example illustrates track-to-track fusion for two sensors $S^{(1)}$ and $S^{(2)}$ using the KF [2].

**Example 13.13** *MSMT Data Fusion: Track-to-Track Fusion* [2, 30] We reanalyze Example 13.12 using the Bar-Shalom-Campo track-to-track fusion algorithm of [2]. If $\boldsymbol{\theta}_k$ denotes the filtered state of the target at time step $k$ and $\mathbf{y}_k^{(m)}$ denotes the measured position of the target at time step $k$ as recorded by the sensor $S^{(m)}$, then the corresponding process and measurement model equations are:

$$\boldsymbol{\theta}_k = \mathbf{F}_k \boldsymbol{\theta}_{k-1} + \mathbf{v}_{k-1} \quad \text{and} \quad \mathbf{y}_k^{(m)} = \mathbf{H}_k^{(m)} \boldsymbol{\theta}_k + \mathbf{w}_k^{(m)} ,$$

where $\mathbf{v}_{k-1} \sim \mathcal{N}(0, \mathbf{Q}_k)$ and $\mathbf{w}_k^{(m)} \sim \mathcal{N}(0, \mathbf{R}_k^{(m)})$.

Let $\boldsymbol{\theta}_k^{(m)} \sim \mathcal{N}\big(\boldsymbol{\mu}_{k|k}^{(m)}, \boldsymbol{\Sigma}_{k|k}^{(m)}\big)$ denote the KF estimate of $\boldsymbol{\theta}_k$ given the sensor measurements $\mathbf{y}_{1:k}^{(m)}$, where $\boldsymbol{\mu}_{k|k}^{(m)}$ and $\boldsymbol{\Sigma}_{k|k}^{(m)}$ are calculated using the KF equations (13.14-13.16). For $M = 2$ we fuse together the two (correlated) estimates $\boldsymbol{\mu}_{k|k}^{(m)}$, $m \in \{1, 2\}$, using the Bar-Shalom-Campo formula:

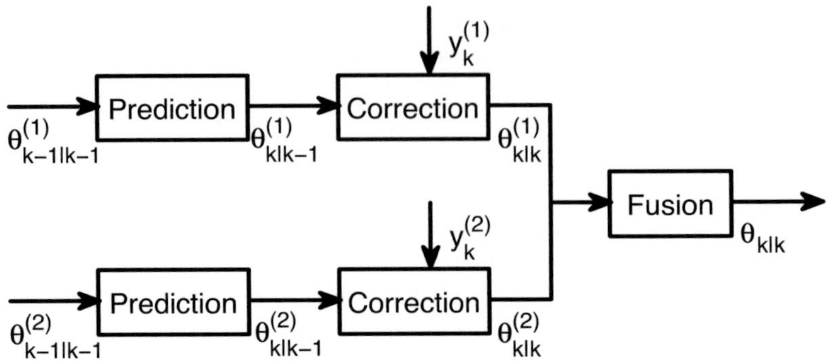


**Fig. 13.6** Illustrates track-to-track fusion process for measurements $\mathbf{y}_{1:k}^{(1)}, \mathbf{y}_{1:k}^{(2)}$ obtained from sensors $S^{(1)}, S^{(2)}$. Each sequence of measurements are separately fused together using a KF. At each time step $k$ the outputs $N(\boldsymbol{\mu}_{k|k}^{(1)}, \boldsymbol{\Sigma}_{k|k}^{(1)})$ and $N(\boldsymbol{\mu}_{k|k}^{(2)}, \boldsymbol{\Sigma}_{k|k}^{(2)})$ are then fused together using the Bar-Shalom-Campo formula

$$\begin{aligned}\widetilde{\boldsymbol{\theta}}_{k|k} = &\big(\boldsymbol{\Sigma}_{k|k}^{(2,2)} - \boldsymbol{\Sigma}_{k|k}^{(2,1)}\big)\big(\boldsymbol{\Sigma}_{k|k}^{(1,1)} + \boldsymbol{\Sigma}_{k|k}^{(2,2)} - \boldsymbol{\Sigma}_{k|k}^{(1,2)} - \boldsymbol{\Sigma}_{k|k}^{(2,1)}\big)^{-1}\boldsymbol{\mu}_{k|k}^{(1)} + \\ &\big(\boldsymbol{\Sigma}_{k|k}^{(1,1)} - \boldsymbol{\Sigma}_{k|k}^{(1,2)}\big)\big(\boldsymbol{\Sigma}_{k|k}^{(1,1)} + \boldsymbol{\Sigma}_{k|k}^{(2,2)} - \boldsymbol{\Sigma}_{k|k}^{(1,2)} - \boldsymbol{\Sigma}_{k|k}^{(2,1)}\big)^{-1}\boldsymbol{\mu}_{k|k}^{(2)} ,\end{aligned}$$

where $\widetilde{\boldsymbol{\mu}}_{k|k}$ is the optimum linear estimate, $\boldsymbol{\Sigma}_{k|k}^{(m,m)} \equiv \boldsymbol{\Sigma}_{k|k}^{(m)}$ is covariance matrix of $\mathbf{y}_k^{(m)}$ and $\boldsymbol{\Sigma}_{k|k}^{(m,n)}$ is cross-covariance matrix between $\mathbf{y}_k^{(m)}$ and $\mathbf{y}_k^{(n)}$.

Figure 13.6 shows the track-to-track fusion of observations received from two sensors $S^{(1)}$ and $S^{(2)}$.

Recently the Bar-Shalom-Campo equations have been extended to the case when $M > 2$ [5]. In the following example we give the equations for track-to-track fusion using $M > 2$ sensors.

**Example 13.14** *Track-to-Track Fusion using the Generalized Millman Equations* [30]. Recently the Bar-Shalom-Campo equations have been extended to more than two sensors. We reconsider Example 13.13 except this time we allow $M > 2$. In this case, we contruct the global (subpotimal) estimate $\widetilde{\boldsymbol{\mu}}_{k|k}$ from the local KF estimates $\boldsymbol{\mu}_{k|k}^{(m)}, m \in \{1, 2, \ldots, M\}$, using the generalized Millman formula (GMF):

$$\widetilde{\boldsymbol{\mu}}_{k|k} = \sum\nolimits_{m=1}^{M} \mathbf{C}_k^{(m)} \boldsymbol{\mu}_{k|k}^{(m)} ,$$

[5] See also Sect. 11.7.

where the time-varying matrices $\mathbf{C}_k^{(m)}$, $m \in \{1, 2, \ldots, M\}$, are given by

$$\sum_{m=1}^{M-1} \mathbf{C}_k^{(m)}(\boldsymbol{\Sigma}_{k|k}^{(m,1)} - \boldsymbol{\Sigma}_{k|k}^{(m,M)}) + \mathbf{C}_k^{(M)}(\boldsymbol{\sigma}_{k|k}^{(M,1)} - \boldsymbol{\Sigma}_{k|k}^{(M,M)}) = 0 ,$$
$$\sum_{m=1}^{M-1} \mathbf{C}_k^{(m)}(\boldsymbol{\Sigma}_{k|k}^{(m,2)} - \boldsymbol{\Sigma}_{k|k}^{(m,M)}) + \mathbf{C}_k^{(M)}(\boldsymbol{\Sigma}_{k|k}^{(M,2)} - \boldsymbol{\Sigma}_{k|k}^{(M,M)}) = 0 ,$$
$$\vdots$$
$$\sum_{m=1}^{M-1} \mathbf{C}_k^{(m)}(\boldsymbol{\Sigma}_{k|k}^{(m,M-1)} - \boldsymbol{\Sigma}_{k|k}^{(m,M)}) + \mathbf{C}_k^{(M)}(\boldsymbol{\Sigma}_{k|k}^{(M,M-1)} - \boldsymbol{\Sigma}_{k|k}^{(M,M)}) = 0 ,$$

and

$$\sum_{m=1}^{M} \mathbf{C}_k^{(m)} = \mathbf{I} ,$$

$$\boldsymbol{\Sigma}_{k|k}^{(m,n)} = \begin{cases} (\mathbf{I} - \mathbf{K}_k^{(m)}\mathbf{H}_k^{(m)})\boldsymbol{\Sigma}_{k|k-1}^{(m)} & \text{if } n = m , \\ (\mathbf{I} - \mathbf{K}_k^{(m)}\mathbf{H}_k^{(m)})(\mathbf{F}_k \boldsymbol{\Sigma}_{k-1|k-1}^{(m,n)} \mathbf{F}_k^T + \mathbf{Q}_k) & \\ \quad \times (\mathbf{I} - \mathbf{K}_k^{(n)}\mathbf{H}_k^{(n)})^T & \text{otherwise} . \end{cases}$$

Track-to-track fusion is widely used and many different variants of the basic scheme have been described. The following example illustrates one such variant.

**Example 13.15** *Modified Gao-Harris track-to-track fusion algorithm* [13]. In the Gao and Harris algorithm [13] we calculate a single predicted pdf $p(\boldsymbol{\theta}_k|\mathbf{y}_{1:k-1}^{(1:M)}, I)$ using the *fused* pdf $p(\boldsymbol{\theta}_k|\mathbf{y}_{1:k}^{(1:M)}, I)$. Figure 13.7 shows a schematic implementation of the Gao-Harris algorithm.

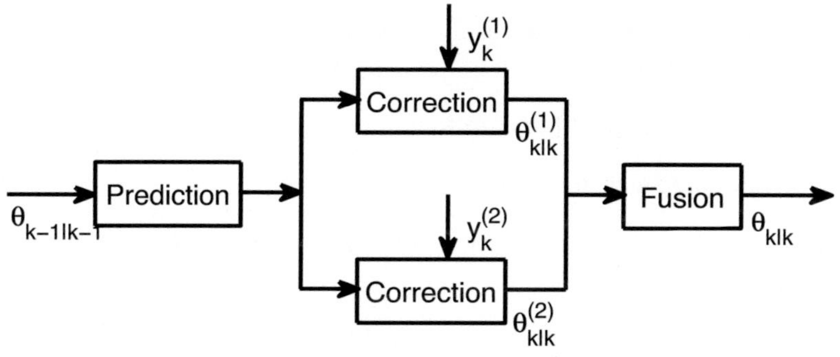


**Fig. 13.7** Illustrates the Gao-Harris track-to-track fusion process for the measurements $\mathbf{y}_{1:k}^{(1)}, \mathbf{y}_{1:k}^{(2)}$ obtained from sensors $S^{(1)}$, $S^{(2)}$. Each sequence of measurements are separately fused together using a KF. However, the KF's use a *common* predictor which is based on the fused output $N(\boldsymbol{\mu}_{k-1|k-1}, \boldsymbol{\Sigma}_{k-1|k-1})$ obtained at previous time step

## 13.8 Software

**MATLAB**. MATLAB SENSOR FUSION AND TRACKING toolbox includes m-files on Kalman filters (including EKF and UKF), particle filters and multi-object tracking.

**PYTHON**. MOTRACKERS and OPENCV have python modules for Kalman filters and multi-object tracking. A python library for KalmanNet tracking is available.

## 13.9 Further Reading

Chen [6] is a very readable, comprehensive survey of Bayesian filtering. Additional modern references on Bayesian filters include [1, 19]. References on specific topics in sequential Bayesian inference include: [23] (unscented Kalman and particle filters), [1] (particle filters), [3, 24] (multi-target tracking). Ref. [11] is an up-to-date survey of multi-sensor fusion in target tracking.

Developments in the KalmanNet have been rapid since its appearance in 2021 [27]. Among the developments are: tracking high-dimensional signals [4], providing a tracking uncertainty measure [10] and adaptive network which adapts to changes in noise statistics [26].

## Problems

**13.1** Define sequential Bayesian inference. Differentiate between smoothing, filtering and forecasting.

**13.2** The Kalman filter assumes a linear Gaussian process and measurement model. Explain what these models are.

**13.3** Describe the Kalman filter equations.

**13.4** Describe some maneuver models.

**13.5** Compare and contrast the Robust, Kriged, Augmented, Extended, Unscented and Switching Kalman filters.

**13.6** Compare and contrast measurement and track-to-track fusion.

**13.7** Define the data association problem. Compare and contrast the following data association algorithms: nearest neighbour, strongest neighbour, covariance union and probabilistic data association.

## References

1. Arulampalam MS, Maskell S, Gordon N, Clapp T (2002) A tutorial on particle filters for online nonlinear/non-Gaussian Bayesian tracking. IEEE Trans Sig Process 50:174–188
2. Bar-Shalom Y, Campo L (1986) The effect of the common process noise on the two sensor fused track covariance. IEEE Trans Aero Elect Sys 22:803–805
3. Blackman SS, Popoli RF (1999) Design and analysis of modern tracking Systems. Artech Pub
4. Buchnik I, Revach G, Steger D et al (2024) Latent-KalmanNet: learned Kalman filtering for tracking from high-dimensional signals. IEEE Trans Sig Proc 72:352–367
5. Challa S, Morelande MR, Musicki D, Evans RJ (2011) Fundamentals of object tracking. Published by Cambridge University Press
6. Chen Z (2003) Bayesian filtering: from Kalman filters to particle filters, and beyond. Adaptive Systems Laboratory. Communications Research Laboratory, McMaster University, Canada
7. Chang K, Saha R, Bar-Shalom Y (1997) On optimal track-to-track fusion. IEEE Trans Aero Elect Sys 33:1271–1276
8. Choi G, Park J, Shlezinger N et al (2023) Split-KalmanNet: a robust model-based deep-learning approach for state estimation. IEEE Trans Vehic Tech 72:12326–12331
9. Cipra T, Romera R (1997) Kalman filter with outliers and missing observations. Test 6:379–395
10. Dahan Y, Revach G, Dunik J et al (2025) Bayesian KalmanNet: quantifying uncertainty in deep learning augmented Kalman filter. IEEE Trans Signal Proc 99:1–15
11. Duncan S, Singh S (2006) Approaches to multisensor data fusion in target tracking: a survey. IEEE Trans Know Data Eng 18:1696–1710
12. Flament M, Fleury G, Davoust M-E (2004) Particle filter and Gaussian-mixture filter efficiency evaluation for terrain-aided navigation. Euro Sig Proc Conf EUSIPCO
13. Gao JB, Harris CJ (2002) Some remarks on Kalman filters for the multisensor fusion. Inf Fusion 3:191–201
14. Huber PJ (1981) Robust Statistics. Published by Wiley
15. Julier S, Uhlmann J (2001) General decentralized data fusion with covariance intersection (CI). In: Hall D, Llians J (eds) Handbook of Multidimensional Data Fusion. CRC Press
16. Julier SJ, Uhlmann JK, Nicholson DN (2004) A method for dealing with assignment ambiguity. Proc 2004 Am Control Conf
17. Lian G, Lai J, Zheng W-S (2011) Spatial-temporal consistent labeling of tracked pedestrians across non-overlapping camera views. Patt Recogn 44:1121–1136
18. Ma K, Zhang H, Wang R et al (2017) Target tracking system for multi-sensor data fusio. IEEE ITNEC Conf 1768–1772
19. Maskell S (2004) Sequentially structured Bayesian solutions. PhD thesis, Cambridge University
20. Galanis G, Anadranistakis M (2002) A one-dimensional Kalman filter for the correction of near surface temperature forecasts. Meteorological Appl 9:437–441
21. Han X, Ding L, Peng C et al (2023) Multi-model KalmanNet for maneuvering target tracking. IET Int Radar Conf (IRC) 399–405
22. Li X, Zhi X (1996) PSNR: a refined strongest neighbor filter for tracking in clutter. Proc 35th IEEE Conf Dec Control
23. van der Merwe R (2004) Sigma-point Kalman filters for probabilistic inference in dynamic state-space models. PhD thesis. Oregon Health and Science University
24. Moore JR, Blair WD (2000) Practical aspects of multisensor tracking. In: Bar-Shalom Y, Blair WD (eds) Multitarget-Multisensor tracking: applications and advances. Pub, Artech
25. Moore M, Wang J (2003) An extended dynamic model for kinematic positioning. J Navig 56:79–88
26. Ni X, Revach G, Shlezinger N (2024) Adaptive KalmanNet: data-driven Kalman filter with fast adaption. In: Proceedings of IEEE international conference acoustics, speech signal processing
27. Revach G, Shlezinger N, van Sloun RJG et al (2022) KalmanNet: data-driven Kalman filtering. Proc IEEE ICASSP 3905–3909
28. Revach G, Shlezinger N, Xiaoyong N et al (2022) KalmanNet: neural network aided Kalman filtering for partially known dynamics. IEEE Trans Sig Proc 70:1532–1547

29. Senel N, Kefferputz K, Doycheva K et al (2023) Processes 11:501
30. Shin V, Lee Y, Choi T-S (2006) Generalized Millman's formula and its application for estimation problems. Sig Proc 86:257–266
31. Siriani ALR, Kodaira V, Mehdizadeh SA et al (2022) Detection and tracking of chickens in low-light images using YOLO network and Kalman filter. Neural Comput App 34:21987–21997
32. Song J, Mei W, Xu Y et al (2024) Practical implementation of KalmanNet for accurate data fusion in integrated navigation. IEEE Sig Proc Lett 31:1890–1894
33. Wang C, Kim J-H, Byun K-Y, Ni J, Ko S-J (2009) Robust digital image stabilization using the Kalman filter. IEEE Trans Consum Elec 55:6–14
34. Zhang L, Sidoti D, Bienkowski A et al (2020) On the identification of noise covariances and adaptive Kalman filtering: a new look at a 50 year old problem. IEEE Access 8:59362–59388
35. Zhang T, Zhang X, Yang Y et al (2020) Efficient golf ball detection and tracking based on convolutional neural networks and Kalman filter. IEEE Int Cong Sys Man Cybernet 2808–2813

# Chapter 14
# Bayesian Decision Theory

**Abstract** The subject of this chapter is Bayesian decision theory and more specifically its use in data fusion. To keep the discussion focused we shall concentrate on the problem of classifying an unknown object into one of $K$ possible classes. We start by considering the curve of dimensionality and how this impinges on the number of features to be used and their selection. We then consider the Naive Bayes classifier and its variants. These include the following Bayesian classifiers: tree augmented, homologous, boosted, committee and model average. We then generalize these classifiers to multiple classes. Finally we briefly consider non-Bayesian classifiers. The chapter includes several detailed examples.

## 14.1 Introduction

The subject of this chapter is *Bayesian decision theory* and its use in data fusion. To make our discussion concrete we shall concentrate on the pattern recognition problem [16] in which an unknown pattern, or object, $O$ is to be assigned to one of $K$ possible classes $\{c_1, c_2, \ldots, c_K\}$. In this chapter we shall limit ourselves to using a single logical sensor, or classifier, $S$. We shall then remove this restriction in Chap. 15 where we consider multiple classifier systems.

In many applications Bayesian decision theory represents the primary fusion algorithm in a multi-sensor data fusion system. In Table 14.1 we list some of these applications together with their classification.

## 14.2 Pattern Recognition

Formally, the pattern recognition problem deals with the optimal assignment of an object $O$ to one of $K$ possible classes, $\{c_1, c_1, \ldots, c_K\}$. We suppose $O$ is characterized by $L$ measurements, or features, $y^{(l)}, l \in \{1, 2, \ldots, L\}$. Then, mathematically, the classifier defines a mapping between a class variable $C$ and the feature vector $\mathbf{y} = (y^{(1)}, y^{(2)}, \ldots, y^{(L)})^T$:



H. B. Mitchell, *Data Fusion*, https://doi.org/10.1007/978-3-662-71023-4_14

**Table 14.1** Applications in which bayesian decision theory is primary fusion Alg

| Class | | Description |
|---|---|---|
| Inter-DeO | Example 1.12 | Automatic speech recognition: preventing catastrophic fusion |
| | Example 3.6 | Tire pressure monitoring device |
| | Example 3.8 | Multi-Modal Biometric Identification Scheme |
| | Example 7.5 | Speech-Recognition in an embedded-speech recognition system |
| | Example 7.7 | Word-Spotting in historical manuscripts |
| | Example 14.3 | Plug-In MAP bayes' classifier versus naive bayes' classifier |
| | Example 14.4 | Naive bayes' classifier: correlated features |
| | Example 14.9 | Gene selection: an application of feature selection in DNA microarray experiments |
| | Example 14.10 | Wrapper feature selection |
| | Example 14.8 | Selective neighbourhood naive bayes' classifier |
| Late-DeO | Example 14.12 | Naive bayes' classifier committee (NBCC) |
| | Example 14.13 | Multi-class naive bayes' classifier |
| | Example 15.1 | Bayesian model averaging of the naive bayes' classifier |

$$\mathbf{y} \rightarrow C \in \{c_1, c_2, \ldots, c_K\} . \tag{14.1}$$

If we wish to minimize the number of classification errors, we use a zero-one loss function (see Sect. 11.2) and assign $O$ to the class with the largest a posteriori probability:

$$C = c_{\text{MAP}} \quad \text{if } p(C = c_{\text{MAP}}|\mathbf{y}, I) = \max_k (p(C = c_k|\mathbf{y}, I)) , \tag{14.2}$$

where $I$ denotes all relevant background information we have regarding $O$ and its measurement vector $\mathbf{y}$.

**Example 14.1** *Confusion Matrix.* A useful tool for recording the performance of a given classifier is the *confusion matrix* $\mathbf{\Omega}$. This is a $K \times K$ matrix in which the $(l, k)$th element $\Omega^{(l,k)}$ records the a posteriori probability $P(C_{\text{true}} = c_l|C = c_k)$ where $C_{\text{true}}$ and $C$ are, respectively, the true and predicted class variables.

In practice the elements $\Omega^{(l,k)}$ are approximated as follows. Let $\mathcal{D}$ denote a training set of $N$ objects $O_i$, $i \in \{1, 2, \ldots, N\}$, then

$$\Omega^{(l,k)} = \frac{n(l,k)}{\sum_{l=1}^{K} n(l,k)} ,$$

where $n(l, k)$ is the number of objects $O_i$ which belong to the true class $C_{\text{true}} = c_l$ but which are assigned by classifier, $S$, to the predicted class $C = c_k$.

Equation (14.2) is known as the maximum a posteriori (MAP) rule. The computation of the a posteriori pdf, $p(C = c_{\mathrm{MAP}}|\mathbf{y}, I)$, is performed by applying Bayes' Theorem:

$$
\begin{aligned}
c_{\mathrm{MAP}} &= \arg\max_k p(C = c_k|\mathbf{y}, I) \ , \\
&= \arg\max_k \left( \frac{\pi(C = c_k|I) p(\mathbf{y}|C = c_k, I)}{p(\mathbf{y}|I)} \right) ,
\end{aligned} \tag{14.3}
$$

where $p(\mathbf{y}|I)$ serves as a normalization constant. This constant can be ignored for classification purposes, giving,

$$
c_{\mathrm{MAP}} = \arg\max_k \pi(C = c_k|I) p(\mathbf{y}|C = c_k, I) \ . \tag{14.4}
$$

For *known* probability densities the MAP rule is the best any classifier can do with a zero-one loss function. However, in practice the pdf's in Eq.(14.4) are unknown. In this case we use the following empirical, or *plug-in*, MAP rule:

$$
c_{\mathrm{PLUG\text{-}IN}} = \arg\max_k \hat{\pi}(C = c_k|I) \hat{p}(\mathbf{y}|C = c_k, I) \ , \tag{14.5}
$$

where $\hat{\pi}(C = c_k|I)$ and $\hat{p}(\mathbf{y}|C = c_k, I)$ denote *estimated* pdf's.

Direct calculation of the a priori probability $\hat{\pi}(C = c_k|I)$ is straightforward: Let $\mathcal{D}$ denote a training set containing $N$ objects $O_i$, $i \in \{1, 2, \ldots, N\}$, of which $N_k$ belong to the class $C = c_k$. If $\mathcal{D}$ is a representative sample of the underlying true population, then an unbiased maximum likelihood estimate of the true a priori probability is

$$
\hat{\pi}(C = c_k|I) = N_k/N \ . \tag{14.6}
$$

However, in many applications, $\mathcal{D}$ is not a random sample of the true population. In this case, and in the absence of any specific information, we often simply let $\hat{\pi}(C = c_k|I) = \frac{1}{K}$.

Direct calculation of the likelihood function, $\hat{p}(\mathbf{y}|C = c_k, I)$, is more difficult, especially for high-dimensional data [30]. The main difficulty is that the required size of $\mathcal{D}$ increases *exponentially* with $L$ [27]. This is known as the *curse of dimensionality* [8] and it affects all aspects of classification theory. For example, the curse of dimensionality implies that for a given number of objects $O_i$ in $\mathcal{D}$, there is a maximum number of features above which the performance of Bayes' classifier will degrade rather than improve. This is illustrated in the following example.

**Example 14.2** *The Small Sample Problem* [32]. We consider a two-class classification problem with equal a priori probabilities. Each class is characterized by an $L$-dimensional Gaussian distribution: $\mathcal{N}(\mathbf{y}|\boldsymbol{\mu}, \boldsymbol{\Sigma})$ and $\mathcal{N}(\mathbf{y}| -\boldsymbol{\mu}, \boldsymbol{\Sigma})$, where

$$\boldsymbol{\mu} = \left(1, 1/\sqrt{2}, 1/\sqrt{3}, \ldots, 1/\sqrt{L}\right)^T \quad \text{and} \quad \boldsymbol{\Sigma} = \text{diag}(\overbrace{1, 1, \ldots, 1}^{L}) .$$

The features are statistically independent and the discriminating power of the successive features decrease monotonically with the first feature providing the maximum discrimination between the two classes. If $\boldsymbol{\mu}$ is *known* then the probability of a misclassification error $P_\epsilon(L)$ decreases with $L$ as shown in Fig. 14.1. However, if $\boldsymbol{\mu}$ is *unknown*, then we estimate $\boldsymbol{\mu}$ and $\boldsymbol{\Sigma}$ from a training set $\mathcal{D}$ containing $N$ labeled objects $O_i, i \in \{1, 2, \ldots, N\}$, with instantiated measurements $\mathbf{y}_i$. In this case, the probability of misclassification, $P_\epsilon(N, L)$, is a function of both $L$ and $N$: For a given value of $N$, $P_\epsilon(N, L)$ initially decreases with $L$ but eventually it increases and takes on the value of $\frac{1}{2}$.

An explanation of this phenomena is: For a fixed value of $N$, the reliability of the likelihood functions decrease with an increase in $L$. As a consequence, the performance of the classifiers also fall with an increase in $L$.

From this perspective, the main effort in Bayesian classification theory is an attempt to make accurate estimates of the likelihoods, $\hat{p}(\mathbf{y}|C = c_k, I)$, while using a small number of labeled objects $O_i$. One way of doing this is to impose a given structure on $\hat{p}(\mathbf{y}|C = c_k, I)$. This idea underlies the naive Bayes' classifier (NBC) which is probably the most widely used classifier used today.

To keep notation simple we shall not, in general, differentiate between the estimated pdf $\hat{p}(\mathbf{y}|C = c_k, I)$ and the true pdf $p(\mathbf{y}|C = c_k, I)$. In this regard, the pdf's are often estimated using a ML or MAP estimators.

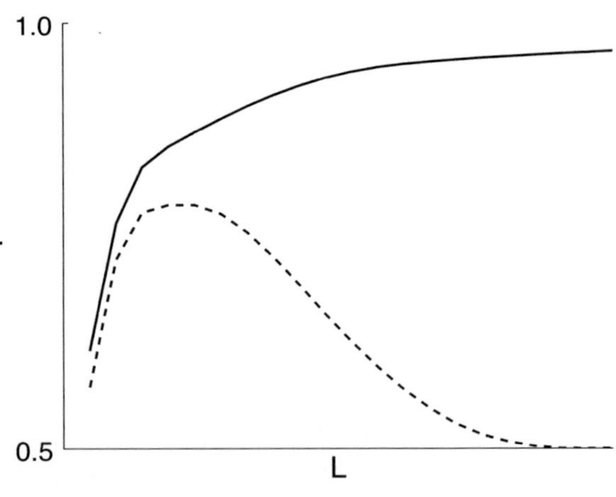


**Fig. 14.1** Shows the probability of correct classification $P = 1 - P_\epsilon$ as a function of the number of features $L$. The solid curve shows $P$ calculated assuming $\boldsymbol{\mu}$ is known. The dashed curve shows $P$ calculated assuming $\boldsymbol{\mu}$ is *estimated* from the $N$ labeled objects in $\mathcal{D}$

## 14.3 Naive Bayes' Classifier (NBC)

In NBC we reduce the impact of the curse of dimensionality, by approximating $p(\mathbf{y}|C = c_k, I)$ as a product of $L$ 1-D likelihood functions $p(y^{(l)}|C = c_k, I)$:

$$p(\mathbf{y}|C = c_k, I) = \prod_l p(y^{(l)}|C = c_k, I) \; . \tag{14.7}$$

In this case, the required number of objects $O_i$ in $\mathcal{D}$ increases only *linearly* with $L$.

**Example 14.3** *Plug-in MAP Classifier vs. NBC.* Consider a classifier with $L$ features $y^{(1)}, y^{(2)}, \ldots, y^{(L)}$, whose instantiated values are limited to a discrete set of $M_l$ values $\alpha_m^{(l)}, m \in \{1, 2, \ldots, M_l\}$. If we assume a minimum of one object $O_i$ for each combination of feature values, then the plug-in MAP classifier requires $\prod_{l=1}^L M_l$ objects, while the NBC only requires $\sum_{l=1}^L M_l$ objects.

Equation (14.7) is clearly not the only approximation which can be used and in some applications the following approximations [20] are used instead:

$$p(\mathbf{y}|C = c_k, I) = \begin{cases} \min_l p(y^{(l)}|C = c_k, I) \; , \\ \frac{1}{L} \sum p(y^{(l)}|C = c_k, I) \; , \\ \text{median}_l \, p(y^{(l)}|C = c_k, I) \; , \\ \max_l p(y^{(l)}|C = c_k, I) \; . \end{cases} \tag{14.8}$$

However, in general, (14.7) is the preferred approximation.[1]

### *14.3.1 Representation*

The NBC is a highly restricted Bayesian network in which all of the features $y^{(l)}, l \in \{1, 2, \ldots, L\}$, have one and the same independent class variable $C$. Graphically, a NBC has the form shown in Fig. 14.2a, in which all arcs are directed from the class variable $C$ to the features $y^{(l)}, l \in \{1, 2, \ldots, L\}$. If $pa(y^{(l)})$ represents the *parents* of a given feature $y^{(l)}$, then by definition the NBC has $pa(C) = 0$ and $pa(y^{(l)}) = C$.

[1] Equation (14.7) may be "justified" as follows: Let $\{\pi(C = c_k|I), p(y^{(1)}|C = c_k, I), \ldots, p(y^{(L)}|C = c_k, I)\}$ denote the set of "low-order" estimates used by the NBC. Then the product approximation contains the smallest amount of information (i.e. the maximum entropy) of all possible approximations to $p(\mathbf{y}|C = c_k, I)$ that use the same set of low-order estimates [34].

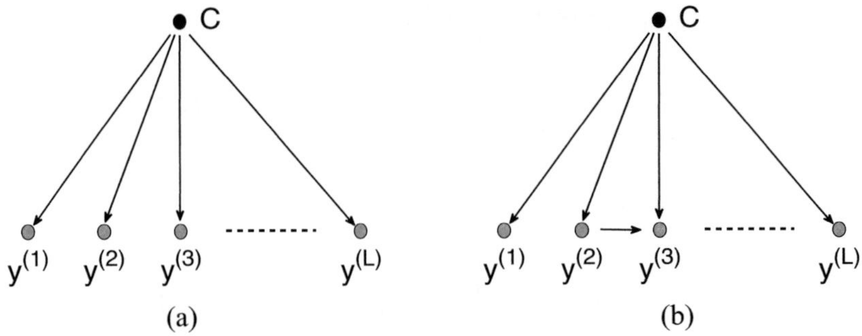


**Fig. 14.2** **a** Shows a NBC with one class variable $C$ and $L$ features $y^{(l)}, l \in \{1, 2, \ldots, L\}$. The NBC has $pa(C) = 0$ and $pa(y^{(l)}) = C$. **b** Shows a Tree Augmented Naive (TAN) Bayes' classifier with one class variable $C$ and $L$ features $y^{(l)}, l \in \{1, 2, \ldots, L\}$. The parents of $y^{(l)}$ are $C$ for all $l$, apart from $l = 3$. The parents of $y^{(3)}$ are $C$ and $y^{(2)}$

## 14.3.2 Performance

The NBC rule corresponding to Eq. (14.7) is

$$c_{NBC} = \arg\max_k \pi(C = c_k|I) \prod_l p_l(y^{(l)}|C = c_k, I) , \tag{14.9}$$

This equation assumes the features $y^{(l)}$ are independent. At first sight, we may believe that the NBC is optimal, i.e. has the smallest number of classification errors, when the $y^{(l)}$ are independent, and perhaps close to optimal when the $y^{(l)}$ are weakly dependent. However, in practice the classifier has a good performance in a wide variety of domains, including many where there are clear dependencies between the $y^{(l)}$. The following example provides a plausible explanation of this phenomenom.

**Example 14.4** *NBC: Correlated Features* [5]. We consider the classification of an object $O$ into two classes $c_1$ and $c_2$. The object $O$ is described by the feature vector $\mathbf{y} = (y^{(1)}, y^{(2)}, y^{(3)})^T$, where we assume $y^{(1)}$ and $y^{(3)}$ are independent, and $y^{(1)}$ and $y^{(2)}$ are completely dependent, i.e. $y^{(1)} = y^{(2)}$. For simplicity we further assume that $\pi(C = c_1|I) = \frac{1}{2} = \pi(C = c_2|I)$. In this case, plug-in MAP and NBC, classification rules are:

$$\begin{aligned} c_{\text{PLUG-IN}} &= \arg\max_k p(y^{(1)}|C = c_k, I) p(y^{(3)}|C = c_k, I) , \\ &= \begin{cases} c_1 & \text{if } B > 1 - A , \\ c_2 & \text{otherwise} , \end{cases} \end{aligned}$$

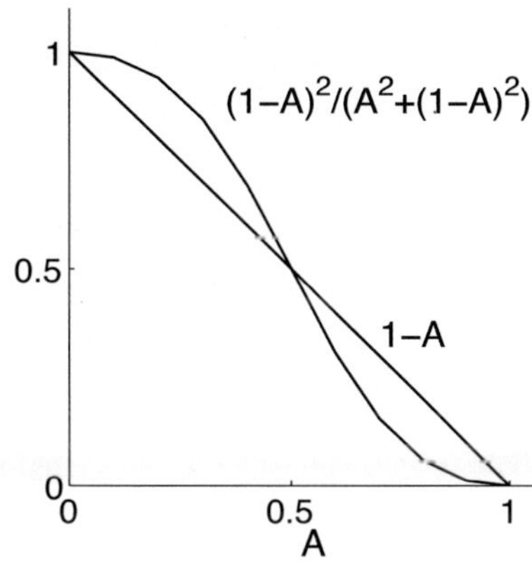


**Fig. 14.3** Shows the curves $B = 1 - A$ and $B = (1 - A)^2/(A^2 + (1 - A)^2)$ as a function of $A$ for the plug-in MAP classifier and the NBC. The only region where the classifiers give different results are regions where one curve is above the other. Everywhere else the two classifiers give identical results

and

$$c_{\text{NBC}} = \arg\max_k p(y^{(1)}|C = c_k, I)^2 p(y^{(3)}|C = c_k, I) \; .$$
$$= \begin{cases} c_1 & \text{if } B > (1 - A)^2/(A^2 + (1 - A)^2) \; , \\ c_2 & \text{otherwise} \; , \end{cases}$$

where $A = p(y^{(l)}|C = c_1, I)$ and $B = p(y^{(l)}|C = c_2, I)$. The two curves, $B = 1 - A$ and $B = (1 - A)^2/(A^2 + (1 - A)^2)$, are shown in Fig. 14.3. We observe that, although the independence assumption is decisively violated ($y^{(2)} = y^{(1)}$), the NBC disagrees with the plug-in MAP classifier only in the two narrow regions that are above one of the curves and below the other; everywhere else the two classifiers give identical classification results.

### Deep learning and Naive Bayes Classifier

The following example illustrates a novel application combining deep learning with NBC.

**Example 14.5** *Autonomous Crack Detection on Metallic Surfaces* [4]. For nuclear power plants regular inspection to detect early cracks in metallic surfaces is critical. In this application, a video of the metal surface is inspected automatically for the presence of cracks. Each frame is divided into overlapping patches of size $d \times d$ and each patch is separately analyzed for the presence of a crack using a CNN (see Fig. 14.4). *Note*: In an earlier work [3] the authors obtained similar results using a hand-crafted LBP texture operator (see Example 4.15).

Since the length of a crack is typically larger than $d$, several overlapping patches are grouped together to form a bounding box of the crack. To enhance the accuracy and robustness of the procedure, each bounding box is tracked in the nearby frames and thus obtain a set of bounding boxes for each crack in successive frames. Each bounding box has a score $s_i$ that the patch is successfully tracked in the $i$th frame.

We use Bayesian decision fusion to compute the probability that the detected crack is an *actual crack* and not an artifact, as follows.

Assume a crack is tracked in $N$ nearby frames. Let $P(C_{\mathrm{AC}}|s_1, s_2, \ldots, s_N)$ and $P(C_{\mathrm{NC}}|s_1, s_2, \ldots, s_N)$ represent a posteriori probabilities of the crack being an actual crack (AC) and a non-crack (artifact) (NC). We assume the scores $s_i$ are independent. In this case, the a posteriori probability that the crack is an actual crack is given by Naive Bayes rule:

$$P(AC|s_1, s_2, \ldots, s_N) = \frac{P(AC)\prod_i P(s_i|AC)}{P(AC)\prod_i P(s_i|AC) + P(NC)\prod_i P(s_i|NC)} ,$$

where $P(s_i|AC)$ and $P(s_i|NC)$ are the tracking accuracies of crack and no-crack respectively; $P(AC)$ is the probability that a certain region of a frame includes an actual crack, $P(NC) = 1 - P(AC)$ and the products are taken over all $i$, $i \in \{1, 2, \ldots, N\}$. $P(AC)$ is obtained by counting the areas of ground truth bounding boxes. Using training videos with ground truth crack bounding boxes, $P(s_i|AC)$ and $P(s_i|NC)$ are obtained by applying the frame-based crack detection and crack bounding box tracking on the video frames.

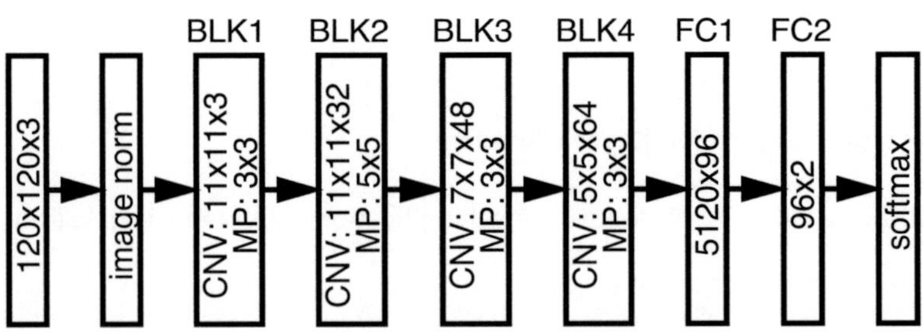


**Fig. 14.4** Shows the CNN architecture for crack detection. Image patches of size $120 \times 120 \times 3$ are normalized and then fed to the network which has 4 blocks (BLK1-BLK4), each block consists of: convolutional layer (plus batch normalization and exponential activation layer) (CNV) and max pooling (MP). Output of BLK4 is fed to fully-connected layers FC1, FC2 and then to softmax layer. Output from softmax has two outputs: crack and no-crack. Each max pooling layer has a stride of 2 and the layers FC1 and FC2 have, respectively, 96 and 2 nodes

**Table 14.2** Principal methods used for estimating the likelihood $p(\mathbf{y}|C = c_k, I)$.

| Method | Description |
|---|---|
| Standard parametric function | Fit $p(y^{(l)}|C = c_k, I)$ to a 1D parametric form using a maximum likelihood, or maximum a posteriori, algorithm |
| Gaussian mixture model (GMM) | Fit $p(y^{(l)}|C = c_k, I)$ to mixture of 1D-Gaussian functions using EM algorithm: $\hat{p}(y^{(l)}|C = c_k, I) \approx \sum_{m=1}^{M} \alpha_{km}^{(l)} \mathcal{N}(\mu_{km}^{(l)}, \Sigma_{km}^{(l)})$, where $\alpha_{km}^{(l)} > 0$, $\sum_{m=1}^{M} \alpha_{km}^{(l)} = 1$. See Sect. 10.4.4 |
| Histogram Binning | Convert the feature values $y^{(l)}$ into a finite set of discrete values $\alpha_m, m \in \{1, 2, \ldots, M_l\}$. See Sect. 6.4.1 |
| Kernel density estimation | Calculate $p(y^{(l)}|C = c_k, I)$ using 1D kernel density estimation procedure. See Sect. 6.4.2 |

### *14.3.3 Likelihood*

Several methods are available for estimating the 1D likelihoods $p(y^{(l)}|C = c_k, I)$ from a training dataset $\mathcal{D}$ (see Table 14.2).

The most common method for estimating the 1D likelihoods $p(y^{(l)}|C = c_k, I)$ is to model them using a standard parametric distribution. Traditionally, we use a 1D Gaussian pdf, $\mathcal{N}(\mu_k^{(l)}, \Sigma_k^{(l)})$, for this purpose, where the parameters, $\mu_k^{(l)}$ and $\Sigma_k^{(l)}$, are estimated using the training dataset $\mathcal{D}$ (which contains $N$ labeled objects $O_i, i \in \{1, 2, \ldots, N\}$, with instantiated values $y_i^{(l)}$). Thus for each class $c_k, k \in \{1, 2, \ldots, K\}$, the unbiased ML estimates of $\mu_k^{(l)}$ and $\Sigma_k^{(l)}$ are given by

$$\mu_k^{(l)} = \frac{1}{n_k} \sum_i \delta_{ik} y_i^{(l)} \quad \text{and} \quad \Sigma_k^{(l)} = \frac{1}{n_k - 1} \sum_i \delta_{ik} \left(y_i^{(l)} - \mu_k^{(l)}\right)^2 , \tag{14.10}$$

where $\delta_{ik} = 1$ if the object $O_i$ belongs to class $C = c_k$, otherwise $\delta_{ik} = 0$ and $n_k = \sum_i \delta_{ik}$.

If the likelihood cannot be modeled using a standard parametric distribution we use a non-parametric method e.g. using a histogram or a kernel density estimator. The following example describes modeling the likelihood with a histogram.

**Example 14.6** *Naive Bayes in Information Retrieval* [1]. The NBC is widely used in information retrieval applications: Given the document class and length, we assume the probability of occurrence of a word does not depend on its position or other words in the document. In spite of being completely unrealistic, this assumption greatly simplifies classifier training. In particular, estimation of the likelihoods reduces to a simple normalization of word counts. Let $\{y^{(1)}, y^{(2)}, \ldots, y^{(L)}\}$ denote a set of $L$ specific words, then

$$p(y^{(l)}|C = c_k, I) = \frac{N_{k,l}}{\sum_{h=1}^{L} N_{k,h}} ,$$

where $N_{k,l}$ is the number of occurrences of the word $y^{(l)}$ in training data from class $C = c_k$. To avoid null estimates we often add a "pseudo-count" $\delta$, $\delta > 0$, to every $N_{k,l}$:

$$p(y^{(l)}|C = c_k, I) = \frac{N_{k,l} + \delta}{\sum_{h=1}^{L}(N_{k,h} + \delta)} ,$$

The default value for $\delta$ is $\delta = 1$ and is referred to as the *Laplace correction*.

Alternatively, instead of using a pseudo-count $\delta$ we may avoid null estimates by "discounting": We discount a small constant $b$, $0 < b < 1$, from every occurence of a word. The probability mass gained by discounting is then distributed among all the words. In this case,

$$p(y^{(l)}|C = c_k, I) = \max\Big(0, \frac{N_{k,l}(1-b)}{\sum_{h=1}^{L} N_{k,h}(1-b)}\Big) + \beta ,$$

where

$$\beta = \frac{1}{L}\sum_{l} \max\Big(0, \frac{N_{k,l}(1-b)}{\sum_{h=1}^{L} N_{k,h}(1-b)}\Big) .$$

The preferred method for non-parametric likelihood estimate is kernel density as described in the following example.

**Example 14.7** *Kernel Density Estimation for NBC* [22]. Different non-parametric density estimates were compared experimentally for use in a NBC. The best classification performance was obtained using a kernel density estimate although surprisingly the most accurate kernel density estimators did *not* give the best classification performance. The bandwidth which gave the best classification performance was:

$$H = \frac{R}{2\big(1 + \log_2(N)\big)} ,$$

where $N$ is the number of training samples in $\mathcal{D}$ and $R$ is the range of the values in $\mathcal{D}$.

In estimating the likelihoods we have assumed that we have used all of the training samples. Recently, [38] suggested using only a subset of the training data in a "lazy" learning approach: Given an object $O$ we construct a NBC using a set of *weighted* training samples, where the weights fall as we move away from $O$ in feature space. The idea behind the lazy NBC is that by locally learning the likelihoods $\hat{p}(y^{(l)}|C = c_k, I)$ we may reduce the effect of any feature-feature correlations.

The following example illustrates a simple, but very effective, lazy NBC.

**Example 14.8** *SNNB: Selective Neighbourhood NBC* [36]. SNNB is a lazy learning algorithm which works as follows. Let $D$ denote training set consisting of $N$ objects $O_i, i \in \{1, 2, \ldots, N\}$. For a given test object $O_{\mathrm{TEST}}$, we define a subset $S_l$ of the $l$ objects $O_i$ in $\mathcal{D}$ which are most similar to $O_{\mathrm{TEST}}$. For each value of $l, l \in \{1, 2, \ldots, N\}$, we learn a NBC (denoted as $H_l$) only on the $l$ objects which belong to $S_l$. We use the NBC with the highest accuracy (as estimated on $\mathcal{D}$) to classify the test object $O_{\mathrm{TEST}}$.

### 14.3.4 Feature Selection

As we mentioned in Sect. 14.2 it is often beneficial to limit the number of features in the NBC. The following example illustrates the feature selection in a DNA experiment (see also Example 14.2)

**Example 14.9** *Gene Selection: An Application of Feature Selection in DNA Microarray Experiments* [15]. DNA microarray experiments, generating thousands of gene expression measurements, are used to collect information from tissue and cell samples regarding gene expression differences that can be useful for diagnosis disease, distinction of the specific tumor type etc. One important application of gene expression microarray data is the classification of samples into known categories.

From a classification point of view, the biological samples can be seen as objects and the genes as features used to describe each object. In a typical microarray dataset the number of samples is small (usually less than 100) but the number of genes is several thousand, with many of the genes being correlated or irrelevant. For diagnostic purposes it is important to find small subsets of genes that are sufficiently informative to distinguish between cells of different types. Studies show that most genes measured in a DNA microarray experiment are not relevant for an accurate distinction among the different classes of the problem and simple classifiers with few genes (less than 15–20) achieve better accuracies. To avoid the "curse of dimensionality" feature selection is required to find the gene subset with the best classification accuracy for a given classifier.

**Table 14.3** Feature selection methods

| Method | Description |
|---|---|
| Exhaustive Search | Evaluate all possible subsets. Guaranteed to find the optimal subset. Not feasible for even moderately large values of $L'$ and $L$ |
| Best Individual Features | Evaluate all $m$ features individually. Select best $L'$ individual features. Computationally simple but not likely to lead to an optimal subset |
| Sequential Forward Selection | Select best single feature. Then add one feature at a time which in combination with the selected features maximizes the criterion function. Computationally attractive. Main drawback is that once a feature is selected it cannot be discarded |
| Sequential Backward Selection | Start with all the $L$ features and successively delete one feature at a time. Computationally attractive although requires more computation than sequential forward selection. Main drawback: once a feature is deleted it cannot be brought back into optimal subset |
| Sequential Forward/ Backward Search | First enlarge the feature subset by $m$ features using forward selection. Then delete $n$ features using backward selection. Technique attempts to eliminate drawbacks of sequential forward and backward selection. Provides close to optimal solution at an affordable computational cost |

In feature selection our goal is to find an optimal subset of the input features that optimizes some criteria $\mathcal{J}$. The simplest feature selection approach is to evaluate each feature individually and select the $L'$ features with the highest scores. Unfortunately this approach ignores feature correlations and it will rarely find an optimal subset. A brute force approach would be to evaluate *all* possible subsets of $L'$ features and select the global optimum, but the number of combinations $\binom{L}{L'}$ becomes impractical even for moderate values of $L'$ and $L$. A compromise between these two approaches is to selectively examine a limited number of feature subsets. Some ways of doing this are listed in Table 14.3.

Given a subset of features two different strategies are available to evaluate the degree to which the subset is optimal. In the first strategy we evaluate the degree to which a subset of features is optimal *indirectly*: it is based on a filter in which we evaluate the optimality of a feature subset on the basis of its "information content" $\Phi$. See Sect. 4.6. The second strategy ("wrapper") follows a more direct approach and evaluates the optimality of a feature subset on the basis of their estimated classification accuracy, or error probability, $P_\epsilon$ [2, 19].

The following example illustrates the wrapper feature selection strategy.

**Example 14.10** *Wrapper Feature Selection* [2, 19]. In wrapper based feature selection we select feature subsets on the basis of their estimated classification accuracy, or probability of error, $P_\epsilon$ (see Sect. 14.5).

Although theoretically optimal, the procedure is liable to *overfit*. This happens when a feature subset has a low $P_\epsilon$ value (as measured on the training data) but which generalizes poorly on the test data [19, 23]. Some effective ways of preventing overfitting are given in [24].

### Feature Extraction

In feature selection we select a small number of the input features for use in the NBC. In *feature extraction* we follow a slightly different approach. We first create a set of new features and then select a small number of them. Mathematically, we define feature extraction as creating new features, $y^{(l)}, l \in \{1, 2, \ldots, L'\}$, from the original set of features $x^{(l)}, l \in \{1, 2, \ldots, L\}$. Our aim is to extract a small number of new features which contain the maximal information concerning the class variable $C$ (see Sect. 4.7.2).

As in the case of feature selection, we may also perform feature extraction by directly optimizing the classifier performance, i.e. by minimizing the estimated probability of classification error $P_\epsilon$.

**Example 14.11** *Feature Extraction Based on Mutual Information (MI)* [21]. We perform feature extraction by searching for a set of linear combinations of the original features, whose mutual information (MI) with the output class is maximized. The MI between the extracted features and the output class is calculated by using a kernel pdf estimate (Sect. 6.4.2).

## 14.4 NBC Variants*

The NBC has a good performance over a wide variety of domains, including many domains where there are clear dependencies between the attributes. The NBC is also robust to noise and irrelevant attributes. As a result, the NBC is widely used in many practical applications. It has also attracted much attention from many researchers who have introduced various modifications in an effort to improve its performance. Some of these variants are listed in Table 14.4.

### Tree-Augmented Naive (TAN) Bayes' Classifer

In TAN Bayes' classifier [9, 18, 28]. we relax the independence assumption as used in the NBC. In TAN we allow each feature in the NBC to depend on $C$, and at most, *one* additional feature. Mathematically, TAN has: $pa(C) = 0$ and $|pa(y^{(l)})| \leq 2$. Graphically, the TAN is a restricted Bayesian network which uses a tree-structure imposed on the NBC structure (Fig. 14.2(b)). The corresponding TAN classifier rule is

$$c_{\text{TAN}} = \arg\max_k \pi(C = c_k|I) \prod_l p_l(y^{(l)}|pa(y^{(l)}), I) .$$

**Table 14.4** Naive bayes classifier variants

| Method | Description |
|---|---|
| Tree-Augmented NBC | Relax NBC assumption of feature independence [9, 18, 28] |
| APM | Adjusted Probability Model. Give different weights to each feature [13] |
| Homologous NBC | Classify multiple objects which belong to same (unknown) class [14] |
| NBC Committee | Use an ensemble of $K$ NBC's. Each NBC is trained on a different subset of features. Final classification is obtained by allowing the NBC's to vote, each classifier having an equal weight [37] |
| Boosted NBC | Use ensemble of $K$ NBC's, in which each classifier is trained on a boosted training set. Final classification is obtained by allowing the NBC's to vote, each classifier having a weight according to the quality of its predictions [6, 29]. See Sect. 15.9 |
| BMA-NBC | Bayesian model averaging—Naive Bayes' Classifier. Similar to NBCC except classifier has a weight according to Bayesian evidence that it is the true (correct) classifier. See Example 15.1 |

Many different methods are available for building the TAN classifier. One method is the super-parent algorithm [18, 25]. Here a "super-parent" is a node with arcs pointing to all other nodes without a parent except for the class label.

The super-parent algorithm consists of three steps:

1. Search for a "super"-parent that has the best estimated classification performance.[2] A super-parent is a node with arcs pointing to all other nodes without a parent except for the class label.
2. Determine one favourite child for the super-parent chosen in the first step, based on its estimated classification accuracy.
3. Add the appropriate arc to the tree structure.

The entire process is repeated until no improvement is achieved, or $L - 1$ arcs are added into the tree.

### Adjusted Probability Model (APM)

In the Adjusted Probability Model (APM) [13] we allow the features to have different weights, or importances, $w_l, l \in \{1, 2, \ldots, L\}$. The corresponding APM classification rule is

$$c_{\text{APM}} = \arg\max_k \pi(C = c_k|I) \prod_l \left( \frac{p(C = c_k|y^{(l)}, I)}{\pi(C = c_k|I)} \right)^{w_l} .$$

[2] Classification performance is conventionally measured using a cross-validation procedure (Sect. 14.5).

Different authors have restricted the weights in various ways. For example, in an unrestricted APM [13] the authors place no resrictions on the $w_l$, and allow both positive and negative values.

### Homologous Naive Bayes' Classifier (HNB)

In Homologous Naive Bayes (HNB) [14] we consider the problem of classifying multiple objects which belong to the same (unknown) class. The homologous classification rule is

$$c_{\mathrm{HNB}} = \arg\max_k \pi(C = c_k|I) \prod_i \prod_l p_l(y_i^{(l)}|C = c_k, I) ,$$

where $y_i^{(l)}$ is the instantiated value of the $l$th feature in the $i$th object.

### Naive Bayes' Classifier Committee (NBCC)

In NBCC we use multiple NBC's. This is illustrated in the following example.

**Example 14.12** *Naive Bayes' Classifier Committee (NBCC)* [37]. In NBCC we use an ensemble of $L$ NBC's, where $L$ is the number of independent features. Let $H_0$ denote the first NBC using all $L$ features. If $E_0$ is the corresponding estimated error rate, then we repeatedly create candidate NBC's by randomly selecting $L/2$ features. Assuming the candidate NBC is not already a member of the committee, then it becomes a member of the committee if its estimated error is less than $E_0$.

To classify a test object $O$, each NBC in the committee is invoked to produce the probability that $O$ belongs to a class $C = c_k$. Then for each class $c_k$, the probabilities provided by all committee members are summed up. The optimum classification of $O$ is the class with largest summed probability.

## 14.5 Error Estimation*

The classification error $P_\epsilon$ is the ultimate measure of the performance of a classifier. Apart from using $P_\epsilon$ to measure the performance of a given classifier we may use it as a general optimization criterion. In Sect. 14.4 we used it in selecting an optimum subset of features ("wrapper") and more generally we may use it to select an optimum classifier.

**Table 14.5** Bounds on the two-class probability of error $P_\epsilon$

| Bound | Description |
|---|---|
| Mahalanobis | $P_\epsilon \leq 2/(4+\Delta)$; where $\Delta$ is the Mahalanobis distance $(\boldsymbol{\mu}_1 - \boldsymbol{\mu}_2)^T \boldsymbol{\Sigma}^{-1}$ $(\boldsymbol{\mu}_1 - \boldsymbol{\mu}_2)$ [10] |
| Bhattacharyya | $\frac{1}{2}(1-\sqrt{1-e^{-2B}}) \leq P_\epsilon \leq \frac{1}{2}e^{-B}$; where $B$ is Bhattacharyya distance [10]: $\frac{1}{8}(\boldsymbol{\mu}_1 - \boldsymbol{\mu}_2)^T(\frac{1}{2}(\boldsymbol{\Sigma}_1 + \boldsymbol{\Sigma}_2))^{-1}(\boldsymbol{\mu}_1 - \boldsymbol{\mu}_2) + (\ln\|(\boldsymbol{\Sigma}_1 + \boldsymbol{\Sigma}_2)/2\|)/(2\sqrt{\|\boldsymbol{\Sigma}_1\|\|\boldsymbol{\Sigma}_2\|})$ |
| Chernoff | $P_\epsilon \leq \int \mathcal{N}(\mathbf{y}\|\boldsymbol{\mu}_1, \boldsymbol{\Sigma}_1)^s \mathcal{N}(\mathbf{y}\|\boldsymbol{\mu}_2, \boldsymbol{\Sigma}_2)^{1-s} d\mathbf{y}$ [10], $s \in [0, 1]$ is a parameter |

Mathematically, the classification error of a Bayesian classifier is defined as

$$P_\epsilon = 1 - \sum_{k=1}^{K} \int_{A_k} \pi(C = c_k|I) p(\mathbf{y}|C = c_k, I) d\mathbf{y} \ , \tag{14.11}$$

where $A_k$ is the region where the class $C = c_k$ has the highest a posteriori probability. In general, it is often very difficult to perform the integration in (14.11) and obtain a closed-form expression for $P_\epsilon$. However, for the case of two classes $C = c_1$ and $C = c_2$ with equal a priori probabilities and Gaussian likelihoods, $\mathcal{N}(\boldsymbol{\mu}_1, \boldsymbol{\Sigma}_1)$ and $\mathcal{N}(\boldsymbol{\mu}_2, \boldsymbol{\Sigma}_2)$, closed-form expression for the bounds on $P_\epsilon$ are available (Table 14.5).

In practice, we cannot be sure that the above conditions hold and then the only option is to estimate $P_\epsilon$ from the measurements $\mathbf{y}_i$, $i \in \{1, 2, \ldots, N\}$, as follows. We split the $\mathbf{y}_i$ into training data $\mathcal{D}$ and test data $\mathcal{T}$. We design the classifier on $\mathcal{D}$ and measure its generalization ability, or $1 - P_\epsilon$, on $\mathcal{T}$. If the number of measurements in $\mathcal{D}$ is small the classifier will not be robust and its generalization ability will be low. On the other hand, if the number of measurements in $\mathcal{T}$ is small, then the confidence in $P_\epsilon$ will be low. In Table 14.6 we list several methods for dividing the $\mathbf{y}_i$ into $\mathcal{D}$ and $\mathcal{T}$ and for estimating $P_\epsilon$.

## 14.6 Pairwise Naive Bayes' Classifier

Until now we have placed no restrictions on the number of possible classes $K$. However, in practice, finding features which are effective for all of the classes $c_k$, $k \in \{1, 2, \ldots, K\}$, becomes increasingly difficult as $K$ increases. For this reason, we often decompose a $K$-class pattern recognition problem into $K(K-1)/2$ two-class pattern recognition problems. In each two-class problem, we may use a different number of different features. The final classification is obtained by combining all $K(K-1)/2$ pairwise results. The following example illustrates the combination of $K(K-1)/2$ pairwise NBC's using a simple voting scheme (see Sect. 15.7.1).

**Table 14.6** Error Estimation Methods

| Method | Description |
|---|---|
| Resubstitution | $\mathcal{T}$ and $\mathcal{D}$ are the same. Thus we use all available data for training and for testing. The $P_\epsilon$ is optimistic biased estimate especially when ratio of sample size to sample dimensionality is small |
| Holdout | Half data is used for $\mathcal{T}$; the other half is used for $\mathcal{D}$. Pessimistic biased estimate. Different partitions give different estimates |
| Leave-One-Out | A classifier is designed using $(M-1)$ samples in $\mathcal{D}$ and evaluated on one remaining sample $\mathcal{T}$. This is repeated $M$ times with different training sets of size $(M-1)$. Estimate is unbiased but it has a large variance. When used for model selection it is asymptotically equivalent to Akaike Information Criterion (AIC) when number of observations is large |
| $K$-Fold CV | A compromise between holdout and leave-one-out methods. Divide available samples into $K$ disjoint subsets, $1 \leq K \leq M$. Use $(K-1)$ subsets for training and remaining subset for testing. Estimate has lower bias than holdout method. Recommended value for $K$ is in range $[M/5, M/2]$. When used for model selection it is asymptotically equivalent to Bayesian Information Criterion (BIC) for an appropriate $K$ when number of observations is large |
| Bootstrap | Generate many bootstrap sample sets of size $M$ by sampling all available data with *replacement*. Bootstrap estimates can have lower variance than the leave-one-out method. Useful in small sample situations |

**Example 14.13** *Multi-Class Classification Using Multiple Pairwise NBC.* We consider the classification of an object $O$ into one of $K$ classes $c_k, k \in \{1, 2, \ldots, K\}$. For each pair of classes, $c_k, c_l, l \neq k$, we obtain an optimal pairwise classification, or vote $V_{kl}$, using a pairwise NBC, where

$$V_{kl} = \begin{cases} 1 & \text{if } p(C = c_k|\mathbf{y}, I) > p(C = c_l|\mathbf{y}, I) \ , \\ 0 & \text{otherwise} \ , \end{cases}$$

$$V_{lk} = 1 - V_{kl} \ ,$$

and the optimal multi-class classification is

$$c_{\text{OPT}} = \arg\max_k \sum_{\substack{l=1 \\ l \neq k}}^{K} V_{kl} \ ,$$

## 14.7 Non-Bayesian Classifiers

Thus far, we have discussed the Bayesian classifier and some of its modifications. Apart from this classifier there are, however, a great number of other classifiers. A brief list of some of these classifiers follows:

**Decision tree.** A decision tree is a classification algorithm that splits the train data $\mathcal{D}$ into increasingly specific categories. The algorithm works in a supervised mode and requires a high-quality data set $\mathcal{D}$.

**Deep Learning.** See Chap. 5.

**Support Vector Machine (SVM).** SVM is a simple algorithm that can used for both classification and regression. It works by finding hyperplanes which optimally divides the data $\mathcal{D}$ into two different classes.

***K*-nearest neighbor.** $K$-nearest neighbor (KNN) is a supervised lazy learner algorithm. It stores the train data $\mathcal{D}$ and compares it to the test data by measuring the degree of similarity between them. Although better performing classifiers are available, $K$-NN classifier is still widely used. Reasons for its popularity include its simplicity and efficiency. Furthermore, in recent years, several modified $K$-NN classifiers with improved classification performance have been developed. See Table 14.7.

## 14.8 Software

**MATLAB.** MATLAB STATISTICS AND MACHINE LEARNING toolbox includes m-files on dimension reduction, feature selection and naive Bayes' classifier.

**PYTHON.** SCIKIT-LEARN includes python package for dimension reduction, feature selection and naive Bayes' classifer.

**Table 14.7** Modified $K$-NN classifiers

| Name | Description |
|---|---|
| Original $K$-NN | Train data $\mathcal{D}$ contains $N$ samples $\mathbf{y}_i$, $i \in \{1, 2, \ldots, N\}$ each of dimension $L$. The samples $\mathbf{y}_i$ in $\mathcal{D}$ belong to $M$ different classes $C = [c_1, c_2, \ldots, c_M]$. For each test sample $\mathbf{x}$ find $K$ closest train samples. If $S_m$ is the number of closest samples which belong to class $m$, then assign $\mathbf{x}$ to class $c^* \equiv c_{m^*}$, where $m^* = \arg\max_m \big(S_m\big)$ |
| Collaborative representation NN | Represent test sample $\mathbf{x}$ as a linear combination of all training samples $\mathbf{y}_i$, $i \in \{1, 2, \ldots, N\}$ with coefficients $\alpha_i$. Optimum coefficients are: $\boldsymbol{\alpha}^* = \arg\min_{\boldsymbol{\alpha}} \big(\lvert\mathbf{x} - \mathbf{Y}\boldsymbol{\alpha}\rvert^2 + \gamma \lvert\boldsymbol{\Gamma}\boldsymbol{\alpha}\rvert^2$, where $\gamma$ is a regularization parameter and $\boldsymbol{\Gamma}$ is a diagonal $N \times N$ matrix whose diagonal elements are: $\Gamma_{ii} = \lvert\lvert\mathbf{x}_i - \mathbf{y}_i\rvert\rvert$. In closed-form: $\boldsymbol{\alpha}^* = \big(\mathbf{Y}^T\mathbf{Y} + \gamma^2\boldsymbol{\Gamma}^T\boldsymbol{\Gamma}\big)^{-1}\mathbf{Y}^T\mathbf{x}$ |
| Pseudo Nearest Neighbor | For each test sample $\mathbf{x}$ find $K$ nearest neighbors *per class*. For each class $c_m$ calculate a cluster center from the $K$ nearest neighbors in the class. Assign $\mathbf{x}$ to the class whose cluster center is closest to $\mathbf{x}$ |

## 14.9 Further Reading

Bayesian decision theory and classification are discussed in detail in [10, 12]. Specific references on Bayesian classification with Gaussian processes are given in [26]. Modern reviews of the NBC and its variants and the assumptions which underlie it are discussed in [7, 11, 28, 35]. The NBC has been used in many applications, including text classification, fraud diagnosis [33], fake review detection [17] and other applications [31].

## Problems

**14.1** What is the curse of dimensionality?

**14.2** Define the naive Bayes' classifier. Explain why it is called naive.

**14.3** In the naive Bayes' the likelihood function is written as a product of individual features. Justify this approximation and suggest alternative approximation which could bee used.

**14.4** In the context of NBC compare and contrast feature selection, feature extraction and feature joining.

**14.5** Explain (a) mRMR (b) wrapper feature selection algorithm.

**14.6** Compare and contrast tree augmented naive Bayes (TAN), adjusted probability model (APM) and homologous naive Bayes' classifiers.

## References


1. Andres-Ferrer J, Juan A (2010) Constrained domain maximum likelihood estimation for naive Bayes text classificaion. Patt Anal Appl 13:189–196
2. Bermejo P, Gamez JA, Puerta JM (2014) Speeding up incremental wrapper feature subset selection with naive bayes classifier. Knowl-Based Syst 55:140–147
3. Chen F-C, Jahanshahi MR, Wu R-T (2017) A texture-based video processsing methodology using Bayesian data fusion for autonomous crack detection on metallic surfaces. Comput-Aided Civil Infrastruct Eng 32:271–287
4. Chen F-C, Jahanshahi MR (2018) NB-CNN: deep learning-based crack detection using a convolutional neural network and naive Bayes' data fusion. IEEE Trans Indust Electron 65:4392–4400
5. Domingos P, Pazzani M (1997) On the optimality of the simple Bayesian classifier under zero-one loss. Mach Learn 29:103–130
6. Elkan C (1997) Boosting and naive Bayes' learning. Tech Rept CS97-557, Sept 1997, University of California, San Diego
7. Flores MJ, Gamez JA, Martinez AM (2014) Domains of competence of the semi-naive bayesian network classifiers. Inf Sci 260:120–148

8. Friedman JH (1997) On bias, variance 0/1-loss and the curse of dimensionality. Data Mining Knowl Dis 1:55–77
9. Friedman N, Geiger D, Goldszmidt M (1997) Bayesian network classifiers. Mach Learn 29:131–163
10. Fukunaga K (1990) Introduction to statistical pattern recognition, 2nd ed. Published by Academic Press
11. Hand DJ, Yu K (2001) Idiot's Bayes: not so stupid after all? Int Stat Rev 69:385–398
12. Hastie T, Tibshirani R, Friedman J (2001) The elements of statistical learning. Published by Springer
13. Hong SJ, Hosking J, Natarajan R (2002) Ensemble modeling through multiplicative adjustment of class probability. Proc IEEE Int Conf Data Mining 621–624
14. Huang H-J, Hsu C-N (2003) Bayesian classification for data from the same unknown class. IEEE Trans Sys Man Cybern-Part B: Cybern 32:137–145
15. Inza I, Larranaga P, Blanco R, Cerrolaza AJ (2004) Filter versus wrapper gene selection approaches in DNA microarray domains. Art Intell Med 31:91–103
16. Jain AK, Duin RPW, Mao J (2000) Statistical pattern recognition: a review. IEEE Trans Patt Anal Mach Intell 22:4–37
17. Kalaivani P, Raj VD, Madhavan R et al (2023) Fake review detection using naive Bayesian classifier. Int Conf Sustainable Comput Smart Syst (ICSCSS) pp 705–709
18. Keogh EJ, Pazzani MJ (2002) Learning the structure of augmented Bayesian classifiers. Int J Art Intell Tools 11:587–601
19. Kohavi R, John G (1997) Wrappers for feature subset selection. Art Intell 97:273–324
20. Kittler J, Hatef M, Duin RPW, Matas J (1998) On combining classifiers. IEEE Trans Patt Anal Mach Intell 20:226–239
21. Kwak N, Choi C-H (2002) Input feature selection using mutual information based on Parzen window. IEEE Patt Anal Mach Intell 24:1667–1671
22. Liu B, Yang Y, Webb GI, Boughton J (2009) LNCS 5476:302–313
23. Loughrey J, Cunningham P (2004) Overfitting in wrapper-based feature subset selection: the harder you try the worse it gets. In: 24th SGAI international conference on innovative techniques and applications of artificial intelligence, pp 33–43
24. Loughrey J, Cunningham P (2005) Using early-stopping to avoid overfitting in wrapper-based feature selection employing stochastic search. Technical report TCD-CS-2005-37, Department of Computer Science, Trinity College, Dublin, Ireland
25. Qiu C, Jiang L, Li C (2015) Not always simple classification: learning preparation for class probability estimation. Expert Syst Appl 13:5440–5483
26. Rasmussen CE, Williams CKI (2006) Gaussian Processes for machine learning. Published by MIT Press, MA
27. Raudys SJ, Jain AK (1991) Small sample size effects in statistical pattern recognition: recommendations for practitioners. IEEE Trans Patt Anal Mach Intell 13:252–264
28. Ren H, Guo Q (2023) Flexible learning tree augmented naive classifier and its application. Knowl-Based Syst 26:110140
29. Ridgeway G, Madigan D, Richardson T, O'Kane JW (1998) Interpretable boosted naive Bayes classification. In: Proceedings of the 4th international conference on knowledge discovery data mining, New York, pp 101–104
30. Silverman B (1986) Density estimation for statistical data analysis. Published by Chapman and Hall
31. Thomas CS (1999) Classifying acute abdominal pain by assuming independence: a study using two models constructed directly from data. Technical report CSM-153, Department of Computing Science and Mathematics, University of Stirling, Stirling, Scotland
32. Trunk GV (1979) A problem of dimensionality: a simple example. IEEE Trans Patt Anal Mach Intell 1:306–307
33. Viaene S, Derrig R, Dedene G (2004) A case study of applying boosting naive Bayes to claim fraud diagnosis. IEEE Trans Know Data Eng 16:612–619

34. Vilalta R, Rish I (2003) A decomposition of classes via clustering to explain and improve naive Bayes. In: Proceedings of the 14th European conference on machine learning ECML/PKDD
35. Webb GI, Boughton J, Wang Z (2005) Not so naive Bayes: aggregating one-dependence estimators. Mach Learn 58:5–24
36. Xie Z, Hsu W, Liu Z, Lee ML (2002) SNNB: a selective neighbourhood-based naive Bayes for lazy classification. LNCS, vol 2336. Published by Springer, pp 104–114
37. Zheng Z (1998) Naive Bayesian classifier committee. LNCS, vol 1398. Published by Springer, pp 196–207
38. Zheng Z, Webb GI (2000) Lazy learning of Bayesian rules. Mach Learn 41:53–87

# Chapter 15
# Ensemble Learning

**Abstract** The subject of this chapter is Ensemble Learning in which the system is characterized by an ensemble of $M$ models. We shall concentrate on supervised classification using a Multiple Classifier System (MCS). We start by introducing several techniques for creating a diverse set of classifiers, including bagging, boosting, subspace techniques and switching class labels. We then discuss different methods for combining the multiple classifiers including BKS (Behavior-Knowledge Systems), Pairwise Fusion Matrix, Borda Count, Boosting, Stacking, Mixture of Experts and Ensembles of Convolutional Neural Networks.

## 15.1 Introduction

The subject of this chapter is ensemble learning in which our system is characterized by an ensemble of $M$ models. The models may share the same common representational format or each model may have its own distinct common representational format. To make our discussion more concrete we shall concentrate on the (supervised) classification of an object $O$ using a *multiple classifier system* (MCS). Given an unknown object $O$, our goal is to optimally assign it to one of $K$ classes, $c_k, k \in \{1, 2, \ldots, K\}$, using an ensemble of $M$ (supervised) classifiers, $S_m, m \in \{1, 2, \ldots, M\}$. The theory of multiple classifier systems suggests that if the *pattern of errors* made by one classifier, $S_m$, is different from the pattern of errors made by another classifier, $S_n$, then we may exploit this difference to give a more accurate and more reliable classification of $O$. If the error rates of the classifiers are less than $\frac{1}{2}$, then the MCS error rate, $E_{\text{MCS}}$, should decrease with the number of classifiers, $M$, and with the mean diversity, $\overline{\sigma}$, between the classifiers.[1] Mathematically, if $\overline{E}$ is the mean error rate of the $S_m$, then

$$E_{\text{MCS}} \sim \alpha \overline{E} \ , \tag{15.1}$$

where $\alpha \sim 1/(\overline{\sigma} M)$ and $0 \leq \alpha \leq 1$.

[1] The concept of diversity is discussed in Sect. 15.5. For the present we may regard $\overline{\sigma}$ as being equal to the inverse of the mean correlation between the $S_m$.



H. B. Mitchell, *Data Fusion*, https://doi.org/10.1007/978-3-662-71023-4_15

**Table 15.1** Applications in which ensemble learning is primary fusion algorithm

| Class | | Application |
|---|---|---|
| Inter-FeO | Example 8.9 | Image segmentation via the method of cluster ensembles |
| | Example 8.10 | Color image segmentation |
| Inter-DeO | Example 15.3 | Multiple color spaces in DenseNet image recognition |
| Late-DeO | Example 12.14 | Target tracking initialization using the Hough transform |
| | Example 15.7 | Weighted majority vote |
| | Example 15.8 | Image thresholding using an unsupervised weighted majority vote rule |
| | Example 3.8 | Multi-modal biometric identification scheme |
| | Example 3.15 | Audio-visual speech recognition system |
| | Example 11.6 | Non-intrusive speech quality estimation using GMM's |
| | Example 14.12 | Naive Bayes' classifier committee |
| | Example 15.1 | Bayesian model averaging of the naive Bayes' classifier (NBC) |

The theory of ensemble learning deals with exploiting these differences and this will be the main emphasis in this chapter. In Table 15.1 we list some of these applications together with their classification.

## 15.2 Bayesian Framework

In this section we present a general Bayesian framework [12, 27, 29] for combining an ensemble of classifiers $S_m, m \in \{1, 2, \ldots, M\}$. Given an unknown object $O$, we suppose each classifier $S_m$ makes a measurement $\mathbf{y}_m$ on $O$ and returns an estimate of the a posteriori class probability $p(C = c_k|\mathbf{y}_m, I)$:

$$\hat{p}(C = c_k|\mathbf{y}_m, I) = p(C = c_k|\mathbf{y}_m, I) + \epsilon_m ,$$

where $\epsilon_m$ is the error made by $S_m$ in estimating $p(C = c_k|\mathbf{y}_m, I)$. We may improve the classification accuracy of $O$ by estimating the joint a posteriori probability, $p(C = c_k|\mathbf{y}_1, \mathbf{y}_2, \ldots, \mathbf{y}_M, I)$. This we do by combining the approximate $\hat{p}(C = c_k|\mathbf{y}_m, I)$ values.

We distinguish two extreme cases [29]:

**Shared Representation.** In this case, the $S_m$ share the same common representational format. It follows that, if we *ignore the effects of noise*, then the classifiers use essentially the same input data $\mathbf{y}_1 = \mathbf{y}_2 = \ldots = \mathbf{y}_M$. As a result,

$$p(C = c_k|\mathbf{y}_1, I) \approx p(C = c_k|\mathbf{y}_2, I) \approx \ldots \approx p(C = c_k|\mathbf{y}_M, I) ,$$

where $p(C = c_k|\mathbf{y}_m, I)$ is estimated by $\hat{p}(C = c_k|\mathbf{y}_m, I)$. If we assume zero-mean errors $\epsilon_m$ for the noise, we may average the $\hat{p}(C = c_k|\mathbf{y}_m, I)$ to obtain a less error-sensitive estimate. This leads to the mean, or average, rule

$$p(C = c_k|\mathbf{y}_1, \mathbf{y}_2, \ldots, \mathbf{y}_M, I) \approx \frac{1}{M}\sum_m \hat{p}(C = c_k|\mathbf{y}_m, I) . \quad (15.2)$$

We may give each classifier $S_m$ a different weight $w_m$, where $0 \leq w_m \leq 1$ and $\sum w_m = 1$. In this case, the weighted mean, or average, rule is

$$p(C = c_k|\mathbf{y}_1, \mathbf{y}_2, \ldots, \mathbf{y}_M, I) \approx \sum_m w_m \hat{p}(C = c_k|\mathbf{y}_m, I) . \quad (15.3)$$

**Different Representations.** In this case, the $S_m$ use different common representational formats, which we assume are independent (see Sect. 14.3). In this case

$$p(\mathbf{y}_1, \mathbf{y}_2, \ldots, \mathbf{y}_M|C = c_k, I) = \prod_m \hat{p}(\mathbf{y}_m|C = c_k, I) .$$

Assuming equal *a priori* probabilities, $\pi(C = c_k|I)$, and small errors, $\epsilon_m$, this reduces [12, 14] to the product rule:

$$p(C = c_k|\mathbf{y}_1, \mathbf{y}_2, \ldots, \mathbf{y}_M, I) \approx \frac{\prod_{m=1}^{M} \hat{p}(C = c_k|\mathbf{y}_m, I)}{\sum_{l=1}^{K}\prod_{m=1}^{M} \hat{p}(C = c_l|\mathbf{y}_m, I)} . \quad (15.4)$$

In many applications we do not use (15.4) but use one of the following approximations:

$$p(C = c_k|\mathbf{y}_1, \mathbf{y}_2, \ldots, \mathbf{y}_M, I) \sim \begin{cases} \min_m \hat{p}(C = c_k|y_m, I) , \\ \frac{1}{M}\sum \hat{p}(C = c_k|y_m, I) , \\ \mathrm{median}_m \hat{p}(C = c_k|y_m, I) , \\ \max_m \hat{p}(C = c_k|y_m, I) . \end{cases} \quad (15.5)$$

Although these approximations are not very accurate they are often preferred because they are less sensitive to noise and outliers [12, 29]. We may also give each classifier, $S_m$, a different weight $w_m$, where $0 \leq w_m \leq 1$. In this case the weighted product rule is

$$p(C = c_k|\mathbf{y}_1, \mathbf{y}_2, \ldots, \mathbf{y}_M, I) \approx \frac{\prod_{m=1}^{M} \hat{p}(C = c_k|\mathbf{y}_m, I)^{w_m}}{\sum_{l=1}^{K}\prod_{m=1}^{M} \hat{p}(C = c_l|\mathbf{y}_m, I)^{w_m}} . \quad (15.6)$$

In the following example we illustrate the weighted mean rule (15.3), where the weights $w_m$ are calculated using Bayesian model averaging (see Sect. 10.6.2).

**Example 15.1** *Bayesian Model Averaging (BMA) of the NBC* [3, 7]. We consider Bayesian model averaging (BMA) of the Naive Bayes' classifier (NBC) (see Sect. 14.3). Given an unknown object $O$ with a $L$-dimensional measurement, or feature, vector $\mathbf{y} = (y^{(1)}, y^{(2)}, \ldots, y^{(L)})^T$, we generate $M$ NBC's $S_m, m \in \{1, 2, \ldots, M\}$, by using $M$ different subsets of features $y^{(l)}$. Let

$$a_{ml} = \begin{cases} 1 \text{ if } S_m \text{ includes the } l\text{th feature } y^{(l)} \;, \\ 0 \text{ otherwise} \;, \end{cases}$$

then, by definition, the a posteriori classification probability, as estimated by $S_m$, is

$$\hat{p}_m(C = c_k|\mathbf{y}, I) = \frac{\pi(C = c_k|I) \prod_{l=1}^{L} (\hat{p}(y^{(l)}|C = c_k, I))^{a_{ml}}}{\hat{p}(\mathbf{y}|I)} \;.$$

For each $S_m$ we assume an *a priori* probability, $\pi(S_m|I)$, which describes our prior belief in $S_m$. Given a training dataset $\mathcal{D}$, we use Bayes' rule to calculate the a posteriori probability

$$\hat{p}(S_m|\mathcal{D}, I) = \pi(S_m|I)\hat{p}(\mathcal{D}|S_m, I)/\hat{p}(\mathcal{D}|I) \;.$$

In BMA we use $\hat{p}(S_m|\mathcal{D}, I)$ to weight the predictions $\hat{p}_m(C = c_k|\mathbf{y}, I)$. The predicted BMA a posteriori probability that the object $O$ belongs to the class $C = c_k$ is

$$\hat{p}_{\text{BMA}}(C = c_k|\mathbf{y}, I) \propto \sum_m p(S_m|\mathcal{D}, I)\hat{p}_m(C = c_k|\mathbf{y}, \mathcal{D}, I) \;.$$

The object $O$ is assigned to the class $C = c_{\text{OPT}}$, where

$$c_{\text{OPT}} = \arg\max_k \hat{p}_{\text{BMA}}(C = c_k|\mathbf{y}, I) \;.$$

Although useful, the Bayesian framework is restricted to classifiers whose output is an a posteriori probability $\hat{p}_m(C = c_k|\mathbf{y}, I)$.[2] Unfortunately, most classifiers do not directly measure the a posteriori probability $p_m(C = c_k|\mathbf{y}, I)$. Instead they rely on a belief $\mu_m^{(k)}$, that $C = c_k$ is the true assignment. Although (theoretically) it may be possible to convert $\mu_m^{(k)}$ to an a posteriori probability, it is rarely performed in practice (see Sect. 9.8). As a consequence, recent research on the MCS has relied less on a formal Bayesian framework, and more on an empirically-based framework.

[2] The above Bayesian framework is subject to additional restrictions [7]. The most important restriction is that the $S_m$ are required to be mutually exclusive and exhaustive, i.e. the $S_m$ must cover all possibilities concerning how the input data was generated.

## 15.3 Empirical Framework

The empirical framework is based on a wealth of experimental studies which all suggest that it is possible to reduce the classification error in a MCS by a factor which increases with the mean diversity $\overline{\sigma}$. In order to realize this reduction in classification error the empirical framework introduces the following four innovations in the Bayesian framework:

**Diversity Techniques**. Empirical techniques are used to create a diverse ensemble of classifiers.

**Diversity Measures**. Probabilistic and non-probabilistic formulas are used to measure the ensemble diversity.

**Classifier Types**. Classifiers are not restricted to those which generate a probabilistic output.

**Combination Strategies**. Probabilistic and non-probabilistic techniques are used to combine the outputs of the individual classifiers.

We now consider each of these innovations in turn.

## 15.4 Diversity Techniques

At the heart of a MCS is an ensemble of diverse classifiers, each classifier having a "reasonable" performance. One way of creating such an ensemble is to devise a set of $M$ classifiers $S_m, m \in \{1, 2, \ldots, M\}$, which rely on different physical principles. The following example illustrates an important application of this technique.

**Example 15.2** *Thresholding Techniques for the Measurement of Breadcrumb Features* [8]. In bread baking the principal factors which determine the quality of the bread are: water, enzyme activity, starch, protein properties and heat. Recently bread quality has been measured by analyzing the corresponding bread crumb images. For this purpose, the breadcrumb images were thresholded using a global thresholding algorithm (see Sect. 9.6). In [8] seven physically different thresholding algorithms were considered.

The more common way of creating such an ensemble is to train a parametric classifier $S$ on multiple training sets $\mathcal{D}_m, m \in \{1, 2, \ldots, M\}$. If the training sets are diverse, we shall in turn obtain a diverse set of *fixed* classifiers $S_m$ which do not have any adjustable parameters. Some common methods for constructing the training sets $\mathcal{D}_m, m \in \{1, 2, \ldots, M\}$, are

**Color Transformations**. Given a training set $\mathcal{D}_{\text{RGB}}$ of input images in RGB space, we transform the images in $\mathcal{D}_{\text{RGB}}$ to $M$ different color spaces. Each color space defines a training set $\mathcal{D}_m, m \in \{1, 2, \ldots, M\}$.

**Disjoint Partitions.** Sample a training set $\mathcal{D}$ without replacement. If $\mathcal{D}$ contains $N$ labeled objects, then on average each $\mathcal{D}_m$ contains $N/M$ labeled objects. The $\mathcal{D}_m$ share a common representational format. In general, the $S_m$ which are produced will be diverse but will have high error rates (due to the small size of the $\mathcal{D}_m$).

**Different $\mathcal{D}_m$.** Construct the $\mathcal{D}_m$, $m \in \{1, 2, \ldots, M\}$, from different input sources. In this case, the $\mathcal{D}_m$ may, or may not, share a common representational format. In general, the $S_m$ will be diverse.

**Sampling $\mathcal{D}$.** Sample $\mathcal{D}$ with replacement. Two important sampling techniques are "bagging" (also called "bootstrapping") and "boosting". The $\mathcal{D}_m$ share a common representational format and, in general, the $S_m$ will be moderately diverse.

**Subspace Techniques.** Sample $\mathcal{D}$ after feature extraction or after applying a subspace transformation. In many cases, the $\mathcal{D}_m$ will have different common representational formats. In general, the $S_m$ will be diverse.

**Switching Class Labels.** The training sets $\mathcal{D}_m$ are generated using the original training set $\mathcal{D}$ with randomized class labels [18]. The class labels of each sample is switched according to a probability that depends on an overall switching rate (defined as the proportion of training samples that are switched on average) and on the proportions of the different class labels in $\mathcal{D}$.

## 15.4.1 Color Transformations

Given the original training set $\mathcal{D}_{\text{RGB}}$ of $N$ color images in RGB space, we create additional training sets $\mathcal{D}_{\text{HVS}}$, $\mathcal{D}_{\text{Lab}}$, $\mathcal{D}_{\text{YCbCr}}$ etc. by transforming the input images in $\mathcal{D}_{\text{RGB}}$ to the color spaces HVS, Lab, YCbCr, etc.

### Color Spaces in Deep Learning

State-of-the-art image classification have been obtained using very deep CNN's (see Chap. 5). An alternative approach to adding more layers to the network is to use an *ensemble* of networks. The following example illustrates state-of-the-art image classification using an ensemble of CNN's each of which operates in a different color space.

**Example 15.3** *Multiple Color Spaces in DenseNet Image Recognition* [9]. Figure 15.1 shows an ensemble of $K = 7$ denseNet networks $\text{DNet}_k$, $k \in \{1, 2, \ldots, K\}$. Each network operates in a different color space $C_k$. The color spaces used in [9] are: {RGB, Lab, HSV, YUV, YCbCr, HED, YIQ}. Each network $\text{DNet}_k$ was trained on the same set of input images after transformation to the color space $C_k$. Similarly, each network $\text{DNet}_k$ receives the test image $I$ after transformation to color space $C_k$. The output scores of each network are then passed to a fully-connected layer which outputs the final image classification.

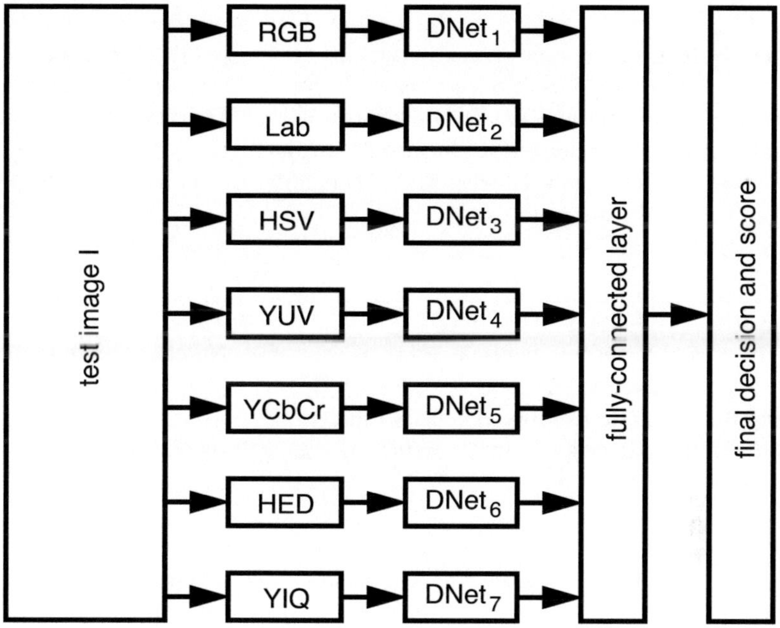


**Fig. 15.1** Shows ensemble of denseNet networks $\text{DNet}_k$, $k \in \{1, 2, \ldots, K\}$, where each network has been trained and operates in a different color space. The test image $I$ is separately transformed into $K = 7$ different color spaces $C_k$ and is then separately processed by $\text{DNet}_k$. The output of each network is passed to a fully-connected layer which outputs the final classification

## 15.4.2 Bagging

Given the original training set $\mathcal{D}$, create $M$ "bootstrapped" training sets $\mathcal{D}_m$, $m \in \{1, 2, \ldots, M\}$, by randomly sampling $\mathcal{D}$ with replacement. On average, each $\mathcal{D}_m$, contains two-thirds of the original samples.

Bagging is widely used in supervised and unsupervised learning algorithms. The following example illustrates the use of bagging in semantic alignment (see also Example 8.13).

**Example 15.4** *Bagging in a DNA microarray cluster algorithm* [4]. An important topic in tumor classification is the identification of new tumor classes using gene expression profiles which are obtained from DNA microarray experiments. Part of the identification process is the accurate partition of the tumor samples into clusters, each cluster corresponding to a different class of tumor. Given the number of clusters $K$ it is important to estimate for each measurement its cluster label and a measure of confidence of the cluster assignment.

In [4] the DNA microarray measurements are bagged (bootstrapped). The cluster algorithm is then applied to each set of bootstrap measurements. For each microarray measurement the final partition is obtained by taking a majority cluster label. The confidence measure of the label $d_n$ is given by the size of the corresponding majority vote. Mathematically, the steps in the procedure are shown in Algorithm 11.

**Algorithm 11:** Bagging in a DNA Microarray Cluster Algorithm

**Input** : DNA measurements $X = \{x_1, x_2, \ldots, x_N\}$. Number of required bootstrapped datasets $X^{(b)}$, $b \in \{1, 2, \ldots, B\}$

**Output** : Final (Bootstrapped) Cluster Label

Form reference decision map $D^* = \{d_1^*, d_2^*, \ldots, d_N^*\}$ by clustering $X$

$b \leftarrow 0$

**while** $b < B$ **do**

 $b \leftarrow b + 1$

 Form bootstrapped dataset $X^{(b)}$ by bootstrapping $X$, where $X^{(b)} = \{x_1^{(b)}, x_2^{(b)}, \ldots, x_N^{(b)}\}$

 Cluster the bootstrapped measurements $X^{(b)}$, where $D^{(b)} = \{d_1^{(b)}, \ldots d_N^{(b)}\}$, denotes corresponding cluster, or decision, map.

 Perform semantic alignment on $D^{(b)}$ by permuting cluster labels assigned to $X^{(b)}$, so there is maximum overlap with reference map $D^*$

**end**

Use a majority vote algorithm (see Eq. 15.10) to assign optimal cluster label to each measurement $x_n$, $n \in \{1, 2, \ldots, N\}$. Relative size of majority vote is a measure of the confidence for each cluster label.

## 15.4.3 Boosting

Boosting is a complete classification system and is discussed in Sect. 15.9. However, for the present we may regard boosting as simply a method of creating an ensemble of training sets $\mathcal{D}_m$ by sampling $\mathcal{D}$ [25]. In boosting we create the $\mathcal{D}_m$, $m \in \{1, 2, \ldots, M\}$, one at a time by sampling $\mathcal{D}$ as follows. Let $S_m$ be a classifier which was learnt on $\mathcal{D}_m$. Then we use the classification results obtained with $S_m$ to create $\mathcal{D}_{m+1}$ by sampling $\mathcal{D}$ such that samples which are misclassified by $S_m$ have a higher chance of being chosen than samples which were correctly classified by $S_m$.

Multi-boosting is a variant of boosting [31] in which we combine boosting and bagging.

**Example 15.5** *Multi-Boosting a Common Training Set $\mathcal{D}$* [31]. We consider multi-boosting a common training set, $\mathcal{D}$. We first boost $\mathcal{D}$ generating $M$ training sets $\mathcal{D}_m$ as explained above. We then perform bagging on each $\mathcal{D}_m$.

### *15.4.4 Switching Class Labels*

In the method of switching class labels [18] we generate the training sets $\mathcal{D}_m$ using the original training set $\mathcal{D}$ with randomized class labels. The class labels of each sample is switched according to the transition matrix

$$\begin{aligned} P_{i \to j} &= \omega P_j \ , \\ P_{i \to i} &= 1 - \omega(1 - P_i) \ , \end{aligned}$$

where $P_{i \to j}$ is the probability that an example which belongs to the class $C = c_i$ is placed in the class $C = c_j$, $P_i$ is the proportion of elements of class $C = c_i$ in the training set and $\omega$ is proportional to the switching rate (average fraction of switched examples), $p$

$$\omega = \frac{p}{1 - \sum_j P_j^2} \ .$$

This form of the transition matrix is ensures that the class proportions remain approximately constant. In order for the method to work, the value of the switching rate $p$ should be small enough to ensure that the training error tends to zero as the size of the ensemble grows. In binary classification, this requires $p < P_{\min}$, where $P_{\min}$ is the proportion of examples that belong to the minority class.

## 15.5 Diversity Measures

In principle, the techniques discussed in Sect 15.4 may generate an unlimited number of classifier ensembles. In order to select the most appropriate ensemble we need to compare the diversity of one ensemble with the diversity of another ensemble. In Table 15.2 we list several quantitative diversity measures [14] which are commonly used for this purpose.[3] The simplest diversity measures are pairwise measures, which

[3] In Sect. 15.1 we defined the diversity of an ensemble as the degree to which the pattern of errors made by one classifier, $S_m$, differs from the pattern of errors made by another classifier, $S_n$, $n \neq m$. This definition is not, however, very useful in comparing the diversity of one ensemble with another ensemble. For this purpose, we use a quantitative diversity measure.

Table 15.2 Diversity measures

| Name | Formula |
|---|---|
| Yule statistic $Q$ | $Q = (ad - bc)/(ad + bc)$, where $a, b, c, d$ are, respectively, # objects $O_i$, that are: correctly classified by $S_1 \& S_2$; correctly classified by $S_1$, incorrectly classified by $S_2$; incorrectly classified by $S_1$, correctly classified by $S_2$; incorrectly classified by $S_1 \& S_2$ |
| Correlation Coeff. | $\rho = (ad - bc)/\sqrt{(a+b)(c+d)(a+c)(b+d)}$ |
| Disagreement $D$ | $D = (b + c)/N$ |
| Double Fault $DF$ | $DF = d/N$ |
| Entropy $E$ | $E = \sum_{i=1}^{N} \min(\xi_i, (M - \xi_i))/(N(M - \lceil M/2 \rceil))$; $\lceil x \rceil$ is ceiling operator, $\xi_i$ is # classifiers that misclassify $O_i$ |
| Kohavi-Wolpart | $KW = \sum_{i=1}^{N} \xi_i (M - \xi_i)/(NM^2)$ |
| Difficulty $\theta$ | $\theta = \sum_{i=1}^{N} (\xi_i - \bar{\xi})^2/(NM^2)$, where $\bar{\xi} = \sum_{i=1}^{N} \xi_i / N$ |

Measures $Q$, $\rho$, $D$ and $DF$ are pair measures while $KW$ and $\theta$ are non-pair measures

are defined between a pair of classifiers. If the ensemble contains $M > 2$ classifiers, we may estimate its diversity by averaging all of the $M(M-1)/2$ pairwise diversity measures. Apart from the pairwise measures, there are also non-pairwise measures which work on an entire ensemble.

Given an accurate diversity measure we may select the optimal ensemble by exhaustive enumeration, that is, by comparing the diversity measure of all ensembles and choosing the ensemble with the largest diversity measure. However, this technique is only feasible if the number of ensembles is relatively small. When the number of ensembles is large, we may search for the optimal ensemble using any of the search algorithms listed in Table 14.3.

Alternatively, and in practice this is the preferred method, we may select the optimal ensemble by directly assessing the classification accuracy of each ensemble (as measured on a given dataset using any of the methods listed in Table 14.6).

## 15.6 Classifier Types

Let $O$ denote an unknown object. In the empirical framework we assume that each classifier $S_m$, $m \in \{1, 2, \ldots, M\}$, generates a *belief* vector,

$$\boldsymbol{\mu}_m = \left(\mu_m(1), \mu_m(2), \ldots, \mu_m(K)\right)^T , \tag{15.7}$$

where $\boldsymbol{\mu}_m(k)$ is the belief that $C = c_k$ is the true assignment. The concept of a belief vector is more general than a probability vector. In particular, belief vectors need not sum to one. Instead we only require that $\mu_m(h) > \mu_m(k)$ if $\hat{p}(C = c_h|\mathbf{y}, I) > \hat{p}(C = c_k|\mathbf{y}, I)$, where $\hat{p}(C = c_k|\mathbf{y}, I)$ is the a posteriori probability that $C = c_k$ is the true assignment given the input measurement $\mathbf{y}$ and the background information $I$.

We may identify several types of "belief" vectors:

**Probability Vector $\mathbf{p}_m$.** The belief vector is equal to the probability vector $\mathbf{p}_m$, where $p_m(k) = \hat{p}(C = c_k|\mathbf{y}, I)$ is the a posteriori probability that $C = c_k$ is the true assignment given the input measurement $\mathbf{y}$ and the background knowledge $I$.

**Measurement Vector $\mathbf{s}_m$.** The belief vector is equal to a measurement vector $\mathbf{s}_m$. We often generate the measurement vector as follows. Off-line we create a set of *prototype* measurements $\mathbf{Y}_1, \mathbf{Y}_2, \ldots, \mathbf{Y}_K$ which correspond to the classes $\{c_1, c_2, \ldots, c_K\}$. Then $s_m(k) = S_m(\mathbf{y}_m, \mathbf{Y})$ is similarity between $\mathbf{y}_m$ and $\mathbf{Y}_k$ as measured by degree-of-similarity operator $S_m$.

**Rank Vector $\mathbf{R}_m$.** The belief vector is equal to a vector of ranks $\mathbf{R}_m = \big(R_m(1), R_m(2), \ldots, R_m(K)\big)$, where $R_m(k) = (K - l + 1)$ if $\mu_m(k)$ is the $l$th largest belief value. E.g. $R_m(k) = 1$ if $\mu_m(k)$ is smallest belief value and $R_m(k) = K$ if $\mu_m(k)$ is largest belief value. See Sect. 9.7.

**Identity Vector $\mathbf{I}_m$.** The belief vector is equal to an identity vector $\mathbf{I}_m$, where $I_m(l) = 1$ if $\mu_m(l) = \max\big(\mu_m(1), \mu_m(2), \ldots, \mu_m(K)\big)$, otherwise $I_m(l) = 0$.

## 15.7 Combination Strategies

Let $O$ denote an unknown object. Suppose each classifier $S_m, m \in \{1, 2, \ldots, M\}$, generates a *belief* vector regarding the classification of $O$. Then the MCS generates a new belief vector, $\widetilde{\boldsymbol{\mu}}$, by combining the $\boldsymbol{\mu}_m$:

$$\widetilde{\boldsymbol{\mu}} = f(\boldsymbol{\mu}_1, \boldsymbol{\mu}_2, \ldots, \boldsymbol{\mu}_M)^T , \tag{15.8}$$

where $f$ denotes a MCS combination function. The corresponding optimum classification is:

$$c_{\mathrm{OPT}} = \arg\max_k \widetilde{\boldsymbol{\mu}} = \arg\max_k \big(\widetilde{\mu}(1), \widetilde{\mu}(2), \ldots, \widetilde{\mu}(K)\big) . \tag{15.9}$$

We may broadly divide the combination functions into several types:

**Identity Combiners.** Simple combination functions which are used with identity vectors $\mathbf{I}_m$. A typical identity function is *majority rule*.

**Rank Combiners.** Simple combination functions which are used with rank vectors $\mathbf{R}_m$. A widely used rank function is *Borda count*.

**Belief Combiners.** Simple combination functions which are used with belief vectors $\boldsymbol{\mu}_m$. A typical belief function is *mean vote*.

**Meta-Learners.** A learnable combination function. Meta-functions are often trained off-line on a training dataset. Meta-learners are vulnerable to over-fitting.

**AdaBoosting.** A learnable combination function which is often trained off-line on a training dataset using the boosting function.

In Table 15.3 we list the most common combination functions.

**Table 15.3** Combination strategies

| Name | Description |
|---|---|
| Max, Min, Median, Mean | Used when each classifier $S_m$ produces a belief vector $\boldsymbol{\mu}_m$. The max, min, median or mean vote for the $k$th class are, respectively, $\max_m(\mu_m(k))$, $\min_m(\mu_m(k))$, $\mathrm{median}_m(\mu_m(k))$ or $\frac{1}{M}\sum_m \mu_m(k)$. Class with highest max, min, median or mean belief wins |
| Borda Count (BC) | Used when each classifier $S_m$ produces a rank vector $\mathbf{R}_m$. The BC for the $k$th class is $B(k) = \sum_{m=1}^{M} R_m(k)$. Class with the highest BC wins |
| Majority Vote | Used when each $S_m$ produces an identity vector $\mathbf{I}_m$. See Eq. 15.10 |
| Weighted Combiner | Used when output of each classifier $S_m$ is a weight $w_m$, where $w_m \in [0, 1]$, $\sum w_m = 1$. Weight $w_m$ is often a measure of performance of $S_m$ as estimated on the validation dataset |
| Stacked Generalization | The output of the ensemble of $S_m$, $m \in \{1, 2, \ldots, M\}$, serves as a feature vector to a meta-classifier |
| Mixture of Experts (MoE) | Feature space is partitioned into different regions, with one expert ($S_m$) in the ensemble being responsible for generating the correct output within that region. Experts in the ensemble and the partition algorithm are trained simultaneously |
| Hierarchical MoE | Mixture of experts is extended to a multi-hierarchical structure, where each component is itself a mixture of experts |
| Boosting Algorithms | Multiple weak classifiers are adaptively trained and fused together |

### 15.7.1 Identity Combiners

The identity combiner only uses class information contained in the identity vectors $\mathbf{I}_m$, $m \in \{1, 2, \ldots, M\}$. Popular identity combiners are majority vote and weighted majority vote.

#### Majority Vote

In the majority vote we find the optimal class $c_{\mathrm{OPT}}$ by selecting the class which is chosen by at least half of the classifiers. Mathematically, the optimal class is

$$c_{\mathrm{OPT}} = k \quad \mathrm{if}\Big(\sum I_m(k)\Big) > M/2 \tag{15.10}$$

For $K > 2$ we often discard the restriction that the optimum class is chosen by at least half of the classifiers. In this case, the majority vote is identical to max vote:

$$c_{\mathrm{OPT}} = \arg\max_k \sum_m I_m(k) . \tag{15.11}$$

**Example 15.6** *Ensemble GMM for Robust Background Subtraction (BS)* [16]. BS is used in a variety of computer vision applications, including video surveillance and human activity recognition (see Examples 4.14 and 6.7). BS aims to segment an input video frame into foreground and background regions, where foreground represents moving objects and background represent a static scene. Several different methods have been used for BS including Gaussian mixture models (GMM) (see Sect. 10.5). The output of a BS algorithm is a binary image $B(i, j)$, where $B(i, j) = 1$ if pixel $(i, j)$ is classified as background pixel, otherwise $B(i, j) = 0$. In [16] several GMM-based BS algorithms were used. The corresponding binary images $B_m(i, j), m \in \{1, 2, \ldots, M\}$, were then fused together using majority vote:

$$\widetilde{B}(i, j) = \begin{cases} 1 & \text{if}\left(\sum_{m=1}^{M} B_m(i, j)\right) > M/2 \ , \\ 0 & \text{otherwise} \end{cases}$$

## Weighted Majority Vote

A variant of the majority voter is the *weighted majority voter*. This is a majority voter in which each vector $\mathbf{I}_m$ is given a weight, $w_m$, which is often learnt on a training dataset $\mathcal{D}$.

**Example 15.7** *Weighted Majority Vote* [14, 24]. Each classifier $S_m, m \in \{1, 2, \ldots, M\}$ is given a weight $w_m$ in proportion to its estimated performance. Then the optimal class is

$$c_{\text{OPT}} = \arg\max \frac{\sum_m w_m I_m(k)}{\sum_m w_m} \ .$$

If the $S_m$ are class independent and there are only $K = 2$ classes, then the optimal weights are

$$w_m \sim \begin{cases} \log \frac{p_m}{1-p_m} & \text{if} p_m > \frac{1}{2} \ , \\ 0 & \text{otherwise} \ . \end{cases}$$

where $p_m$ is the probability of correct classification if we were to use $S_m$ by itself. This weighting scheme is used in AdaBoost (Sect. 15.9).

In some applications the weights are *not* learnt on a training data set $\mathcal{D}$. The following example illustrates one such application.

**Example 15.8** *Image Thresholding Using an Unsupervised Weighted Majority Vote Rule* [20]. The effectiveness of a thresholding algorithm is strongly dependent on the input image characteristics. A given thresholding algorithm may appear to be the best for one image, while it may fail completely for another image. This makes it difficult to choose an appropriate algorithm to binarize a given image. One way to solve this problem is to binarize the input image using $M$ different thresholding algorithms and then fuse the thresholded images using an *unsupervised* weighted majority vote rule. Suppose $t_m$ is the optimal threshold found by the $m$th thresholding algorithm. If $g(i, j)$ denotes the gray-level of pixel $(i, j)$, then $I_m(i, j) = 1$ if the $g(i, j) \geq t_m$, otherwise $I_m(i, j) = 0$. For each pixel, we form the weighted sum

$$\phi(i, j) = \frac{\sum_m I_m(i, j) w_m(i, j)}{\sum_m w_m(i, j)} ,$$

where $w_m(i, j)$ denotes the weight associated with the $m$th thresholding algorithm when applied to pixel $(i, j)$. The final binary image $B(i, j)$ is

$$B(i, j) = \begin{cases} 1 \text{ if } \phi(i, j) \geq \frac{1}{2} , \\ 0 \text{ otherwise} . \end{cases}$$

The formula for the weights $w_m(i, j) = 1 - \exp(-\gamma |g(i, j) - t_m|)$ is based on the idea that the larger the difference $|g(i, j) - t_m|$, the higher the degree of confidence in the decision of the $m$th algorithm, where $\gamma$ is a real positive constant which controls the steepness of the weight function.

### Behavior-Knowledge Space

The behaviour-knowledge space (BKS) [11] is a supervised identity-level combiner which takes into account inter-classifier correlations. In the BKS combiner we record how often the classifiers $S_m, m \in \{1, 2, \ldots, M\}$, produce a given combination of class labels. If there are $K$ classes, then altogether there are $K^M$ different combinations of class labels. The true class for which a particular combination of class labels is observed most often, is chosen every time that a given combination of class labels occurs during testing. Let $n(k, l)$ denote the number of objects $O_i, i \in \{1, 2, \ldots, N\}$, in $\mathcal{D}$ which belong to class $C = c_k$ and which have the $l$th combination of class labels $\omega_l$. Suppose the test object is found to belong to a given combination $\omega_{l^*}$. Then we classify it as belong to the class

$$c_{\text{OPT}} = \arg\max_k \Big( n(k, l^*) \Big) .$$

BKS provides good performance if large and representative training datasets are available. Furthermore, BKS makes no assumptions regarding inter-classifier correlations. In particular, it does *not* assume conditional independence between the $S_m$. Nevertheless, BKS has significant drawbacks:

**Large Training Sets.** Large and representative training sets are required. Given $K$ classes and $M$ classifiers we have $K^M$ label combinations $\omega_l, l \in \{1, 2, \ldots, K^M\}$. If each combination requires a minimum number $n_{\min}$ of training samples, then the BKS algorithm requires at least $K^M \times n_{\min}$ training samples.

**Overfitting.** BKS suffers from the small sample size problem. If training set is not sufficiently large overfitting will occur leading to large errors.

**Ambiguous combinations.** BKS produces high errors for "ambiguous" combinations, i.e. label combinations $\omega_l$ in which the probability of the most representative class is not very high. This is an intrinsic limitation of the BKS method.

The following example illustrates the use of BKS for credit fraud detection. Here enormous training datasets are available and thus it is feasible to use BKS.

**Example 15.9** *Credit card fraud detection* [22]. Credit card fraud leads to enormous financial losses every year and both merchants and consumers are significantly affected by the consequences of fraud. Capturing irregular transaction patterns is a vital step in fraud detection. Multiple classifications algorithms $S_m, m \in \{1, 2, \ldots, M\}$, were designed to capture irregular patterns. The BKS is then leveraged to combine the predictions from the $S_m$.

### Pairwise Fusion Matrix

A modification of BKS which requires much less training data is pairwise fusion matrix (PMF) [13]. In PMF we divide the $M$ classifiers $S_m, m \in \{1, 2, \ldots, M\}$, into an ensemble $\mathbf{P} = (P_1, P_2, \ldots, P_N)^T$ of all possible pairs of classifiers, where $P_1 = (S_1, S_2),\ P_2 = (S_1, S_3), \ldots, P_N = (S_{M-1}, S_M)$ and $N = M(M-1)/2$. For each pair of classifiers we perform a BKS-type classification where each $P_i$ has only $K^2$ label combinations $\omega_l, l \in \{1, 2, \ldots, K^2\}$, associated with it. Finally the set of $N$ classifications are fused together using an appropriate combiner e.g. majority rule.

## 15.7.2 Rank Combiners

Rank combiners are used to combine multiple rank vectors $\mathbf{R}_m, m \in \{1, 2, \ldots, M\}$. These combiners are used when we do not believe the a posteriori probability distributions generated by a classifier but we do believe the relative ordering of the distributions. The Borda count is probably the most popular rank-level combiner. It is defined as the sum of ranks. Thus

$$\widetilde{R}(k) = \sum_m R_m(k) , \tag{15.12}$$

and the optimum class is defined as the class with the largest Borda count:

$$c_{OPT} = \arg\max_k \widetilde{\mathbf{R}} = \arg\max_k \sum_m R_m(k) . \tag{15.13}$$

**Example 15.10** *Face Sketch Recognition Using Borda Count* [17]. Face sketch recognition deals with matching a hand-drawn face sketch ($I_{TEST}$) to a training set of $K$ face images. In this application, the aim is to find the person whose image is most similar to the hand-drawn face-sketch $I_{TEST}$. The method for doing this is to first transform the face images into a gallery of $K$ synthesized sketches $I_n, n \in \{1, 2, \ldots, K\}$, using face sketch synthesis (see Example 4.24). We then compute the similarity between $I_{TEST}$ and each synthesized sketch $I_n$ using the $M = 4$ image quality assessments (IQA):

- Structual similarity index
- Visual information fidelity
- Feature similarity index
- Gradient magnitude similarity deviation.

Each IQA returns a rank vector $\mathbf{R}_m = (R_m(1), R_m(2), \ldots, R_m(K))^T, m \in \{1, 2, \ldots, M\}$, where $R_m(k)$ is the similarity (expressed as a rank) between $I_{TEST}$ and $I_k$ for the $m$th IQA. The $\mathbf{R}_m, m \in \{1, 2, \ldots, M\}$, are combined using the Borda count:

$$\widetilde{\mathbf{R}} = \sum_m \mathbf{R}_m ,$$

Finally we identify the person whose appearance is most similar to the hand-drawn sketch $I_{TEST}$ as

$$k^* = \arg\max_k \widetilde{R}(k) ,$$

Some variants of rank fusion are listed in Table 15.4 [1].

The following example illustrates a weighted Borda Count in which we weigh the different classifiers according to the confidence we have in the classifiers themselves.

**Table 15.4** Rank fusion variants

| Name | Description |
|---|---|
| Borda count (BC) | BC for $k$th class is $\widetilde{R}(k) = \sum_{m=1}^{M} R_m(k)$, where $R_m(k)$ is rank of the class $C = c_k$ as determined by $m$th classifier. Class with highest BC wins |
| Weighted BC | Weighted BC for $k$th class is $\widetilde{R}(k) = \sum_{m=1}^{M} w_m R_m(k)$, where $w_m$ is confidence of $m$th classifier. Class with highest weighted BC wins |
| Exponential BC | Exponential BC for $k$th class is $\widetilde{R}(k) = \sum_{m=1}^{M} \exp\big(w_m \times R_m(k)\big)$. Class with highest exponential BC wins |
| Logarithm BC | Logarithm BC for $k$th class is $\sum_{m=1}^{M} \log\big(1 + w_m \times R_m(k)\big)$. Class with highest logarithm BC wins |
| Cross-entropy Fusion | Optimization based rank fusion. See Example 15.12 |

**Example 15.11** *Confidence-based Rank Fusion for Audio-Visual Identification* [2]. Let $\mathbf{F}$ denote a set of audio and video features which belong to an unknown ("test") speaker. Then in audio-speaker speaker identification our aim is to find whether $\mathbf{F}$ belongs to one of the $K$ subjects registered in the database or not. The system identifies the speaker if there is a match, rejects otherwise. Let $S_m, m \in \{1, 2, \ldots, M\}$, denote a set of $M$ classifiers which operate on $\mathbf{F}$. For the test speaker we obtain a $M \times K$ belief matrix:

$$\boldsymbol{\mu} = \begin{pmatrix} \mu_1(1) & \mu_1(2) & \ldots & \mu_1(K) \\ \mu_2(1) & \mu_2(2) & \ldots & \mu_2(K) \\ \vdots & & & \\ \mu_M(1) & \mu_M(2) & \ldots & \mu_M(K) \end{pmatrix},$$

where $\mu_m(k)$ is the belief value generated by the $m$th classifier $S_m$ assuming the identity of the test speaker is, in fact, the $k$th registered speaker.

Sort each row of $\boldsymbol{\mu}$. This gives us the rank matrix $\mathbf{R}$ corresponding to $\boldsymbol{\mu}$:

$$\mathbf{R} = \begin{pmatrix} R_1(1) & R_1(2) & \ldots & R_1(K) \\ R_2(1) & R_2(2) & \ldots & R_2(K) \\ \vdots & & & \\ R_M(1) & R_M(2) & \ldots & R_M(K) \end{pmatrix}.$$

In each row of $\boldsymbol{\mu}$ we record the highest belief value and the mean of the remaining belief values:

$$\begin{pmatrix} \mu_1^* \\ \mu_2^* \\ \vdots \\ \mu_M^* \end{pmatrix} \quad \text{and} \quad \begin{pmatrix} \overline{\mu}_1 \\ \overline{\mu}_2 \\ \vdots \\ \overline{\mu}_M \end{pmatrix}$$

where $\mu_m^* = \max_k \big(\mu_m(1), \mu_m(2), \ldots, \mu_m(K)\big)$ and $\overline{\mu}_m = \frac{1}{K-1}\big(\sum_{k=1}^{K} \mu_m(k) - \mu_m^*\big)$. Then the confidence for the $m$th classifier $S_m$ is

$$w_m = \frac{|\mu_m^* - \overline{\mu}_m|}{\overline{\mu}_m} \ .$$

In this case the corresponding confidence-weighted Borda count is

$$\widetilde{R}(k) = \sum_m w_m R_m(k) \ ,$$

The last entry in Table 15.4 is *cross-entropy rank fusion*. This is a recent rank-level fusion method [1] in which fusion is formulated as an optimization problem.

**Example 15.12** *Multimodal Biometric Identification Using Cross-Entropy Rank Fusion* [1]. Let $P_1, P_2, \ldots, P_M$ be various biometric traits used to identify a person. Let $Q_m(k)$ be the matching score associated with each biometric trait $P_m$ for the $k$th person. Rank vectors are obtained for each trait $P_m, m \in \{1, 2, \ldots, M\}$, by sorting the scores $\{Q_m(1), Q_m(2), \ldots, Q_m(K)\}$. The result is a set of $M$ rank vectors $\mathbf{R}_1, \mathbf{R}_2, \ldots, \mathbf{R}_M$ where the rank vector $\mathbf{R}_m$ corresponds to the $m$th biometric trait $P_m$. A combination of these $M$ rank vectors generates a new rank vector $\widetilde{\mathbf{R}}$:

$$\widetilde{\mathbf{R}} = f(\mathbf{R}_1, \mathbf{R}_2, \ldots, \mathbf{R}_M) \ ,$$

The objective of cross-entropy rank fusion is to generate a vector $\widetilde{\mathbf{R}}$ which has a minimum distance from the input rank lists $\mathbf{R}_1, \mathbf{R}_2, \ldots, \mathbf{R}_M$. Mathematically,

$$\widetilde{\mathbf{R}} = \arg\min \Big( \sum_m w_m d(\widetilde{\mathbf{R}}, \mathbf{R}_m) \Big) \ ,$$

where $w_m$ is a weight associated with the rank vector $\mathbf{R}_m$ and $d(\widetilde{\mathbf{R}}, \mathbf{R}_m)$ is an appropriate distance function.

### 15.7.3 Belief Combiners

The belief combiners use all of the information contained in the belief vectors $\boldsymbol{\mu}_m, m \in \{1, 2, \ldots, M\}$. The *mean*, or average, vote is a popular belief combiner. It finds the optimal class $c_{OPT}$ by selecting the class with the highest average value. Mathematically, the optimal class is

$$c_{OPT} = \arg\max_k \frac{1}{M} \sum_m \mu_m(k) , \quad (15.14)$$

where $\mu_m(k)$ represents the belief as generated by the $m$th classifier for the $k$th class.

A variant of the mean vote is the *weighted mean (average) vote* in which each belief vector is given a weight $w_m$ which is usually learnt.

**Example 15.13** *Weighted Mean Vote*. The weighted mean approach is similar to the mean vote, however, the outputs of the various classifiers are multiplied with a weighting factor or

$$c_{OPT} = \arg\max_k \left( \frac{\sum_m w_m \mu_m(k)}{\sum_m w_m} \right) \quad (15.15)$$

The weights $w_m, m \in \{1, 2, \ldots, M\}$, can be trained by minimizing the error of the different classifiers on the training dataset.

In principle the belief vector $\boldsymbol{\mu}_m$ can be transformed into an a posteriori pdf. In this case we can interpret the weights $w_m$ as the a posteriori probabilities $p(S_m|\mathbf{y}, I)$ and use the Bayesian model averaging technique (see Example 15.1).

## 15.8 Meta-Learners

In the meta-learners the combination function is a meta-classifier which we denote by $\widetilde{S}$. Two widely used meta-learners are the *stacked generalization model* and the *mixture of experts model* which we now consider.

#### Stacked Generalization Model

In the stacked generalization (SG) model [14] the classifiers $S_m, m \in \{1, 2, \ldots, M\}$, are trained on the training sets $\mathcal{D}_m$. The outputs of these classifiers and the corresponding true classes are then used as input/output training pairs for the meta-classifier

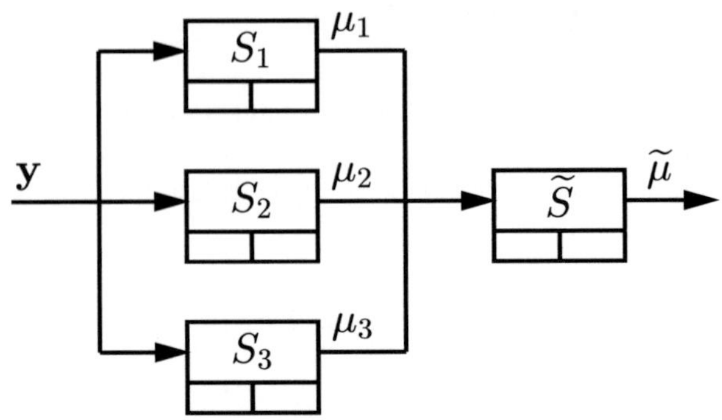


**Fig. 15.2** Shows SG with $M = 3$ classifiers $S_m, m \in \{1, 2, 3\}$, and a meta-classifier $\widetilde{S}$. Input to $\widetilde{S}$ are belief vectors $\boldsymbol{\mu}_m, m \in \{1, 2, 3\}$ generated by $S_m$. Output from $\widetilde{S}$, is a new belief vector $\widetilde{\boldsymbol{\mu}}$, where in some cases, $\widetilde{\boldsymbol{\mu}}$ is an identity vector $\mathbf{I}$

$\widetilde{S}$ (see Fig. 15.2). However, by training the meta-classifier on the same training data used to train the pairwise classifiers, $\widetilde{S}$ may suffer from overfitting. One way of reducing the problem of overfitting is to train the meta-classifier $\widetilde{S}$ via cross-validation. This is explained in Example 15.14. *Note*. In place of cross-validation bootstrap may be used.

**Example 15.14** *Pairwise Classification Using Stacking* [28]. We consider the multi-class classification of an object $O$ into one of $K$ classes using $K(K-1)/2$ pairwise classifiers. The outputs of the pairwise classifiers and the corresponding true classes are used as input/output training pairs for the stacked generalization model.

We partition the training data $\mathcal{D}$ into $L$ separate training sets $\mathcal{D}_1, \mathcal{D}_2, \ldots, \mathcal{D}_L$. Then we define $\mathcal{D}_l^*$ as the training set obtained when we remove $\mathcal{D}_l$ from $\mathcal{D}$. We train each binary pairwise classifier on $\mathcal{D}_l^*, l \in \{1, 2, \ldots, L\}$. We thus obtain $M = LK(K-1)/2$ pairwise classifiers $S_m, m \in \{1, 2, \ldots, M\}$, which are then used as input to the stacked meta-classifier $\widetilde{S}$.

## Mixture of Experts Model

The mixture of experts (MoE) model is another popular meta-learner.

**Example 15.15** *Mixture of Experts (MoE) Model* [19]. The MoE model (see Fig. 15.3) is conceptually similar, but different, to the stacked generalization model. In the MoE model, the input to $\widetilde{S}$ is the measurement vector $\mathbf{y}$ and the output is a weight vector $\mathbf{w} = (w_1, w_2, \ldots, w_M)^T$. The final output of the MoE model is a belief vector, $\widetilde{\boldsymbol{\mu}}$, which is formed by combining the belief vectors $\boldsymbol{\mu}_m, m \in \{1, 2, \ldots, M\}$, with $\mathbf{w}$ using a simple combiner $f$. The MoE

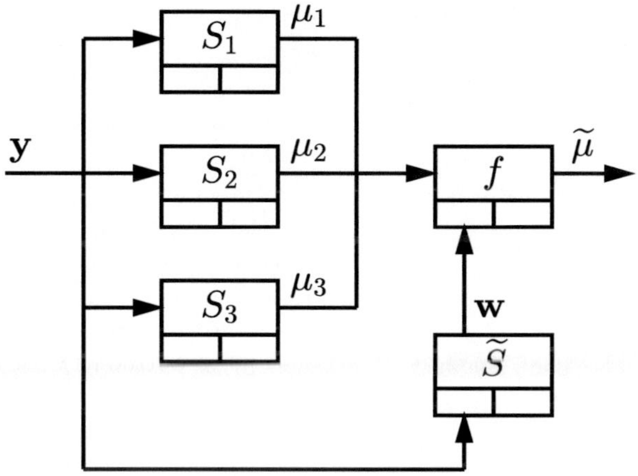


**Fig. 15.3** Shows the mixture of experts model. This consists of an ensemble of $M = 3$ classifiers $S_m, m \in \{1, 2, 3\}$, and a meta-classifier $\widetilde{S}$. The meta-classifier outputs a weight vector $\mathbf{w} = (w_1, w_2, w_3)^T$. The final output is formed by combining the belief vectors $\boldsymbol{\mu}_m, m \in \{1, 2, 3\}$, with the weight vector $\mathbf{w}$, using a simple combiner $f$

model is often interpreted as a classifier selection algorithm in which individual classifiers are regarded as experts in some portion of the feature space. In this case, for each measurement vector $\mathbf{y}$, $\widetilde{S}$ selects the most appropriate classifier, $S_n$, by selecting a weight vector $\mathbf{w} = (w_1, w_2, \ldots, w_M)^T$, where $w_n = 1$ and $w_m = 0, m \neq n$.

The following example illustrates deep learning MoE for fall detection.

**Example 15.16** *Fall detection using mixtures of CNN's* [10]. Falls are a major public health issue all around the world. After traffic accidents it is the second highest cause of unintentional injury mortality. World-wide more than half a million people die from falls every year. Reference [10] describe fall detection using a MoE model in which each model is a CNN.

## 15.9 Boosting

Another meta-learner is "boosting" [5, 21] which is a very effective ensemble learning algorithm. Some common variants of the boosting algorithm are listed in Table 15.5.

We first introduced the concept of boosting in Sect. 15.4.3, where we described it as "a technique for generating an ensemble of training sets $\mathcal{D}^{(m)}, m \in \{1, 2, \ldots, M\}$". Boosting is, however, much more than this. It is a *complete* ensemble learning tech-

**Table 15.5** Boosting algorithms [5, 21]

| Name | Description |
|---|---|
| AdaBoost | The most popular boosting algorithm |
| Real AdaBoost | Generalization of AdaBoost. Often regarded as basic boosting algorithm |
| Modest AdaBoost | A regularized version of AdaBoost |
| Gentle AdaBoost | A more robust and stable version of Real AdaBoost |
| FloatBoost | An improved algorithm for learning a boosted classifier |
| SmoteBoost | A boosting algorithm which improves the performance of classes which are under-represented in the training set $\mathcal{D}$ |
| MadaBoost, EtaBoost, LogitBoost | Various robust versions of AdaBoost |
| SpatialBoost | An algorithm which adds spatial reasoning to AdaBoost |

nique and includes a method for generating multiple training sets $\mathcal{D}^{(m)}$ and an algorithm for combining the corresponding classifiers $S^{(m)}$.[4]

We start by discussing "Adaboost" for *two-class* classification. In adaboost we iteratively build a set of classifiers $S^{(m)}$, $m \in \{1, 2, \ldots, M\}$ that are combined to yield a more powerful ensemble classifier $\widetilde{S}$. At each iteration $m$, a new classifier $S^{(m)}$ is generated and the training samples are re-weighted according to the classification results: the weights of samples which are classified incorrectly are increased while the weights of samples which are correctly classified are decreased. The new weights are used to generate the next classifier $S^{(m+1)}$.

Mathematically, adaboosting works as follows. Let $\mathcal{D}$ denote a common training set containing $N$ samples $\mathbf{x}_i$, $i \in \{1, 2, \ldots, N\}$. In the $m$th iteration, we create a training set $\mathcal{D}^{(m)}$, by randomly sampling $\mathcal{D}$ with replacement, using a non-uniform probability distribution $p_i$, where $p_i = w_i / \sum w_i$ and $w_i$ is the weight given to the $i$th sample.

We use $\mathcal{D}^{(m)}$ to learn the parameters of $S^{(m)}$. The result is a classifier $S^{(m)}$. In boosting we adjust the weights $w_i$, $i \in \{1, 2, \ldots, N\}$, in accordance with the performance of the previous classifier, $S^{(m-1)}$, which was trained on the previous training set, $\mathcal{D}^{(m-1)}$. By increasing the weights of misclassified samples we cause the learning algorithm to focus on different samples and thus lead to different classifiers $S^{(m)}$. The sequence of classifiers $S^{(m)}$, $m \in \{1, 2, \ldots, M\}$, are then fused together to form the "boosted" classifier.

The following algorithm (Algorithm 12) shows the pseudo-code for creating a set of fixed classifiers $S^{(m)}$, $\in \{1, 2, \ldots, M\}$, by boosting a parametric classifier $S$.

[4] From this point of view, we may regard bagging as complete ensemble learning techniques, since it too contains a method for creating multiple training sets $\mathcal{D}^{(m)}$, $m \in \{1, 2, \ldots, M\}$, on which the classifiers $S^{(m)}$ are learnt, as well as an algorithm (majority vote) for combining the $S^{(m)}$.

**Algorithm 12:** AdaBoosting Algorithm

**Input**: Training set $D$ containing $N$ Objects $O_i, i \in \{1, 2, \ldots, N\}$ with features $\mathbf{x}_i$ and labels $y_i = \{0, 1\}$.
**Output**: Boosted Classifier

**Initialize:** $w_i^{(1)} = 1, i \in \{1, 2, \ldots, N\}$
**for** $m \leftarrow 1$ ***to*** $M$ **do**
    Create training set $\mathcal{D}^{(m)}$ by randomly sampling $\mathcal{D}$ with replacement according to non-uniform probability distribution $p_i^{(m)} = w_i^{(m)} / \sum_{i=1}^{N} w_i^{(m)}$
    Learn classifier $S^{(m)}$ on $\mathcal{D}^{(m)}$
    Calculate normalized error $\epsilon^{(m)}$ for $S^{(m)}$ using error function $\delta_i = 1 - \frac{1}{2}|y_i - \hat{y}_i|$ where $\hat{y}_i^{(m)}$ is class of $O_i$ predicted by $S^{(m)}$.
    $\epsilon^{(m)} = \sum_{i=1}^{N} p_i^{(m)} \delta_i$
    Update weights according to: $w_i^{(m+1)} = w_i^{(m)} \left(\frac{\epsilon^{(m)}}{1-\epsilon^{(m)}}\right)^{(1-\delta_i)}$
**end**
Let $p_m(C = c|\mathbf{x}_i, I)$ denote probability $O_i$ belongs to $C = c$ as estimated by $S_m$. Then calculate $r(\mathbf{x}_i)$: $r(\mathbf{x}_i) = \frac{1}{\sum_{m=1}^{M} \ln(1/\beta_m)} \sum_{m=1}^{M} p_m(C = c|\mathbf{x}_i, I) \ln(1/\beta_m)$
Output final (boosted) *a posteriori* probability $p(C = c|\mathbf{x}_i, I)$ that $O_i$ belongs to class $C = c$: $p(C = c|\mathbf{x}_i, I) = \frac{1}{1+\prod_{m=1}^{M} \beta_m^{2r(\mathbf{x}_i)-1}}$
**return**

### Mutual Boosting

The basic idea behind adaboosting is to build a strong classifier from a set of weak classifiers. At each round, we select the weak classifier with the lowest weighted classification error and adjusts the weights based on this error. Weight adjustment aims to alter the distribution of training samples such that the weak classifier selected at current round $m$ is "uncorrelated" with the class label in the next round $m + 1$. However the classifier selected at round $m + 1$ could be similar to the one selected in a previous round $m', 0 < m' < t$. As a result, many classifiers selected by the adaboost algorithm might be similar, and are redundant. A better solution is to adjust the weights in such a way that all of the $M$ selected weak classifiers are "uncorrelated" with the class label. However, such a $M$-dimensional optimization problem might be practically intractable, and the solution might not even exist.

Instead the mutualboost [26] algorithm uses mutual information (MI) to eliminate non-effective weak classifiers. Before a weak classifier is selected, the MI between the new classifier and each of the previously selected ones is examined to make sure that the information carried by the new classifier has not been captured before.

#### Robust Adaboost

Bosting algorithms are sensitive to outliers in the dataset $\mathcal{D}$. The reason is that the boosting algorithm learns by preferentially choosing examples which are not classified correctly and, by definition, these examples are often outliers.

Robust boosting algorithms have however been developed for these conditions [15].

#### Multi-class Boosting

In going from two-class classification to multi-class classification, we often decompose a $K$-class pattern recognition problem into $K(K-1)/2$ two-class pattern recognition problems. The final classification is then obtained by combining all $K(K-1)/2$ pairwise results. Example 14.13 illustrates this procedure for a NBC.

Alternatively, we may use a multi-class AdaBoost algorithm [32].

## 15.10 Recommendations

The various ensemble generation functions have been compared for a wide variety of different classification problems [23]. The consensus is that boosting usually achieves better generalization than bagging but it is more sensitive to noise and outliers. Regarding the combination functions: In general, the preferred functions are majority vote, Borda count and averaging.

## 15.11 Software

**MATLAB**. MATLAB STATISTICS AND MACHINE LEARNING toolbox includes m-files on: stacked generalization, ensemble algorithms (including bagging, boosting). Other useful toolboxes are: MIXTURE OF EXPERTS (Author: Santarcangelo) and ENSEMBLE LEARNING (Author:Ribeiro).

**PYTHON**. SCIKIT-LEARN and SCIKIT-IMAGE includes python modules on: stacked generalization, ensemble algorithms (including bagging, boosting).

## 15.12 Further Reading

Two comprehensive and modern books on ensemble learning are [14, 24]. Recently boosting has been successfully combined with the naive Bayes' classifier [30]. Recently encouraging results have been obtained in multi-class classification by using meta-classifiers to combine pairwise classifiers [28]. See also [6]. A modern survey of ensemble learning is [21].

## 15.13 Postscript

In writing this book our aim was to provide the reader with an introduction to the concepts and ideas in data fusion. The theories and techniques may be formulated in many different frameworks, but for reasons explained in the preface, we have, by and large, restricted ourselves to the Bayesian framework. The reader should, however, keep in mind that data fusion is a pragmatic activity which is driven by practicalities. When circumstances dictate the engineer may abandon the Bayesian framework and adopt an alternative framework instead.

During the time the different editions of this book was being written we have witnessed a growing interest in data fusion: Many commercial applications now employ data fusion in one form or another. We have tied to convey this growing interest by giving examples drawn from a wide range of real-world applications and the reader is now encouraged to review these examples.

## Problems

**15.1** Describe the basic idea of ensemble learning.

**15.2** Compare and contrast the empirical and Bayesian frameworks for ensemble learning.

**15.3** Compare and contrast two methods for constructing an ensemble of diverse training sets.

**15.4** Compare and contrast the following diversity measures: Yule statistic, correlation coefficient, disagreement measure, double fault measure, entropy, Kohavi-Wolpart variance and measure of difficulty.

**15.5** Compare and contrast the following combination strategies: mean vote, Borda count, majority vote, mixture of experts.

**15.6** Describe the method of boosting.

## References

1. Ahmad S, Pal R, Ganivada A (2022) Rank level fusion of multimodal biometrics using genetic algorithm. Multimedia Tools and Appl 81:40931–40958
2. Alam MR, Bennamoun M, Togneri R et al (2015) A confidence-based late fusion framework for audio-visual biometric identification. Patt Recogn Lett 52:65–71

3. Boulle M (2006) Regularization and averaging of the selective naive Bayes classifier. Proc 2006 Int Joint Conf Neural Netw 2989–2997
4. Dudoit S, Fridlyand J (2003) Bagging to improve the accuracy of a clustering procedure. Bioinformatics 19:1090–1099
5. Ferreira AJ, Figueiredo MAT (2012) Boosting algorithms: a review of methods, theory and applications. Published by Springer, In Ensemble Machine Learning
6. Garcia-Pedrajas N, Ortiz-Boyer D (2011) An empirical study of binary classifier fusion methods for multi-class classification. Inf Fusion 12:111–130
7. Ghahramani Z, Kim H-C (2003) Bayesian classifier combination. University College, University of London, UK, Gatsby Tech Rept
8. Gonzalez-Barron U, Butler F (2006) A comparison of seven thresholding techniques with the k-means clustering algorithm for measurement of bread-crumb features by digital image analysis. J Food Eng 74:268–278
9. Gowda SN, Yuan C (2019) ColorNet: investigating the importance of color spaces for image classification. Lect Notes Comput Sci 11364:581–596
10. Ha TV, Nguyen HM, Thanh SH et al (2024) Fall detection using mixtures of convolutional neural networks. Multimedia Tools App 83:18091–18118
11. Huang YS, Suen CY (1995) A method of combining multiple experts for the recognition of unconstrained handwritten numerals. IEEE Trans Patt Anal Mach Intell 17:90–94
12. Kittler J, Hatef M, Duin RPW et al (1998) On combining classifiers. IEEE Trans Patt Anal Mach Intell 20:226–239
13. Ko AHR, Sabourin R, de Souza Britto A Jr et al (2007) Pairwise fusion matrix for combining classifiers. Patt Recogn 40:2198–2210
14. Kuncheva LI (2014) Combining pattern classifiers, 2nd ed. Published by Wiley
15. Li AH, Bradic J (2018) Boosting in the presence of outliers: adaptive classification with non-convex loss functions. J Am Stat Assoc 113:660–674
16. Liang S, Baker D (2023) Real-time background subtraction under varying lighting conditions. In: International conference on robotics and automation (ICRA 2023), pp 9317–9323
17. Mahfoudi S, Daamouche A, Bengherabi M et al (2022) Hand-drawn face sketch recognition using rank-level fusion of image quality assessment metrics. Bull Polish Academy Sci Tech Sci 70:Art No e143554
18. Martinez-Munoz G, Suarez A (2005) Switching class labels to generate classification ensembles. Patt Recogn 38:1483–1494
19. Masoudnia S, Ebrahimpour R (2014) Mixture of experts: a literature survey. Artif Intell Rev 42:275–293
20. Melgani F (2006) Robust image binarization with ensembles of thresholding algorithms. J Elec Imag 15:023010
21. Mienye ID, Sun Y (2022) A survey of ensemble learning: concepts, algorithms, applications and prospects. IEEE Access 10:99129–99144
22. Nandi AS, Randhawa KK, Chua HS et al (2022) Credit card fraud detection using a hierarchical behavior-knowledge space model. PLoS ONE 17:e0260579
23. Polikar R (2006) Ensemble based systems in decision making. IEEE Circuit Syst Mag 6:21–45
24. Rokach L (2019) Ensemble learning, 2nd ed. Published by World Scientific
25. Schapire RE (2002) The boosting approach to machine learning: an overview. Proceedings of the MSRI workshop nonlinear estimation and classification
26. Shen L, Bai L (2006) MutualBoost learning for selecting Gabor features for face recognition. Patt Recogn Lett 27:1758–1767
27. Shi X, Manduchi R (2007) On the Bayes fusion of visual features. Image Vis Comput 25:1748–1758
28. Shiraishi Y, Fukumizu K (2011) Statistical approaches to combining binary classifiers for multi-class classification. Neurocomput 74:680–686

29. Tax DMJ (2001) One class classification. PhD thesis, Delft University, The Netherlands
30. Viaene S, Derrig R, Dedene G (2004) A case study of applying boosting naive Bayes to claim fraud diagnosis. IEEE Trans Know Data Eng 16:612–619
31. Webb GI (2000) Multiboosting: a technique combining boosting and wagging. Mach Learn 40:159–197
32. Zhu J, Zou H, Rosset S, Hastie T (2009) Multi-class adaboost. Statistics Interface 2:349–360

# Index




H. B. Mitchell, *Data Fusion*, https://doi.org/10.1007/978-3-662-71023-4

MIX
Papier aus verantwortungsvollen Quellen
Paper from responsible sources
FSC® C105338

FSC
www.fsc.org

If you have any concerns about our products,
you can contact us on
**ProductSafety@springernature.com**

In case Publisher is established outside the EU,
the EU authorized representative is:
**Springer Nature Customer Service Center GmbH**
**Europaplatz 3, 69115 Heidelberg, Germany**

Printed by Libri Plureos GmbH
in Hamburg, Germany